Principles of
CHEMICAL ENGINEERING PROCESSES

Nayef Ghasem
Redhouane Henda

CRC Press
Taylor & Francis Group
Boca Raton London New York

CRC Press is an imprint of the
Taylor & Francis Group, an **informa** business

CRC Press
Taylor & Francis Group
6000 Broken Sound Parkway NW, Suite 300
Boca Raton, FL 33487-2742

© 2009 by Taylor & Francis Group, LLC
CRC Press is an imprint of Taylor & Francis Group, an Informa business

No claim to original U.S. Government works
Printed in the United States of America on acid-free paper
10 9 8 7 6 5 4 3 2 1

International Standard Book Number-13: 978-1-4200-8013-1 (Hardcover)

Library of Congress Cataloging-in-Publication Data

Ghasem, Nayef.
    Principles of chemical engineering processes / Nayef Ghasem, Redhouane Henda.
        p. cm.
    Includes bibliographical references and index.
    ISBN 978-1-4200-8013-1 (hardback : alk. paper)
    1. Chemical engineering. I. Henda, Redhouane. II. Title.

TP155.G47 2008
660--dc22   2008024363

Visit the Taylor & Francis Web site at
http://www.taylorandfrancis.com

and the CRC Press Web site at
http://www.crcpress.com

Contents

Preface

Purpose of the Book

The objective of this book is to introduce chemical engineering students to the basic principles and calculation techniques used in the field, and to acquaint them with the fundamentals of material and energy balances as applied to chemical engineering. The book is mainly intended for junior chemical engineers. It exposes them to problems in material and energy balances that arise in relation to problems involving chemical reactors. It also introduces them to numerical methods used to solve these problems with simple software packages.

The focus is toward subjects related to material and energy balances. Subjects that could be covered in other courses such as chemical engineering, thermodynamics, and fluid mechanics are not included to avoid redundancy. The theory is explained in a few lines and in simple language so that it encourages students to get used to the book. Many examples are included and solved in a simple and understandable format that explains the related theory. A few examples are solved using E–Z Solve® and Polymath and this enables students to use such software in solving material and energy balances of complicated problems. The book can be covered in one semester if based on a 4-hour contact course or two semesters on a 2-hour contact course.

Course Objectives

Upon completion of the course using *Principles of Chemical Engineering Processes*, you should be able to

1. Draw and fully label a flowchart for a given process description.
2. Choose a convenient basis of calculation, for single- and multiple-unit processes.
3. Identify possible subsystems for which material and energy balances might be written.
4. Perform a degree of freedom analysis for the overall system and each possible subsystem, formulate and simplify the appropriate material and energy balance equations, and perform the necessary calculations. You should be proficient at performing these analyses for

single- and multiple-unit processes involving recycle, bypass, or purge streams, and for systems involving reactions.

5. Understand and apply the first law of thermodynamics (conservation of energy), calculate energy and enthalpy changes using tabulated data and heat capacities, and construct energy balances on closed and open systems.

6. Be familiar with basic processing terminology (batch, semibatch, continuous, purge, and recycle) and standard operations (reaction, distillation, absorption, extraction, and filtration).

7. Have an appreciation of why these basic skills are required and are important for all chemical engineers, independent of their final career choice.

Course Outcomes

Every course that a student takes should further his or her knowledge, building on what was previously learned. By the end of this course, each student should be able to

1. Explain the concepts of dimensions, units, psychrometry, steam properties, and conservation of mass and energy.

2. Solve steady-state and transient mass and energy balance problems involving multiple-unit processes and recycle, bypass, and purge streams.

3. Solve and understand simple unsteady-state mass and energy balances.

4. Solve more complicated problems using the software appropriate for the problem.

5. Present the solutions to engineering problems in both oral and written form in a clear and concise manner.

Relationship of Course to Program Outcomes (Accreditation Board for Engineering and Technology [ABET] Criteria)

(a) Ability to apply knowledge of mathematics, science, and engineering.

(b) Ability to design a system, component, or process to meet desired needs within realistic constraints such as economic, environmental, social, political, ethical, health and safety, manufacturability, and sustainability.

(c) Ability to identify formulas and solve engineering problems related to materials and energy balances.

(d) Ability to use the techniques, skills, and modern engineering tools necessary for engineering practice.

A Note about the CD-ROM

The book's accompanying CD-ROM contains a summary of all chapters in the book, exercises, property estimation software, PowerPoint presentations that explain the simulations and property estimation using HYSYS, and psychometric chart parameter estimation software.

The CD-ROM also contains POLYMATH software, which the user will be allowed to run 200 times free of charge. For the last 50 uses, a warning will be issued when the software is started to indicate the number of startups remaining and to give information as to where the POLYMATH software can be purchased. The Web link for this software is

http://www.polymath-software.com/ghasem/

More information about POLYMATH software can be obtained at the following link:

http://www.polymath-software.com

Acknowledgments

The contents of this book are based on topics covered by many universities worldwide and not only on topics covered in available textbooks. First, the authors would like to thank Allison Shatkin, Acquiring Editor for this book, and Rachael Panthier, the Project Editor, for their help and cooperation. We also thank those professors who have given us the chance to see their course syllabi and contents. They include Professor Ezzat Chan from the University of Malaya, Professors McAuley and Hutchinson from Queen's University, Professors Rowley and Fletcher from Brigham Young University, Professor Joseph McCarthy from the University of Pittsburgh, Professor Zollars from Washington State University, Professors Prausnitz, Brown, Jessica, Bartling, Koros, and Timothy from the Georgia Institute of Technology, and Professors Palecek, Murphy, Root, and Rawling from the University of Wisconsin–Madison. We extend our thanks to all those who contribute to Wikipedia and, finally, to all professors who are teaching or contributing to this subject.

Nayef Ghasem
Redhouane Henda

Authors

Nayef Ghasem is an associate professor of chemical engineering at United Arab Emirates University, where he teaches chemical process principles and chemical reaction engineering as undergraduate courses along with other courses in chemical engineering. Previously, he taught these courses at the University of Malaya, Malaysia. He has published primarily in the areas of modeling and simulation, bifurcation theory, polymer reaction engineering, and advanced control of polyethylene and polystyrene polymerization processes. He has also authored *Simulation of Unit Operations with HYSYS*, published by University of Malaya Press. Dr. Ghasem is a member and ambassador of the Institute of Chemical Engineers and a member of the American Chemical Society.

Redhouane Henda is an associate professor of chemical engineering at Laurentian University, Ontario, Canada. He has taught chemical engineering principles as an undergraduate course along with other core courses in chemical engineering at both undergraduate and graduate levels. Dr. Henda's professional activities are in the areas of process engineering and advanced materials. He and his students and other colleagues have written a number of journal articles on modeling and simulation of complex chemical systems, and on advanced thin solid films. Dr. Henda is a licensed professional engineer in Ontario and is currently an associate editor of the *Canadian Metallurgical Quarterly*. He is a fellow of the Alexander von Humboldt foundation.

Systems of Units

Systems of units are defined with reference to Newton's second law for a system of constant mass:

$$F = ma \quad (\text{mass–length/time}^2)$$

where $F = $ force required to accelerate a body of mass, m, at a rate a (length/time^2)

System	Length	Time	Mass	Force	g_c
SI	Meter **m**	Second **s**	Kilogram **kg**	Newton **N**	1.0 (kg-m)/(N·s^2)
CGS	Centimeter **cm**	Second **s**	Gram **g**	dyne	1.0 (g-cm)/(dynes^2)
AES	Foot **ft**	Second **s**	Pound mass **lb$_m$**	Pound force **lb$_f$**	32.174 (lb$_m$-ft)/(lb$_f$-s^2)

1.0 N = Force that will accelerate a mass of 1.0 kg by 1.0 m/s^2

1.0 dyne = Force that will accelerate a mass of 1.0 g by 1.0 cm/s^2

1.0 lb$_f$ = Force that will accelerate a mass of 1.0 lb$_m$ by 32.174 ft/s^2

Metric prefixes

10^{12}	T	tera	trillion	10^{-1}	d	deci	tenths
10^{9}	G	giga	billion	10^{-2}	C	centi	hundredths
10^{6}	M	mega	million	10^{-3}	m	milli	thousandths
10^{3}	k	kilo	thousand	10^{-6}	μ	micro	millionths
10^{2}	h	hecto	hundred	10^{-9}	n	nano	billionths
10^{1}	da	deca	ten	10^{-12}	p	pico	trillionths

Acceleration of gravity	$g = 9.8066$ m/s^2 (sea level, 45° latitude)
	$g = 32.174$ ft/s^2
Gas constant	$R = 10.731$ psia-ft^3/lbmol $= 0.7302$ atm-ft^3/lbmol
	$R = 0.082056$ atm-L/mol-K $= 8.3143$ Pa-m^3/mol-K
	$R = 0.08314$ L-bar/mol-K $= 1.987$ Btu/lb mol $= 8314.3$ J/kg mol-K
	$R = 8.3143$ J/mol-K $= 62.36$ L-mmHg/mol-K $= 1.987$ cal/mol-K
Density of water at 4°C	$\rho(H_2O, 4°C) = 1.0$ g/cm^3 $= 1.0$ kg/L $= 10^3$ kg/m^3
	$\rho(H_2O, 4°C) = 8.34$ lb$_m$/gal $= 62.43$ lb$_m$/ft^3
Specific gravity of water	$= 1.0$
Specific gravity of Hg	$= 13.6$

Conversion Factors

Mass	$1\ lb_m = 5 \times 10^{-4}\ t = 0.453593\ kg = 453.593\ g = 16\ oz$
	$1\ kg = 1000\ g = 2.20462\ lb_m = 0.001\ t$
	$1\ t = 2000\ lb_m;\ 1\ t = 1000\ kg$
Length	$1\ ft = 12\ in.;\ 1\ ft = 0.3048\ m = 30.48\ cm;\ 1\ in. = 2.54\ cm;\ 1\ mile = 5280\ ft$
	$1\ m = 10^{10}\ \mathring{A} = 39.37\ in. = 3.2808\ ft = 1.0936\ yd = 0.0006214\ mi$
Volume	$1\ ft^3 = 7.481\ gal = 1728\ in.^3 = 28.317\ L = 28{,}317\ cm^3$
	$1\ gal = 231\ in.^3;\ 1\ in.^3 = 16.387\ cm^3$
	$1\ cc = 1\ cm^3 = 1\ mL;\ 1000\ mL = L$
	$1000\ L = 1\ m^3 = 35.3145\ ft^3 = 220.83\ \text{imperial gallons} = 264.17\ gal = 1056.68\ qt$
	$8\ fl\ oz = 1\ cup;\ 4\ cup = 1\ quart;\ 4\ quart = 1\ gal = 128\ fl\ oz$
Density	$1\ g/cm^3 = 1\ kg/L = 1000\ kg/m^3 = 62.428\ lb/ft^3 = 8.345\ lb_m/gal$
Force	$1\ lb_f = 32.174\ lb_m\text{-}ft/s^2 = 4.448222\ N = 4.4482 \times 10^5\ dynes$
	$1\ N = 1\ kg\text{-}m/s^2 = 10^5\ dynes = 10^5\ g\text{-}cm/s^2 = 0.22481\ lb_f$
Pressure	$1\ bar = 10^5\ Pa = 100\ kPa = 10^5\ N/m^2$
	Pascal (Pa) is defined as $1\ N/m^2 = 1\ kg/m\text{-}s^2$
	$1\ atm = 1.01325\ bar = 14.696\ lb_f/in.^2 = 760\ mmHg\ \text{at}\ 0°C\ (torr) = 29.92\ in\ Hg\ \text{at}\ 0°C$
	$1\ psi = 1\ lb_f/in.^2;$ psia (absolute) = psig (gauge) + 14.696
Temperature	$1\ K = 1.8°R$ (absolute temperatures)
	$T\ (°C) = T\ (K) - 273.15$
	$T\ (°F) = T\ (°R) - 459.67$
	$T\ (°F) = 1.8T\ (°C) + 32$
Energy	$1\ J = 1\ N\text{-}m = 1\ kg\text{-}m^2/s^2 = 10^7\ ergs = 10^7\ dyne\text{-}cm = 2.778 \times 10^{-7}\ kW\text{-}h$
	$= 0.23901\ cal = 0.7376\ ft\text{-}lb_f = 9.486 \times 10^{-4}\ Btu$
	$1\ cal = 4.1868\ J;\ 1\ Btu = 778.17\ ft\text{-}lb_f = 252.0\ cal$
	$1\ Btu/lb_m\text{-}F = 1\ cal/g\text{-}°C$
Power	$1\ hp = 550\ ft\text{-}lb/s = 0.74570\ kW$
	$1\ W = 1\ J/s = 0.23901\ cal/s = 0.7376\ ft\text{-}lb_f/s = 9.486 \times 10^{-4}\ Btu/s$
	$1\ kW = 1000\ J/s = 3412.1\ Btu/h = 1.341\ hp$

1

Introduction

At the End of This Chapter You Should Be Able to

1. Understand what is chemical engineering.
2. Explain the difference between values, units, and dimensions in a given expression.
3. Convert from one set of units to another.
4. Identify an invalid equation, based on dimensional arguments.
5. Compare two quantities using a dimensionless group.
6. Write a value in scientific notation (using the correct number of significant digits).
7. Determine the number of significant digits both in a given value and in a arithmetic result.

1.1 Definition of Chemical Engineering

There has been no universally accepted definition of chemical engineering, to date. A definition found in a standard dictionary states that "Chemical engineering is a branch of engineering which involves the design and operation of large-scale chemical plants, petrochemical refineries, and the like." Another definition from the information highway: "Chemical engineering is concerned with processes that cause substances to undergo required changes in their chemical or physical composition, structure, energy content, or physical state." Despite the lack of consensus among chemical engineers on the definition of their discipline, which may be attributed to the broad nature of the discipline itself, there is no disagreement among chemical engineering practitioners that chemical engineers

- Turn low value materials into high value products
- Involved in product design and development

- Design processes to manufacture products
- Involved with process scale-up, development, and optimization
- Perform economic analysis of the production process
- Operate and control the processes to ensure that product quality satisfies the required specification
- Involved in management of the processes
- Involved in product sales and technical service

Because of the reliance of chemical engineering on principles that are versatile in the natural world, the possibilities offered to and challenges faced by chemical engineers are quite formidable and constantly changing. Without intending to be extensive, the list of chemical engineering applications includes

- Traditional areas, for example, mining, pulp and paper, oil refining, materials (rubber, plastics, etc.), and environment
- Nontraditional areas such as microelectronics (semiconductor manufacturing), biotechnology (pharmaceutical production processes, genetic engineering, etc.), and nanotechnology
- Other areas (e.g., medicine, law, and business)

A similarity among all chemical engineering systems is that they involve processes designed to transform raw materials into desired products. A typical problem in the design of a new process or the modification of an existing process is "given the amount and the properties of the raw materials, calculate the amount of the products and determine their properties or vice versa." In order to answer this question, chemical engineers have powerful tools at their disposal that consist of the principles of material and energy balances.

1.2 Material and Energy Balances

Material balances can be used to describe material quantities as they pass through processing operations. Such balances are statements on the conservation of mass. Similarly, energy quantities can be described by energy balances, which are statements on the conservation of energy. If there is no accumulation, what goes into a process must come out. Material and energy balances are very important in any industry. Material balances are fundamental to the control of processing, particularly in the control of product yield. The first material balances are determined in the exploratory stages of a new process, improved upon during pilot plant experiments when the process is being planned and tested, checked when the plant is commissioned, and then refined and maintained as a control instrument as production continues. The material balances need to be determined when any changes occur in the process. The increasing cost of energy has caused industries to examine means of reducing energy consumption during processing. Energy balances are used in the examination of the various stages of a process, across the whole process, and even extending over the entire production system from the raw material to the finished product.

Material and energy balances can be simple; however, sometimes they can be very complicated, but the basic approach is general. Experience in working with simpler systems such as individual unit operations will develop facilities to extend the methods to more complicated situations, which do arise. The increasing availability of computers has meant that very complex mass and energy balances can be set up and manipulated quite readily, and therefore used in everyday process management to maximize product yields and minimize costs. A balance on a conserved quantity (i.e., mass or energy) in a system may be written generally as

$$\text{Accumulation} = (\text{in} - \text{out}) + (\text{generation} - \text{consumption}) \qquad (1.1)$$

The proper understanding and mastering of material and energy balances is critical to the chemical engineering profession. The formulation of the balances on a process system has far reaching implications as it

- Provides a basis for modeling systems that would be difficult and expensive to study in the lab on paper or by computer simulation (i.e., numerically)
- Assists in the synthesis of chemical processes and evaluation of design processes
- Guides the analysis of physical systems
- Provides a basis for estimating the economic costs and benefits of a project

1.3 Values, Units, and Dimensions

Chemical engineers, like many other engineers, use values, units, and dimensions all the time. For example, a grocery list shows that

- 1 L of milk (value: 1; units: liter; dimensions: volume [length^3])
- 1.5 kg of meat (value: 1.5; units: kg; dimensions: mass)

Another example is to consider buying 2.0 L of drinking water; in this case, the "value" is the numerical quantity (2.0). The "unit" indicates what this quantity represents, in this case 2.0 L. The "dimension" is the measurable

TABLE 1.1

Quantities, Units, and Symbols of SI System

Quantity	Unit (Base Unit)	Symbol
Length	Meter	m
	Centimeter	cm
Mass	Kilogram	kg
	Gram	g
Time	Second	s
	Day	day
Temperature	Celsius	°C
	Kelvin	K

property that the units represent. Liter is a unit of volume (units are a specific example of a dimensional quantity). A measured or counted quantity has a numerical value (e.g., 2.0) and a unit (whatever there are 2.0 of). It is necessary in engineering calculations to report both the value and the unit. A value without its unit is meaningless, for example, the length of a table is 2 m, and other such examples include 17 s, 2.5 kg, 7 gold rings. A dimension is a property that can be measured, such as length, mass, or temperature, or calculated by multiplying or dividing other dimensions such as length/time = velocity, length^3 = volume, and mass/length^3 = density. Measurable units are specific values of dimensions that have been defined by convention, such as grams for mass, seconds for time, and centimeters for length.

Units can be treated like algebraic variables when quantities are added, subtracted, multiplied, or divided. Derived units can be obtained by multiplying or dividing base units (i.e., units of length, mass, and time) (see Tables 1.1 and 1.2).

1.3.1 Systems of Units

There are several systems of units, but two primary systems that engineers use are the International System of Units (SI system) and the American

TABLE 1.2

Units for Mass, Length, and Time

Quantity	Unit	Symbol	In Terms of Base Units
Volume	Liter	L	0.001 m^3
Force	Newton	N	1 (kg·m)/s^2
Energy	Joule	J	1 N·m = 1 (kg·m^2)/s^2
Pressure	Pascal	Pa	N/m^2
Density			g/cm^3
Molecular weight			g/mol

TABLE 1.3

Units Associated with Systems of Units

System	Mass (*m*)	Length (*l*)	Time (*t*)	Temperature (*T*)
SI	Kilogram (kg)	Meter (m)	Second (s)	Kelvin (K)
AES	Pound mass (lb$_m$)	Foot (ft)	Second (s)	Degree Fahrenheit (°F)
CGS	Gram (g)	Centimeter (cm)	Second (s)	Kelvin (K)
FPS[a]	Pound mass (lb$_m$)	Foot (ft)	Second (s)	Degree Fahrenheit (°F)
British	Slug	Foot (ft)	Second (s)	Degree Celsius (°C)

[a] Imperial System units are sometimes referred to as FPS.

Engineering System of Units (AES). Other systems are centimeter gram second (CGS), foot pound second (FPS), and the British System of Units (British). Table 1.3 shows the units associated with these systems of units.

1.4 Unit Conversion

A measured quantity can be expressed in terms of units having the appropriate dimension. For example, velocity may be expressed in terms of foot per second, miles per hour, centimeter per year, or any other ratio of length and time. The numerical value of velocity depends on the units chosen, for example, 20 m/s = 66 ft/s.

To convert a quantity expressed in terms of one unit to its equivalent in terms of another unit, multiply the given quantity by the conversion factor (new unit/old unit).

For example, to convert 36 mg to its equivalent in grams

$$36 \text{ mg} \times \left(\frac{1 \text{ g}}{1000 \text{ mg}}\right) = 0.036 \text{ g}$$

The following generally used conversion units are important.

1.4.1 Time

$$60 \text{ s} = 1 \text{ min}, \quad \frac{60 \text{ s}}{1 \text{ min}} = 1$$

$$\frac{60 \text{ s}}{\text{min}} \bigg| \frac{60 \text{ min}}{\text{h}} \bigg| \frac{24 \text{ h}}{\text{day}} \bigg| \frac{365.25 \text{ days}}{\text{year}}$$

1.4.2 Mass

$$\frac{2000 \text{ lb}_m}{\text{ton}} \bigg| \frac{453.6 \text{ g}}{\text{lb}_m} \bigg| \frac{\text{kg}}{1000 \text{ g}} \bigg| \frac{\text{metric ton}}{1000 \text{ kg}}$$

1.4.3 Length

$$\frac{100 \text{ cm}}{\text{m}} \Big| \frac{\text{in.}}{2.54 \text{ cm}} \Big| \frac{\text{ft}}{12 \text{ in.}} \Big| \frac{\text{mile}}{5280 \text{ ft}}$$

$$1 \text{ m} = 100 \text{ cm} = 1000 \text{ mm} = 10^6 \text{ } \mu\text{m} = 10^{10} \text{ Å}$$

1.4.4 Volume

$$\frac{\text{cc}}{\text{cm}^3} \Big| \frac{\text{cm}^3}{\text{mL}} \Big| \frac{1000 \text{ mL}}{\text{L}} \Big| \frac{1000 \text{ L}}{\text{m}^3}$$

1.4.5 Density

Density of water at 4°C expressed in various units

$$\frac{\text{g}}{\text{cm}^3}; \frac{\text{kg}}{\text{L}}; \frac{10^3 \text{ kg}}{\text{m}^3}; \frac{8.34 \text{ lb}_\text{m}}{\text{gal}}; \frac{62.4 \text{ lb}_\text{m}}{\text{ft}^3}$$

Specific gravity of water $= 1.0$
Specific gravity of Hg $= 13.6$

1.4.6 Force

According to Newton's second law of motion, force is proportional to the product of mass and acceleration (length/time^2), and has units of $(\text{kg·m})/\text{s}^2$ (SI), $(\text{g·cm})/\text{s}^2$ (CGS), and $(\text{lb}_\text{m}\text{·ft})/\text{s}^2$ (AES). For the metric systems, the following derived units are commonly used

$F = m \cdot a$
1 Newton (N) $\equiv 1 \text{ (kg·m)}/\text{s}^2$
1 dyne $\equiv 1 \text{ (g·cm)}/\text{s}^2$

In the AES, the derived force unit is called a pound-force (lb_f) and is defined as the product of a unit mass (1 lb_m) and the acceleration of gravity, which is $32.174 \text{ ft}/\text{s}^2$.

$$\text{lb}_\text{f} = 32.174 \frac{\text{lb}_\text{m}\text{ft}}{\text{s}^2}; \quad \frac{32.174 \frac{\text{lb}_\text{m}\text{ft}}{\text{s}^2}}{\text{lb}_\text{f}}$$

$$\text{N} = \frac{\text{kg·m}}{\text{s}^2}; \quad \frac{\text{N}}{(\text{kg·m})/\text{s}^2}$$

1.4.7 Pressure

Pressure is the ratio of force to the area over which that force acts.

$$1 \text{ pascal (Pa)} = 1 \frac{\text{N}}{\text{m}^2} = \frac{\text{kg}}{\text{ms}^2} \text{ (pressure in base units)}$$

$$\frac{lb_f/in.^2}{psi} \Bigg| \frac{14.696 \text{ psi}}{atm} \Bigg| \frac{atm}{760 \text{ mmHg}}; \frac{atm}{1.01325 \text{ bar}}$$

$$\frac{bar}{100 \text{ kPa}}; \frac{bar}{10^5 \text{ Pa}}; \frac{Pa}{N/in.^2}$$

1.4.8 Energy

$$1 \text{ J} = 1 \text{ N m} = 1 \frac{\text{kg m}^2}{\text{s}^2} \quad \text{(base units of energy)}$$

$$\frac{kJ}{10^3 \text{ J}} \frac{J}{\text{N m}} \Bigg| \frac{\text{calorie}}{4.184 \text{ J}}$$

$$\frac{10^3 \text{ J}}{kJ} \Bigg| \frac{1.05 \text{ kJ}}{\text{Btu}}$$

1.4.9 Power

$$W = \frac{J}{s}; \quad kW = 1.34 \text{ hp}$$

$$\frac{J/s}{W} \Bigg| \frac{10^3 \text{ W}}{kW} \Bigg| \frac{kW}{1.34 \text{ hp}}$$

Gravitational constant

$$g = 9.8066 \frac{m}{s^2} = 32.174 \frac{ft}{s^2}$$

Speed of light: $c = 3 \times 10^8 \dfrac{m}{s}$

1.4.10 Weight

The weight of an object is the force exerted on the object by gravitational attraction:

$$W = m \cdot g$$

The value of g at sea level and 45° latitude for each system of units is

$$g = 9.8066 \text{ m/s}^2 \text{ (SI)}$$
$$= 980.66 \text{ cm/s}^2 \text{ (CGS)}$$
$$= 32.174 \text{ ft/s}^2 \text{ (AES)}$$

1.5 Dimensional Homogeneity

The dimensions on both sides of the "equals" sign must be the same for an equation to be valid. Another way to say this is that valid equations must be dimensionally homogeneous. It is good practice to make units similar (via conversion). It is also good practice to show all units throughout a problem to test equation validity. Identify an invalid equation based on dimensional arguments.

Which of these equations are dimensionally homogeneous?

1. $x(\text{m}) = x_o(\text{m}) + 0.3048(\text{m/ft})v(\text{ft/s})t(\text{s}) + 0.5a(\text{m/s}^2)[t(\text{s})]^2$

2. $P\left(\dfrac{\text{kg}}{\text{ms}^2}\right) = 101325.0\left(\dfrac{\text{Pa}}{\text{atm}}\right)1\left(\dfrac{\text{kg/ms}^2}{\text{Pa}}\right)P_0(\text{atm}) + \rho\left(\dfrac{\text{kg}}{\text{m}^3}\right)v\left(\dfrac{\text{m}}{\text{s}}\right)$

Solution

1. It is sometimes simpler to write the units in vertical fractions to facilitate canceling.

$$x(\boxed{\text{m}}) = x_0(\boxed{\text{m}}) + 0.3048\left(\frac{\boxed{\text{m}}}{\text{ft}}\right)v\left(\frac{\text{ft}}{\text{s}}\right)t(\text{s}) + 0.5a\left(\frac{\boxed{\text{m}}}{\text{s}^2}\right)[t(\text{s}^2)]$$

$$m \quad [=] \quad m \quad + \qquad m \qquad\qquad + \qquad m$$

The above equation is dimensionally homogeneous because each term has the unit of length (m).

2. Simplify the equation by writing the units in vertical fractions to facilitate canceling.

$$P\left(\frac{\text{kg}}{\text{ms}^2}\right) = 101325.0\left(\frac{\text{Pa}}{\text{atm}}\right)1\left(\frac{\text{kg}}{\text{ms}^2\,\text{Pa}}\right)P_0(\text{atm}) + \rho\left(\frac{\text{kg}}{\text{m}^3}\right)v\left(\frac{\text{m}}{\text{s}}\right)$$

$$\frac{\text{kg}}{\text{ms}^2} \quad = \qquad\qquad \frac{\text{kg}}{\text{ms}^2} \qquad\qquad + \quad \frac{\text{kg}}{\text{m}^2\text{s}^2}$$

As can be seen in the above equation, it is not dimensionally homogeneous. A dimensionally homogeneous equation does not mean that it is valid; dimensional considerations act as a first test for only validity. Units are also important to chemical engineers as computing tools. Because all equations must be dimensionally consistent, engineers can exploit that constraint to put things in different units and to find the units of variables. By dimensionally consistent, we mean that equality established by the equals sign requires not only that the value be identical but also that the units be the same on both sides of the equation.

1.6 Significant Figures

The significant figures of a number are the digits from the first nonzero digit on the left to either the last digit (zero or nonzero) on the right if there is a decimal point or the last nonzero digit of the number if there is no decimal point. The number of significant figures is readily seen if scientific notation is used. Scientific notation is generally a more convenient way of representing large numbers (Tables 1.4 and 1.5).

Three significant figures such as

Conventional number: 123,000,000
Scientific notation: 1.23×10^8

represent the precision at which a value is known:
Clock without a second hand → 7:31 AM
Clock with a second hand → 7:31:07 AM
Stop watch → 7:31:07.13 AM

Numbers can be written to show significant figures:

- With decimal: count from nonzero on left to last digit on right
- Without decimal: count from nonzero on left to last nonzero on right
- Scientific notation: typically only significant figures are shown

Examples
40500 (4.0500×10^4) → 5 significant figures
0.0012 (1.2×10^{-3}) → 2 significant figures
0.001200 (1.200×10^{-3}) → 4 significant figures
40500 (4.05×10^4) → 3 significant figures
1.47×10^8 → 3 significant figures

TABLE 1.4

Significant Figures for Numbers
without Decimal Point

Numbers	Significant Figures
2300	2
230001	6
02301	4
023010	4

TABLE 1.5

Significant Figures for Numbers
with Decimal Points

Numbers	Significant Figures
2300.	4
230.001	6
0230.10	5
0.0103	3
0.01030	4

For a number with a decimal point, include all the digits from the leftmost nonzero digit to the last digit, including zeros.

There is an accepted convention regarding the reporting of computed results. It is simply that a final result should be listed to no more significant figures than the least accurate factor in the expression evaluated. For example, the distance traveled by a vehicle moving with velocity 16.223 m/s in 2.0 s is 32 m. The fact that your calculator may list 32.446000 m as the result of the computation is an electronic artifact. It is your job to round the result to an appropriate number of significant figures. Intermediate calculations should be performed with the full precision of the tool you are using (calculator or computer) but the result must reflect the true precision.

1.6.1 Multiplication and Division

The number of significant figures in your answer should equal the lowest number of significant figures of any of the numbers being multiplied or divided.

1.6.2 Addition and Subtraction

Compare the positions of the last significant figure of each number: the position farthest to the left is the position of the last significant figure in your answer.

Example 1.1 Significant figures
How many significant digits are there in 2.04×10^{-3}?

Solution
There are three significant digits in the given scientific number.

Example 1.2 Addition of significant figures
Your car weighs 2100 lb_f; you put an 8 lb_f bag in the trunk. What is the car's weight now?

Solution
The car is still thought of as 2100 lb$_f$. You add the two numbers

$$2100 \text{ lb}_f + 8 \text{ lb}_f = 2108 \text{ lb}_f$$

but can only keep the 100s place value since we do not know the weight of the car (without the bag) to any precision greater than 100s of pound-force.

Example 1.3   Multiplication of significant figures
You measure a sample mass to be 3 kg. Calculate the sample weight (multiply by gravity 9.81 m/s^2). What is your answer?

Solution
The weight of the sample is thought of as 30 N. You multiply the two numbers

$$(3 \text{ kg})(9.81 \text{ m/s}^2) = 29.43 \text{ N}$$

but can only keep one significant figure since our original mass was only accurate to a certain extent (i.e., 30 N).

1.7   Process and Process Variables

- Process flow sheet—it is a sequence of process units connected by process streams. It shows the flow of materials and energy through the process units (Figure 1.1).

- Process—a process is any operation or series of operations that causes physical or chemical changes in a substance or a mixture of substances.

- Process unit—a process unit is an apparatus, equipment in which one of the operations that constitute a process is carried out. Each process unit has associated with it a set of input and output "process streams," which consists of materials that enter and leave the unit.

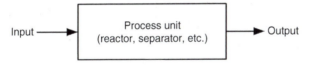

FIGURE 1.1
Overall process flow sheet of a process.

Process streams—a process stream is a line that represents the movement of material to or from process units. Typically, these streams are labeled with information regarding the amount, composition, temperature, and pressure of the components.

1.7.1 Density, Mass, and Volume

Density $(\rho) \equiv$ mass (m) per unit volume (V) of a substance; units are mass/length^3 (kg/m^3, g/cm^3, and lb$_m$/ft^3)

$$\rho = \frac{m}{V} = \frac{\text{mass}}{\text{volume}}$$

Density can be used as a conversion factor to relate the mass and volume of a substance. Densities of pure solids and liquids are essentially independent of pressure and vary slightly with temperature. For most compounds, density decreases with temperature (volume expansion). Water is an exception.

$$\rho\,(H_2O(\text{liquid})) = 0.999868 \text{ g/cm}^3 \text{ at } 0°C$$
$$= 1.00000 \text{ g/cm}^3 \text{ at } 4°C$$
$$= 0.95828 \text{ g/cm}^3 \text{ at } 100°C$$

Assumptions: Solids and liquids are incompressible, which means density is constant with change in pressure. Gases (vapors) are compressible, which means density changes as pressure changes.

Specific volume $(1/\rho) \equiv$ volume occupied by a unit mass of a substance; unit length^3/mass (reciprocal of density).

Specific gravity (SG) $\equiv$ ratio of the density (ρ) of a substance to the density (ρ_{ref}) of a reference substance at a specific condition

$$SG = \frac{\rho}{\rho_{ref}} = \frac{\text{density of substance}}{\text{density of a reference substance}} = [\,] = \text{dimensionless}$$

The reference commonly used for solids and liquids is water at 4°C, which has the following density: $\rho_{ref} = \rho_{H_2O}|_{@4°C} = 1.000 \text{ g/cm}^3 = 1000 \text{ kg/m}^3 = 62.43$ lbm/ft^3.

The following notation signifies that the specific gravity of a substance at 20°C with reference to water at 4°C is 0.6:

$$SG = 0.6 \frac{\rho_{sub} \; 20°C}{\rho_{ref} \; 4°C}$$

1.7.2 Flow Rate

Flow rate is the rate at which a material is transported through a process line (Figure 1.2).

$\dot{m}$ (kg fluid/s) or
$\dot{V}$ (m³ fluid/s)

FIGURE 1.2
Material transportation
through process line.

Mass flow rate $= \dot{m} =$ mass/time
Volume flow rate $= \dot{V} =$ volume/time

where the "dot" above the m and V refers to a flow rate, that is relatively to time.

The density of a fluid can be used to convert a known volumetric flow rate of a process stream to the mass flow rate of that particular stream or vice versa:

$$\text{Density} = \rho = \frac{m}{V} = \frac{\dot{m}}{\dot{V}}$$

1.7.3 Moles and Molecular Weight

Atomic weight is the mass of an atom of an element. *Mole* is the amount of a species whose mass in grams is numerically equal to its molecular weight; 1 mol of any species contains approximately 6.02×10^{23} (Avogadro's number) molecules of that species. *Molecular weight* is the sum of the atomic weights of the atoms that constitute a molecule of the compound (same as molar mass); units are of the form kg/kmol, g/mol, or lb/lb-mol. Molecular weight is the conversion factor that relates the mass and the number of moles of a quantity of a substance.

1.7.4 Mass Fraction and Mole Fraction

Process streams occasionally contain one substance, but more often they consist of mixtures of liquids or gases, or solutions of one or more solutes in a liquid solvent. The following terms may be used to define the composition of a mixture of substances, including a species A:

$$\text{Mass fraction: } x_A = \frac{\text{mass of A}}{\text{total mass}} \left(\frac{\text{kg A}}{\text{kg total}} \text{ or } \frac{\text{g A}}{\text{g total}} \text{ or } \frac{\text{lb}_m \text{ A}}{\text{lb}_m \text{ total}} \right)$$

$$\text{Mole fraction: } y_A = \frac{\text{moles of A}}{\text{total moles}} \left(\frac{\text{kmol A}}{\text{kmol}} \text{ or } \frac{\text{mol A}}{\text{mol}} \text{ or } \frac{\text{lb-mol A}}{\text{lb-mol}} \right)$$

The mass percent of A is $100x_A$, and the mole percent of A is $100y_A$. Mass fractions can be converted to mole fractions or vice versa by assuming a basis of calculation. Remember, fractions are unitless.

1.7.5 Concentration

The concentration of a component of a mixture or solution is the quantity of this component per unit volume of the mixture. Mass concentration is the mass of a component per unit volume of the mixture (g/cm^3, lb_m/ft^3, and $kg/in.^3$). Molar concentration is the number of moles of the component per unit volume of the mixture ($kmol/m^3$ and lb-mol/ft^3). Trace species (species present in minute amounts) in mixtures of gases or liquids are typically expressed in units of parts per million (ppm) or parts per billion (ppb). If y_i is the fraction of component i, then

$$ppm_i = y_i \times 10^6, \quad ppb_i = y_i \times 10^9$$

1.7.6 Pressure

Pressure is the ratio of a force to the area on which the force acts. Pressure units are force units divided by area units (e.g., N/m^2 or pascal (Pa), $dynes/cm^2$, and $lb_f/in.^2$ or psi). The pressure at the base of a vertical column of fluid of density (ρ) and height (h) is called the hydrostatic pressure, and is given by

$$P = P_0 + \rho g h$$

where
P_0 is the pressure exerted on the top of the column
g is the acceleration of gravity

The earth's atmosphere can be considered as a column of fluid with zero pressure at the top. The fluid pressure at the base of this column (e.g., at sea level) is the atmospheric pressure, P_{atm}.

$$P_{atm} = 760 \text{ mmHg} = 1.01325 \times 10^5 \text{ N/m}^2 \text{ (Pa)} = 1 \text{ atm}$$

The absolute pressure, P_{abs}, of a fluid is the pressure relative to a perfect vacuum ($P=0$). Many pressure measurement devices report the gauge pressure of the fluid, which is the pressure of the fluid relative to atmospheric pressure. A gauge pressure of 0 indicates that the absolute pressure of the fluid is equal to the atmospheric pressure. A relationship for converting between absolute and gauge pressure is

$$P_{absolute} = P_{gauge} + P_{atmospheric}$$

Pressure is defined as force per unit area. Pressure can be expressed as a relative pressure or absolute pressure. The units of pressure are

$$\text{SI unit: } \frac{N}{m^2} \text{ (or Pa)}$$

$$\text{CGS unit: } \frac{dynes}{cm^2}$$

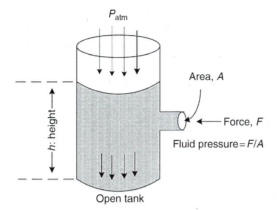

FIGURE 1.3
Fluid pressure on the base of a tank.

$$\text{AES unit: } \frac{\text{lb}_f}{\text{in.}^2}\ (\text{or psi})$$

Chemical engineers are most often interested in pressures that are caused by a fluid:

1. If a fluid is flowing through a horizontal pipe and a leak develops, a force must be applied over the area of the hole that causes the leak. This pressure is called the fluid pressure (the force applied must be divided by the area of the hole). This is schematically shown in Figure 1.3.
2. If a vertical container contains a fluid, the mass of the fluid will exert a force on the base of the container. This pressure is called the hydrostatic pressure. Hydrostatic pressure is the pressure caused by the mass of a fluid. This is shown schematically in Figure 1.4.

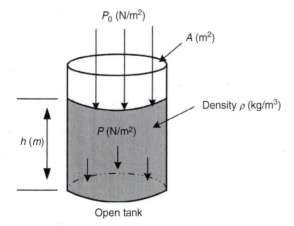

FIGURE 1.4
Pressure at the base of fluid column (hydrostatic pressure).

1.7.7 Types of Pressures

(a) Atmospheric pressure, P_{atm}, is the pressure caused by the weight of the earth's atmosphere. Often atmospheric pressure is called Barometric Pressure.

(b) Absolute pressure, P_{abs}, is the total pressure. An absolute pressure of 0.0 is a perfect vacuum. Absolute pressure must be used in all calculations unless a pressure difference is used.

(c) Gauge pressure, P_{gauge}, is pressure relative to atmospheric pressure.

(d) Vacuum pressure, P_{vac}, is a gauge pressure that is a pressure below atmospheric pressure. It is used so that a positive number can be reported.

Absolute pressure, gauge pressure, and the atmospheric pressure are related by the following expression:

$$P_{absolute} = P_{gauge} + P_{atmospheric}$$

Standard atmosphere is defined as the pressure equivalent to 760 mmHg at sea level and at 0°C. The unit for the standard pressure is the atmosphere (atm). Pressure equivalents to the standard atmosphere (atm) are

$$1 \text{ atm} = 760 \text{ mmHg} = 76 \text{ cmHg}$$

$$= 1.013 \times 10^5 \frac{N}{m^2} \text{ (or Pa)} = 101.3 \text{ kPa} = 1.013 \text{ bar}$$

$$= 14.696 \text{ psi} \left(\text{or } \frac{lb_f}{in.^2} \right) = 29.92 \text{ in Hg} = 33.91 \text{ ft } H_2O$$

The units psi and atm often carry a suffix "a" or "g" to indicate that the pressure is either absolute or gauge pressure. Thus, by psig, we mean gauge pressure in psi, and by psia, we mean absolute pressure in psi. So is the meaning of atma or atmg, if nothing is noted otherwise.

Example 1.4   Fundamentals of pressure

Consider the manometer in Figure E1.4. If $h = 10$ in. and the manometer fluid is mercury ($\rho = 13.6$ g/cm^3), then what is the gauge and absolute pressure?

Solution

Gauge pressure = absolute pressure – atmospheric pressure
$\rho = 13.6$ g/cm^3 (or 13.6 times greater than H_2O)

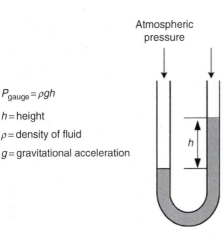

Atmospheric
pressure

$P_{gauge} = \rho g h$

h = height

ρ = density of fluid

g = gravitational acceleration

FIGURE E1.4
Pressure measurement using a manometer.

Then gauge pressure $= P_{gauge} = \rho \dfrac{g}{g_c} h$

$$= \left(13.6 \times \frac{62.4\,\text{lb}}{\text{ft}^3}\bigg|\frac{1\,\text{ft}^3}{(12)^3\,\text{in.}^3}\right)\left(\frac{32.174\,\text{ft}/\text{s}^2}{32.174\,\text{lb}\,\text{ft}/\text{lb}_f\,\text{s}^2}\right)(10\,\text{in.})$$

$$= 4.9\,\frac{\text{lb}_f}{\text{in.}^2} = 4.9\ \text{psig}$$

The absolute pressure $= P_{gauge} + P_{atm} = 4.9 + 14.7 = 19.6$ psia.

Example 1.5   Pressure measurements

Atmospheric pressure decreases with increased elevation above the surface of the earth. The temperature of the atmosphere also decreases with increasing elevation up to about 10 km where the temperature begins to increase again. The temperature decreases linearly with elevation up to 10 km given by the constant $C = 6.5$ K/km or $6.5°C/km$. The pressure at any elevation may be calculated by the following equation:

$$\frac{P}{P_0} = \left(1 - \frac{B}{T_0} z\right)^{\frac{Mg}{BR}}$$

where

P = pressure at elevation

P_0 = pressure at sea level

B = 6.5 K/km

T_0 = temperature at sea level

z = elevation above sea level

M = 29 g/mol = molecular weight of air

g = 9.80665 m/s² = gravitational constant

R = 8.314 J/mol K = gas constant

(a) Satisfy yourself that both the contents of the parenthesis and the exponent are unitless values.

(b) Calculate the pressure (in atm) at the top of Mount Everest (29,028 ft) using a sea level temperature of 20°C and average gravitational constant of 9.8066 m/s².

Solution
(a) Substitute the units as shown below

$$\frac{P}{P_0} = \left(1 - \frac{6.5\,\cancel{K}}{\cancel{km}(293.15\,\cancel{K})} 29028\,\cancel{ft}\, \frac{12\,\cancel{in.}}{\cancel{ft}}\, \frac{0.0254\,\cancel{m}}{\cancel{in.}}\, \frac{\cancel{km}}{10^3\,\cancel{m}}\right)^{\left\{\frac{29\,\frac{g}{mol}\,\frac{kg}{10^3\,g}\,9.8066\,\frac{m}{s^2}}{6.5\frac{K}{km}\,\frac{km}{10^3 m}\,8.314\,\frac{J}{mol\,K}\,\frac{kgm^2/s^2}{J}}\right\}}$$

(b) $\dfrac{P}{P_0} = 0.804^{5.2625}$, $\dfrac{P}{P_0} = 0.317$, $P_0 = 1$ atm $\Rightarrow P = 0.317$ atm

Example 1.6 Pressure gauge in a tank
A pressure gauge on a tank reads 20 psi. What is the gauge pressure?

Solution
The gauge reads the gauge pressure directly. Therefore, we must interpret the reading as gauge pressure and to avoid confusion, we should report the pressure as 20 psig instead of 20 psi.

Example 1.7 Absolute pressure
A gauge on a tank reads 15 psi. What is the absolute pressure in the tank?

Solution

$$P_{absolute} = P_{gauge} + P_{atm}$$

Since atmospheric pressure was not given, we must assume it is 14.69 psi. Thus, we find the absolute pressure in the tank as

$$P_{abs} = 14.69 + 15 = 30 \text{ psia}$$

Example 1.8 Absolute pressure from vacuum pressure
The pressure gauge on a tank reads 20 cmHg vacuum. What is the absolute pressure in the tank?

Solution
The gauge reads a vacuum gauge pressure directly. The pressure is below atmospheric, or is −20 cmHg relative to atmospheric pressure. Thus, the absolute pressure is

$$P_{abs} = P_{atm} + P_{gauge} = 76 + (-20) = 56 \text{ cmHg}$$

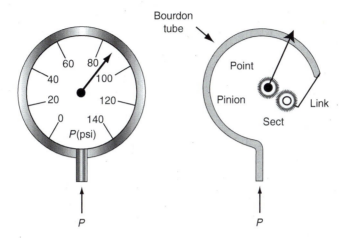

FIGURE 1.5
Bourdon gauge.

Types of pressure-sensing devices include the Bourdon gauge (Figure 1.5),
diaphragm capsule, capacitance sensor, column of fluid, manometer, barometer, silicon diaphragm, and semiconductor strain gauges, the latter used
for better response and increased sensitivity. A manometer is a U-shaped
device that uses a fluid having greater density than other fluids in the
process unit. Manometer operation is based on the fact that hydrostatic
pressure at the same level in the same fluid must be the same in each leg.
To understand how a manometer works, we must understand how to
determine the hydrostatic pressure caused by a mass of a column of fluid.

By definition of pressure

$$P = \frac{F}{A} = \frac{mg}{A} = \frac{\rho V g}{A} = \frac{\rho A h g}{A} = \rho g h$$

Whenever we need to determine the hydrostatic pressure caused by a mass
of fluid, it is simply

$$P = \rho g h$$

Example 1.9   Pressure caused by a column of fluid

What is the pressure (in kPa) caused by a 25.0 cm column of fluid (SG = 6.43)
at sea level and ambient conditions?

Solution

$$P = \rho g h$$

In SI units: $h = 25$ cm $= 0.250$ m; $\rho_{ref} = 1000 \dfrac{kg}{m^3}$; $g = 9.8 \dfrac{m}{s^2}$

$$P = \rho g h = (\rho_{ref} \times SG)gh$$

$$= \left(1000\ \frac{kg}{m^3}\right)(6.43)\left(9.8\ \frac{m}{s^2}\right)(0.250\ m)\left(\frac{1\,N}{1\frac{kg \cdot m}{s^2}}\right)\left(\frac{1\,kPa}{1000\ \frac{N}{m^2}}\right) = 15.8\ kPa$$

Example 1.10   Pressure caused by a column of mercury

What is the pressure (in psi) caused by a 6.34 ft column of mercury at ambient temperature where $g = 29.7\ \dfrac{ft}{s^2}$?

Solution

The specific gravity for Hg is 13.546.

The density of water (reference fluid): $\rho_{ref} = 62.4\ \dfrac{lb_m}{ft^3}$

$$P = \rho g h = \rho_{ref} \times (SG) \times g$$

$$= \left(62.4\ \frac{lb_m}{ft^3}\right)(13.546)\left(29.7\ \frac{ft}{s^2}\right)(6.34\,ft)\left(\frac{lb_f \cdot s^2}{32.174\,lb_m \cdot ft}\right)\left(\frac{1\,ft}{12\,in.}\right)^2 = 34.4\,psi$$

1.7.8   Manometers for Pressure and ΔP Measurement

Three different arrangements of manometers are shown in Figure 1.6. These manometers can be used to measure pressure using a column of a dense liquid.

- a. An open-end manometer can give the gauge pressure.
- b. A differential manometer gives ΔP between two points. Note that the pressure decreases in the direction of flow.
- c. A closed-end manometer gives absolute pressure. A manometer that has one leg sealed and the other leg open measures atmospheric pressure. It is called a barometer.

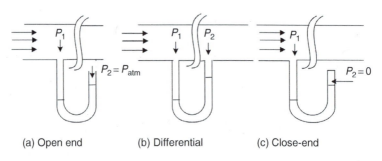

(a) Open end   (b) Differential   (c) Close-end

FIGURE 1.6
Arrangement of manometers.

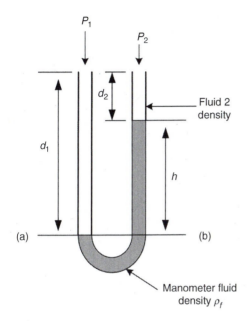

Fluid 2
density

Manometer fluid
density ρ_f

FIGURE 1.7
Manometer variables.

The basic variables that need to be considered in a manometer are the ones that measure pressure or pressure difference as shown in Figure 1.7. The line $a - b$ is at the interface between the manometer fluid and the higher-pressure fluid. The hydrostatic pressure on each leg is the same at that point. It becomes our reference point. We do a pressure balance by equating the pressures on each leg.

Applying the above allows us to develop the general manometer equation as

$$\sum \text{Pressure on leg } 1 = \sum \text{pressure on leg } 2$$

$$\text{i.e., } P_1 + \rho_1 g d_1 = P_2 + \rho_2 g d_2 + \rho_f g h$$

Example 1.11 Pressure drop across orifice
Determine the pressure drop across the orifice meter as shown in Figure E1.11.

Solution
This is a differential manometer. Note that the hydrostatic pressure above the 32 mark is the same on both sides: it cancels out. The reference line to select is at the 10 mark. The manometer equation is

$$P_1 + \rho_1 g (d + l) = P_2 + \rho_f g d + \rho_1 g l$$

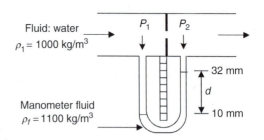

FIGURE E1.11
Pressure drop across orifice meter.

This is simplified to

$$P_1 + \rho_1 gd = P_2 + \rho_f gd$$

We now can write the pressure difference across the orifice from the above equation as

$$P_1 - P_2 = \Delta P = gd(\rho_f - \rho_1)$$

Substituting the appropriate known quantities, we get

$$\Delta P = \left(9.8\,\frac{m}{s^2}\right)(0.022\,m)(1100 - 1000)\frac{kg}{m^3}\left(\frac{1N}{1\frac{m \cdot kg}{s^2}}\right) = 21.6\frac{N}{m^2} = 21.6\,Pa$$

Manometers are simple, inexpensive, and accurate devices used to measure fluid pressure. Because of the low density of the gas, the gas pressure will be virtually uniform within the container. We see then that the differential height of the fluid in the column (Figure E1.12) is a direct measurement of the gauge pressure of the gas.

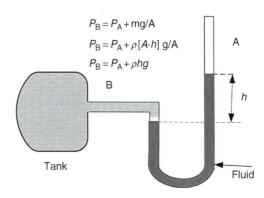

FIGURE E1.12
Pressure calculation using manometer.

Example 1.12   Pressure in manometer

Consider Figure E1.12. Assume that the manometer fluid is oil with a specific gravity of 0.87 and that we can read elevation differences of 0.001 m (1 mm).

Solution

$$P_{gauge} = \rho g h$$
$$P_{gauge} = (0.87 \times 1000 \, kg/m^3)(9.81 \, m/s^2)(0.001 \, m)$$
$$= 8.53 \, Pa$$
$$= 8.42 \times 10^{-5} \, atm$$
$$= 1.24 \times 10^{-3} \, psi$$

1.7.9   Temperature Measurement

Temperature is a measure of the average kinetic energy of a substance. We can measure temperature using physical properties of a substance that change as a function of temperature.

- Volume of a fluid (thermometer)
- Resistance of a metal (resistance thermometer)
- Voltage at the junction of two dissimilar metals (thermocouple)
- Spectra of emitted radiation (pyrometer)

Common temperature scales are the following:

- Fahrenheit: Freezing point of water is 32 and boiling point is 212.
- Celsius: Freezing point of water is 0 and boiling point is 100.
- Rankine: Absolute zero (when all kinetic energy vanishes) is 0, increments = Fahrenheit ($459.67°R = 0°F$).
- Kelvin: Absolute zero (when all kinetic energy vanishes) is 0, increments = celsius ($273.15 \, K = 0°C$).

1.7.10   Converting Temperatures

$$T \, (K) = T \, (°C) + 273.15$$
$$T \, (°R) = T \, (°F) + 459.67$$
$$T \, (°R) = 1.8T \, (K)$$
$$T \, (°F) = 1.8T \, (°C) + 32$$

Example 1.13   Temperature conversion
Perform the following temperature unit conversions:

(a) If temperature outside is 32°C, what will it be in units of °F, °R, and K?

(b) If the temperature outside drops from 72°F to −40°F, what will it be in units of °C, K, and °R?

(c) What is the temperature change (ΔT) in °F, °R, °C, and K?

Solution

(a) The temperature outside is 32°C. T the temperature in °F, °R, and K is calculated as

$$\circ F = 32°C\frac{1.8°F}{°C} + 32°F = 89.6°F = 90°F$$

$$\circ R = (90°F + 460°F)\frac{°R}{°F} = 550°R$$

$$K = (32°C + 273.15°C)\frac{K}{°C} = 305\ K$$

(b) The temperature outside drops from 72°F to −40°F. The temperature in °C, K, and °R is calculated as

$$\frac{(-40°F - 32°F)}{1.8\dfrac{°F}{°C}} = -40°C$$

$$(-40°C + 273.15°C)\frac{K}{°C} = 233\ K$$

$$(-40°F + 460°F)\frac{1°R}{1°F} = 420°R$$

(c) The temperature change (ΔT) in °F, °R, °C, and K is calculated as

$$\Delta T = -(72°F - (-40°F)) = -112°F$$

$$\Delta T = 112°F\frac{1°R}{1°F} = -112°R$$

$$\Delta T = 112°F\frac{1°C}{1.8°F} = -62°C$$

$$\Delta T = 112°F\frac{1\ K}{1.8°F} = -62\ K$$

1.7.11   Ideal Gas Law

The ideal gas law relates the molar quantity and volume of a gas to temperature and pressure. The law is derived from the kinetic theory of gases, which assumes that gas molecules have negligible volume, exert no forces on one another, and collide elastically with the walls of their container. The equation usually appears in the form

$$PV = nRT \text{ or } P\dot{V} = \dot{n}RT, \text{ valid only at low pressure}$$

where
P is absolute pressure of a gas
$V(\dot{V})$ is volume (volumetric flow rate) of the gas
$n(\dot{n})$ is number of moles (molar flow rate) of the gas
R is the gas constant, the value depending on units of $P, V, n,$ and T
T is absolute temperature of the gas

Alternatively, in the following form

$$P\hat{V} = RT$$

where $\hat{V} = V/n$ (the specific molar volume of the gas).

Example 1.14   Volumetric flow rate calculations

In Figure E1.14, nitrogen flows at a rate of 6 lb$_m$/min from a tank at $-350°F$ through a compressor. It is then heated in a heater. The gas leaves the heater at 150°F and 600 psia.

(a) Calculate the volumetric flow rate and the specific volume of gas leaving the heater. Assume ideal gas law.
(b) The volumetric flow rate at standard conditions.
(c) Weekly gas flow rate to the heater (SG = 0.81).

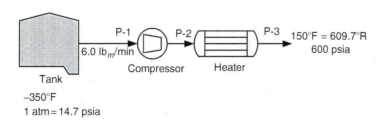

FIGURE E1.14
Flow sheet of nitrogen liquefactions.

Solution

(a) Volumetric flow rate of ($\dot{V}$) gas leaving the heater
N_2: MW 28 g/mol
28 lb_m/lb_m-mol

$$\dot{V} = \frac{\dot{n}RT}{P}$$

$$\dot{V} = \left(\frac{6\ lb_m}{min}\right)\left(\frac{lb\ mole}{28\ lb_m}\right)\left(\frac{609.7°R}{1}\right)\left(\frac{1}{600\ psia}\right)\left(\frac{10.73\ ft^3\ psia}{lb\ mole°R}\right) = 2.34\ \frac{ft^3}{min}$$

$$\hat{V} = \frac{2.34}{6/23} = 10.92\ ft^3/lb_m$$

(b) Flow rate in Standard Cubic Feet per Minutes (SCFM)
Standard temperature and pressure (STP) 492°R
1 atm = 14.7 psia

$$\dot{V} = \frac{\dot{n}RT}{P} = \left(\frac{6\ lb_m}{min}\right)\left(\frac{lb\text{-}mol}{28\ lb_m}\right)\left(\frac{10.73\ ft^3\ psia}{lb\text{-}mol\ °R}\right)\left(\frac{492°R}{14.7\ psia}\right) = 76.95\ \frac{ft^3}{min}$$

$$= 76.95\ SCFM$$

(c) Weekly gas flow rate

$$\left(\frac{6\ lb_m}{min}\right)\left(\frac{60\ min}{h}\right)\left(\frac{24\ h}{day}\right)\left(\frac{7\ day}{week}\right)\left(\frac{ft^3}{0.81(62.4\ lb_m)}\right) = 1196.6\ \frac{ft^3}{week}$$

1.7.12   Standard Temperature and Pressure

Standard temperature and pressure are used widely as a standard reference point for expression of the properties and processes of ideal gases.

$$\text{Standard temperature} = T_s = 0°C \Rightarrow 273\ K$$

$$\text{Standard pressure} = P_s = 1\ atm = 760\ mmHg = 101.3\ kPa$$

$$\text{Standard molar volume} = \hat{V}_s = 22.4\ \frac{m^3(STP)}{kmol} \Leftrightarrow 22.4\ \frac{L(STP)}{mol} \Leftrightarrow 359\ \frac{ft^3(STP)}{lb\text{-}mol}$$

One molecule of an ideal gas at 0°C and 1 atm occupies 22.4 L.

Example 1.15   Maintaining dimensional consistency

Suppose that it is known that the composition C varies with time, t, in the following manner:

$$C = 0.03\, e^{(-2.00t)}$$

where C [=] (has units of) kg/L and t [=] s.

(a) What are the units associated with 0.03 and 2.00?
(b) Now suppose that during a particular experiment the data are obtained in units of C [=] lb_m/ft^3 and t [=] h. Change the above equation so that the data can be directly applied for these units.

Solution

(a) Because the argument of an exponential function must be dimensionless, 2.00 t must be unitless. Therefore, 2.00 must have units of s^{-1}.

(b) What we will do is write a new expression for the variables C and t using the previous units and the new variables C and t using the new units. Thus,

$$t\,[\text{s}] = t'\left[\cancel{\text{h}}\right] \cdot \left|\frac{3600\,\boxed{\text{s}}}{\text{h}}\right| = 3600 \cdot t'$$

$$C\,[\text{kg/L}] = C'\left[\frac{\cancel{lb_m}}{ft^3}\right] \cdot \left[\frac{453.59\,\cancel{g}}{\cancel{lb_m}}\left|\frac{1\,\boxed{\text{kg}}}{1000\,\cancel{g}}\right|\frac{35.3145\,\cancel{ft^3}}{1000\,\boxed{\text{L}}}\right] = 0.0160 \cdot C'$$

So, in terms of the new variables we have

$$0.0160C' = 0.03\,[-(2.0)\,(3600)t']$$

Finally,

$$C'\left[\frac{\text{lbm}}{ft^3}\right] = 1.9e^{(-7200t'\,[\text{h}])}$$

Example 1.16   Unit conversion

Engineers use dimensionless groups. What are the units of h, knowing that $h/(C_pG)$ is unitless?

$$\left(\frac{h}{C_pG}\right)\left(\frac{C_p\mu}{k}\right)^{2/3} = \frac{0.23}{(DG/\mu)}$$

where

C_p [=] Btu/(lb$_m$ °F)
μ [=] lb$_m$/(h ft)
k [=] Btu/(h ft °F)
D [=] ft
G [=] lb$_m$/(h ft^2)

Solution

Note that you do not need to work through all of the equations to get the units of h. Since each individual group must be unitless or dimensionless, all that we need to do is work from a single group containing h. Thus,

$$h[=]C_pG[=]\left[\frac{\text{Btu}}{\text{lb}_m\cdot{}^\circ\text{F}}\right]\left[\frac{\text{lb}_m}{\text{h}\cdot\text{ft}^2}\right][=]\frac{\text{Btu}}{(\text{h}\cdot\text{ft}^2\cdot{}^\circ\text{F})}$$

Example 1.17 Dimensional analysis

Find the dependence of the diffusion coefficient on molecular size σ, intermolecular attraction ε, and molecular mass m of the particles.

Solution

Dimensional analysis is a very powerful tool to develop relationships between properties if the independent variables are known. For example, in this case, we know that the only independent variables are the molecular parameters (σ, ε, and m). We know that whatever correlation we use must be dimensionally consistent, and so what we do is a three-step process.

(1) First we write the dependent variable as a product of the independent variables raised to an unknown power.

(2) Next, we substitute in the units for each of the independent variables and collect terms of like units.

(3) Then we match the power of each separate unit on both sides of the equation in order to maintain dimensional consistency.

(4) Finally, we solve for the unknown powers. This is illustrated in the example.

Step 1: Write the dependent variable as a product of independent variables raised to unique powers.

$$D = D(\varepsilon, \sigma, m) \text{ or } D[=]\varepsilon^x\sigma^y m^z$$

Step 2: Substitute in all of the units.

$$\varepsilon[=]\text{ergs}, \sigma[=]\text{cm}, m[=]\text{g}, D[=]\text{cm}^2/\text{s}$$

$$D[=]\text{cm}^2/\text{s}[=]\varepsilon^x\sigma^y m^z[=](\text{g}\cdot\text{cm}^2/\text{s}^2)^x(\text{cm})^y(\text{g})^z[=]\text{g}^{x+z}\ \text{cm}^{2x+y}\ \text{s}^{-2x}$$

Step 3: Match powers on each unit on both sides of the equation.

$$\text{g: } x+z=0$$
$$\text{cm: } 2x+y=2$$
$$\text{s: } -2x=-1$$

Step 4: Solve for the unknown powers.

Solving for x, y, and z gives $x = 1/2$, $y = 1$, and $z = -1/2$.
Thus, $D [=] D [\sigma (\varepsilon/m)^{0.5}]$.

Without knowing any theory about the diffusion coefficient, we can see that D depends on the molecular diameter σ linearly. Thus, for molecules with twice the diameter, the diffusion coefficient will be twice as large. Similarly, all other things being equal, a molecule that is four times as massive will have half the value of D. These are not the only properties that will affect the diffusion coefficient, but if the others are all constant, this is the way the molecular properties σ, ε, and m would affect D.

1.8 Process Classification

Before writing a material balance you must first identify the type of process in question.

Batch: In batch processes no material is transferred into or out of the system over the period of interest (e.g., heating a sealed bottle of milk in a water bath).

Continuous: A material is transferred into and out of the system continuously (e.g., pumping liquid at a constant rate into a distillation column and removing the product streams from the top and bottom of the column).

Semibatch: Any process that is neither batch nor continuous (e.g., slowly blending two liquids in a tank).

Steady state: Process variables (i.e., T, P, V, flow rates) do not change with time.

Transient: Process variables that change with time.

1.9 Problems

1.9.1 Process Classification

Classify the following processes as batch, continuous, or semibatch, and transient or steady state.

(a) A balloon is being filled with air at a steady rate of 2 g/min. (Answer: Semibatch.)
(b) A bottle of milk is taken from the refrigerator and left on the kitchen table. (Answer: Batch, transient heat.)

(c) Carbon monoxide and steam are fed into a tubular reactor at a steady rate and react to form carbon dioxide and hydrogen. Products and unused reactants are withdrawn at the other end. The reactor contains air when the process is started up. The temperature of the reactor is also constant, and the composition and flow rate of the entering reactant stream are independent of time. Classify the process initially (answer: continuous, transient) and after a long period has elapsed (answer: continuous, steady state).

1.9.2 Types of Processes

Water ($\rho = 1000$ kg/m^3) enters a 2.00 m^3 tank at a rate of 6.00 kg/s and is withdrawn at a rate of 3.00 kg/s. The tank is initially half-full.

(a) Is the process continuous, batch or semibatch? (Answer: Continuous.)

(b) Is it transient or steady state? (Answer: Transient.)

(c) Use the general material balance equation to determine how long it will take the tank to overflow. (Answer: 1000 kg/(3 kg/s).)

1.9.3 Unit Conversion

How many seconds are in a year? (3600 s/h × 24 h/day × 365 days/yr)

1.9.4 Flow Rate through Horizontal Pipe

Natural gas is flowing through a horizontal pipe at a flow rate of 50 ft^3/s. What is this flow rate in m^3/s, gal/h? (Use conversion table.)

1.9.5 Molar Flow Rate

Propane gas (C_3H_8) flows to a furnace at a rate of 1450 m^3/h at 15°C and 150 kPa (gauge), where it is burned with 8% excess air. Calculate the molar flow rate (mol/s) of propane gas entering the furnace. (Hint: Assume ideal gas and use $P\dot{V} = \dot{n}RT$.)

1.9.6 Dimensional Homogeneity

A quantity k depends on the temperature T in the following manner:

$$k\left(\frac{\text{Mol}}{\text{cm}^3 \cdot \text{s}}\right) = A \ \exp\left(-\frac{Ea}{R\,T}\right)$$

The units of the quantity Ea are cal/mol, and T is in K (Kelvin). What are the units of A and R. (Hint: Note that the exponent is of unitless values; $Ea/RT = [\ ]$.)

1.9.7 Calculation of Mass for Specific Gravity and Volume

The specific gravity of gasoline is approximately 0.7. What is the mass (kg) of 10 L of gasoline? (Answer: 7 kg.)

1.9.8 Conversion of Equation to Other Units

The heat capacity of ammonia, $C_p = 2.5$ J/g·°C. Convert the expression for C_p to Btu/(lb$_m$·°F). (Hint: Note that the °C and °F here is the difference in temperature; $\frac{1°C}{1.8°F}$; $\frac{1 K}{1.8°R}$; $\frac{1°C}{1 K}$; $\frac{1°C}{1 K}$.)

Further Readings

1. Reklaitis, G.V. (1983) *Introduction to Material and Energy Balances*, John Wiley & Sons, New York.
2. Felder, R.M. and R.W. Rousseau (1999) *Elementary Principles of Chemical Processes*, 3rd edn, John Wiley, New York.
3. Himmelblau, D.M. (1974) *Basic Principles and Calculations in Chemical Engineering*, 3rd edn, Prentice-Hall, New Jersey.
4. Whirwell, J.C. and R.K. Toner (1969) *Conservation of Mass and Energy*, Blaisdell, Waltham, Massachusetts.
5. Cordier, J.-L., B.M. Butsch, B. Birou, and U. von Stockar (1987) The relationship between elemental composition and heat of combustion of microbial biornass. *Appl. Microbiol. Biotechnol. 25*, 305–312.
6. Atkinson, B. and F. Mavituna (1991) *Biochemical Engineering and Biotechnology Handbook*, 2nd edn, Macmillan, Basingstoke.
7. Scott Fogler, H. (1999) *Elements of Chemical Reaction Engineering*, 3rd edn, Prentice Hall Inc., New Jersey.

2

Process Units and Degree of Freedom Analysis

At the End of This Chapter You Should Be Able to

1. Draw and fully label a process flow sheet.
2. Perform a degree of freedom analysis (DFA).
3. Calculate the number of degree of freedom in a problem to ascertain that a unique solution exists for a problem using the given data.
4. Decide what equation to use if you have redundant equations.
5. Prepare a material flow diagram and translate the problem into a material balance.
6. Understand the function of most frequently used unit operations in chemical engineering processes.

2.1 Degree of Freedom Analysis

When attempting to solve a material balance problem, two questions that one may ask are

1. How many equations do I need?
2. Where do they come from?

A DFA is used to answer these two questions. DFA is a highly useful tool for systematic analysis of block flow diagrams. DFA provides a rapid means for determining if a specific problem is "solvable," that is, if the information available is sufficient, and provides a structured method for determining which equations to solve, and in which order to solve them. Briefly, one simply counts the number of independent variables and then the number of equations. To do the analysis draw a flow diagram, label each stream with the components that are present in that stream, and make a list of additional information such as known flow rates, compositions, ratios, and conversions. There are two main points here.

The first has to do with drawing "balance boundaries," that is, the number of systems where you can write the material balance equation.

There are three rules for drawing system boundaries:

   a. Draw a boundary around each process unit.
   b. Draw a boundary around junction points.
   c. Draw a boundary around the entire process (unless there is only one boundary).

The second point has to do with how many equations you can write for each drawn boundary. You can write as many equations as there are unique components passing through the boundary. However, understanding when you can write equations is important when you have to solve a system of equations or when you are dealing with reactions. Counting unknowns is simple; just look at your carefully drawn flowchart. As you should remember from algebra, the number of equations necessary is equal to the number of unknowns.

In drawing the flowchart, one must know the following to fully specify a stream:

   1. Total amount of the flow within the stream.
   2. Composition of the stream (label what you do not know with variables).

DFA = number of unknowns + number of independent reactions − number of independent material balance equations − number of useful auxiliary relations.

2.1.1 Possible Outcomes of the DFA

   a. DFA = 0. The system is completely defined. You get a unique solution.
   b. DFA > 0. The system is underdefined (underspecified). There are an infinite number of solutions.
   c. DFA < 0. The system is overdefined (overspecified). There are too many restrictions. Check if you wrote too many equations or have too many restrictions. Overdefined problems cannot be solved to be consistent with all equations.

In general, DFA can be determined using the following equations for molecular species balance, the extent of the reaction method, and atomic balance (more details on DFA for reactive processes will be discussed in Chapter 5).

Molecular species balances and extent of reaction:

$$\text{Number of degrees of freedom} = \left\{\begin{array}{l} \text{Number} \\ \text{of} \\ \text{unknowns} \end{array}\right\} + \left\{\begin{array}{l} \text{Number of} \\ \text{independent} \\ \text{chemical reactions} \end{array}\right\}$$

$$- \left\{\begin{array}{l} \text{Number of} \\ \text{independent molecular} \\ \text{species balances} \end{array}\right\}$$

$$- \left\{\begin{array}{l} \text{Number of other} \\ \text{equations relating} \\ \text{variables} \end{array}\right\} \qquad (2.1)$$

Degree of freedom equation for molecular species balances and extent of reaction technique are the same.

Atomic species balances:

$$\text{Number of degrees of freedom} = \left\{\begin{array}{l} \text{Number} \\ \text{of} \\ \text{unknowns} \end{array}\right\} - \left\{\begin{array}{l} \text{Number of} \\ \text{independent atomic} \\ \text{species balances} \end{array}\right\}$$

$$- \left\{\begin{array}{l} \text{Number of molecular} \\ \text{balances or independent} \\ \text{nonreactive species} \end{array}\right\}$$

$$- \left\{\begin{array}{l} \text{Number of other} \\ \text{equations relating} \\ \text{variables} \end{array}\right\} \qquad (2.2)$$

2.2 Sources of Equations

Sources of equations that relate unknown process variables include:

1. Material balances for a nonreactive process, usually but not always, the maximum number of independent equations that can be written equals the number of chemical species in the process.
2. Energy balances.
3. Process specifications given in the problem statement.
4. Physical properties and laws: e.g., density relation, gas law.
5. Physical constraints: mass or mole fractions must add to unity.
6. Stoichiometric relations: systems with chemical reactions.

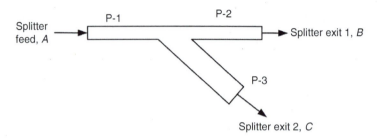

FIGURE 2.1
Splitter flow lines.

2.3 Process Units: Basic Functions

Typical operations involving the transfer of mass are illustrated in the following.

2.3.1 Divider/Splitter

The schematic diagram of a splitter is shown in Figure 2.1:

- The total mass balance is $A = B + C$ (see Figure 2.1).
- Composition of streams A, B, and C is the same.
- There is only one independent material balance since all compositions are equal.
- Mass flow rates of A, B, and C may be different.

2.3.2 Mixer (Blender)

The mixing process flow sheet is shown in Figure 2.2.
   The mixing process has the following characteristics:

- There are two or more entering streams (see Figure 2.2).
- There is only one exit stream, a "mixed" stream.
- The streams can be any phase, that is, gas, liquid, or solid.

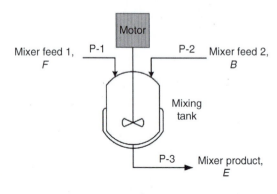

FIGURE 2.2
Flow sheet of a mixing process.

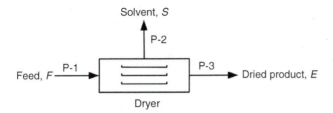

FIGURE 2.3
Flow sheet of a drying process.

2.3.3 Dryer (Direct Heating)

Drying is a mass transfer process resulting in the removal of water moisture or moisture from another solvent, by evaporation from a solid, semisolid, or liquid (hereafter product) to end in a solid state. To achieve this, there must be a source of heat, and a sink of the vapor is thus produced. The flow sheet of the drying process is shown in Figure 2.3.

Drying processes have the following characteristics:

- Solvent stream leaves as a pure vapor and is free of solids (see Figure 2.3).
- Exit dried solids are in the solid phase.
- Dried solids may not be solvent free.
- Feed can be solid, slurry, or solution.

2.3.4 Filter

Filtration is a technique used either to remove impurities from an organic solution or to isolate an organic solid (see Figure 2.4).

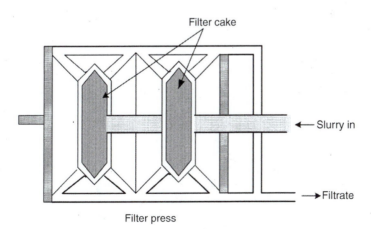

FIGURE 2.4
Schematic diagram of a filter press.

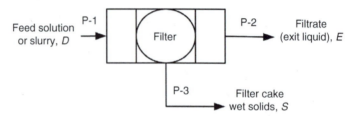

FIGURE 2.5
Flow sheet of a filtration process.

Filtration processes have the following characteristics:

- Filtrate, the exit liquid, is free of solids.
- Filtrate is saturated with soluble components.
- The filter cake leaves with some liquid attached (see Figure 2.5).
- Concentration of stream E and liquid attached to the filter cake is the same.

2.3.5 Distillation Column

Distillation is a method of separating chemical substances based on differences in their volatilities (see Figure 2.6). Distillation usually forms part of a larger chemical process, and is thus referred to as a unit operation.

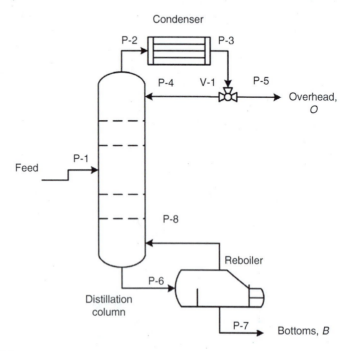

FIGURE 2.6
Schematic diagrams of a distillation column.

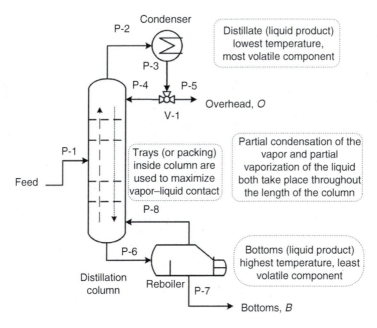

FIGURE 2.7
Distillation column separation mechanisms.

The distillation column (Figure 2.7) has the following characteristics:

- More volatile components are in the distillate.
- Less volatile components are in the bottom.
- Separation is accomplished by boiling.
- Perfect separation is not possible.

An industrial distillation column is shown in Figure 2.8a. Internal trays (or packing) are used to enhance component separations (Figure 2.8b). Each tray accomplishes a fraction of the separation task by transferring the more volatile species to the gas phase and the less volatile species to the liquid phase. Material and energy balances can be performed on an individual tray, the column, bottom reboiler, or top condenser, and the entire system.

2.3.6 Evaporator

Evaporation remains a basic process for the concentration of solutions in spite of significant competition from different baro-membrane processes. The processes of evaporation are used in the different branches of industry, from food to chemicals, in which evaporation is a widespread process. Theoretically, the multiple-effect evaporators (see Figure 2.9) allow decreased consumption of energy for a concentration almost proportionally equal to the number of effects. However, evaporators being expensive

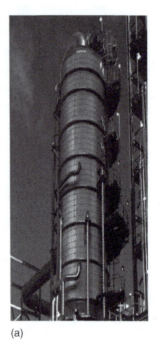
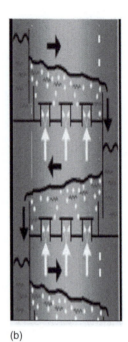

FIGURE 2.8
(a) Industrial distillation column and (b) inside the distillation column.   (a)   (b)

require the reduction in the number of evaporators. Hence, to determine optimal number of effects the exact calculation results are required. The specifications of an evaporator are similar to those of a dryer, except that both process streams are liquids for an evaporator (Figure 2.10).

2.3.7 Dehumidification

A dehumidifier with internal cooling and heating coils is shown in Figure 2.11. It is a device that reduces the level of humidity in air. The dehumidification

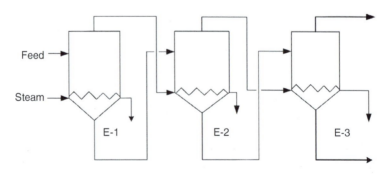

FIGURE 2.9
Multiple-effect evaporators.

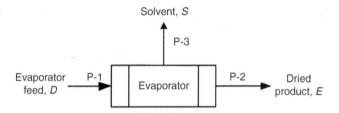

FIGURE 2.10
Evaporator process flow sheet.

process flowchart is shown in Figure 2.12. Dehumidification processes have the following characteristics:

- Feed stream contains a condensable component and a noncondensable component.
- Condensate is a liquid with the condensable component only.
- The "dry gas" exit stream is saturated with the condensable component at T and P of the process.

2.3.8   Humidifier

Humidifier is a device that increases the amount of moisture in indoor air or a stream of air. It operates by allowing water to evaporate from a pan or a wetted surface, or by circulating air through an air-washer compartment that contains moisture (Figure 2.13).

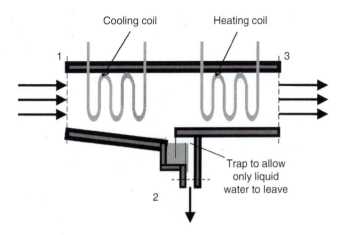

FIGURE 2.11
Dehumidifier process with internal cooling and heating coils.

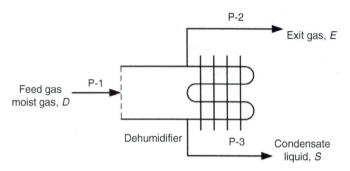

FIGURE 2.12
Dehumidifier process flow sheet.

Humidifier processes have the following characteristics:

- Feed gas is not saturated.
- Liquid is evaporated in the process unit.
- Vapor exit product may or may not be saturated.

2.3.9 Leaching and Extraction

Leaching is the removal of materials by dissolving them from solids. The chemical process industries use leaching but the process is usually called extraction. Leaching of toxic materials into groundwater is a major health concern (see Figure 2.14).

Leaching processes have the following characteristics:

- Two liquid solvents must be immiscible.
- They must have different specific gravity.

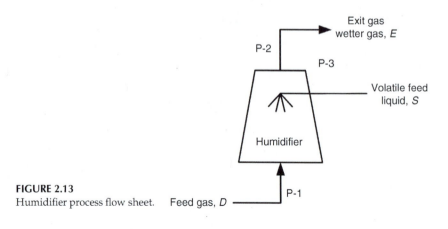

FIGURE 2.13
Humidifier process flow sheet.

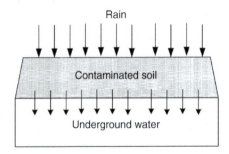

FIGURE 2.14
Leaching process.

- At least one component is transferred from one solvent to the other by difference in solubility.
- The process is often called liquid–liquid extraction.
- If one of the feed streams is a solid, the process is called leaching or liquid–solid extraction (see Figure 2.15).

2.3.10 Absorption (Gas Absorption) and Desorption

In gas absorption a soluble component is absorbed by contact with a liquid phase in which the component is soluble (Figure 2.16). This system is used for scrubbing gas streams of components such as sulfur dioxide, carbon dioxide, and ammonia. In this experiment, water, amines, and potassium carbonate can be used to remove CO_2 from air or natural gas.

Absorption processes have the following characteristics:

- Purpose of the unit is to have the liquid absorb a component from the feed gas. An absorber is often called a scrubber (Figure 2.17).
- Liquid stream flows down through the tower by gravity.

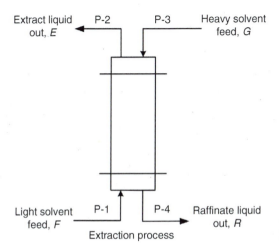

FIGURE 2.15
Flow sheet of an extraction process.

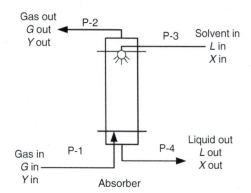

FIGURE 2.16
Flow sheet of an absorber column.

- Gas stream is pumped upward through the tower.
- No carrier gas is transferred to the liquid.
- Generally, no liquid solvent is transferred to the gas stream (check this assumption).
- Desorption is the same process as gas absorption except that the component transferred leaves the liquid phase and enters the gas phase. The absorber is sometimes called a stripper.

2.3.11 Partial Condenser

A partial condenser partly (not completely) condenses a vapor stream (Figure 2.18).

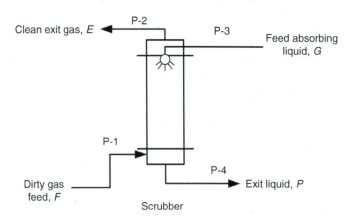

FIGURE 2.17
Flow sheet of a scrubber column.

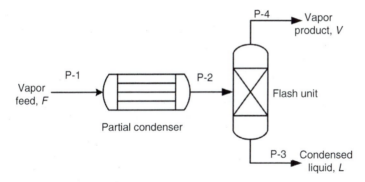

FIGURE 2.18
Flow sheet of a partial condenser with flash unit.

Partial condensers have the following characteristics:

- Feed stream contains only condensable vapor components.
- Exit streams, L and V, are in equilibrium.
- Condensation is caused by cooling or increasing pressure.

2.3.12 Flash Vaporizer and Flash Distillation

Flash vaporization/distillation splits a liquid feed into vapor and liquid-phase products (Figure 2.19).
Flash units have the following characteristics:

- Process flow sheet is the same as a partial condenser except the feed is a liquid.
- Vaporization is caused by reducing the pressure or by heating.
- Vapor and liquid streams are in equilibrium.

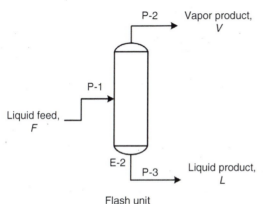

Flash unit

FIGURE 2.19
Flow sheet of a flash unit separator.

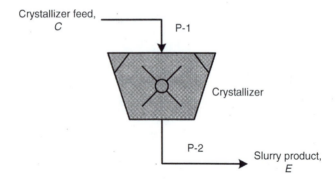

FIGURE 2.20
Flow sheet of a crystallization process.

2.3.13 Crystallizer

Solid crystals are formed in the unit by a change in temperature. The flow sheet for a crystallizer is a combination crystallizer–filter so as to separate solid crystals from solution (Figure 2.20).

2.3.14 Reactors (Chemical Reactor, Combustor, Furnace, and Reformer)

Figure 2.21 shows a typical reactor that has two reactant feed streams and a recycle stream.

In reactive processes,

- If a single reaction takes place, put the conversion in the box.
- A reactor is often named by the reaction taking place.
- A reactor is sometimes preceded by a fictitious mixer if the combined reactor feed is specified or must be determined.
- Multiple exit streams are shown to remind you to watch for exit streams that separate because of their different phases.

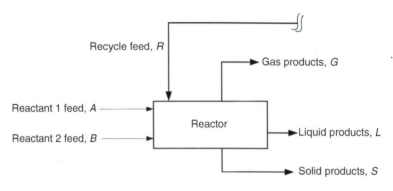

FIGURE 2.21
Flow sheet of a reactor.

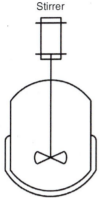

Stirrer

FIGURE 2.22

Batch reactor   Schematic diagram of a batch reactor.

A chemical reactor carries out a chemical reaction that converts molecular species in the input to different molecular species in the output where a chemical compound looses its identity. There are various types of reactors used in industry; the most common ones are batch reactor (Figure 2.22), plug flow reactor (PFR) (Figure 2.23a), packed bed reactor (PBR) (Figure 2.23b), continuous stirred tank reactor (CSTR), and fluidized bed reactor.

2.3.14.1 Batch Reactor

The batch reactor is shown in Figure 2.22.

2.3.14.1.1 Key Characteristics of Batch Reactor

The key characteristics of a batch reactor are unsteady-state operation (by definition), no spatial variation of concentration or temperature, that is, lumped parameter system, mainly used for small-scale operations, suitable for slow reactions, mainly (not exclusively) used for liquid-phase reactions, and large charge-in/cleanup times.

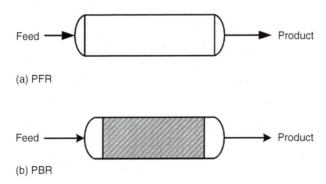

Feed → Product

(a) PFR

Feed → Product

(b) PBR

FIGURE 2.23

Schematic of (a) PFR and (b) PBR.

2.3.14.2   Plug Flow and Packed Bed Reactor

The PFR is shown in Figure 2.23a.

2.3.14.2.1   Key Characteristics of PFR

The key characteristics of a PFR are steady-state operation, spatial variation but no temporal variation, suitable for fast reactions and mainly used for gas-phase reactions, difficult temperature control, and no moving parts.

2.3.14.3   Continuous Stirred Tank Reactor and Fluidized Bed Reactor

Figure 2.24a shows the diagram of a CSTR where continuous inlet feed and continuous outlet product exist. Fluidized bed reactors are some times treated as stirred tank reactors (Figure 2.24b).

2.3.14.3.1   Key Characteristics of CSTR

The key characteristics of a CSTR are steady-state operation, use in series, good mixing leading to uniform concentration and temperature, used for liquid-phase reactions, and suitable for viscous liquids.

Example 2.1   Binary mixture separation process

100 lb_m/h of a mixture of 40% benzene and 60% toluene is separated completely. What are the flow rates of the product streams?

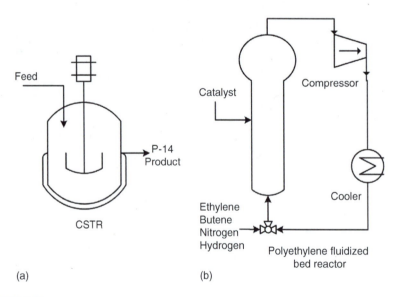

FIGURE 2.24
Schematic diagram of (a) CSTR and (b) fluidized bed reactor for.

Solution

Approach: Read the whole problem, draw a process flow sheet, label all streams, choose a basis, choose a system, make a table, write equations, using variables when needed (total balances, component balances), and solve.

a. After reading the whole problem carefully, the following process flow sheet is constructed (Figure E2.1):

Basis: 100 lb_m/h feed stream mass flow rate
System: overall unit
Total mass balance: $F = O + B$
Toluene balance: $0.6F = B$
Benzene balance: $0.4F = O$

Two of these three equations are independent material balance equations because the total mass balance equation can be generated by adding the two component balance equations as shown below:

$$0.6\,F = B$$
$$+0.4\,F = O$$
$$\overline{F = B + O}$$

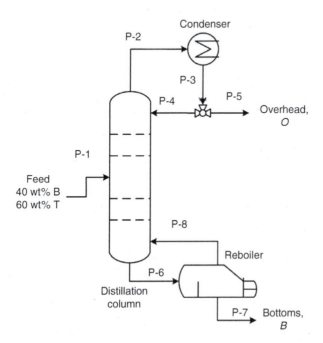

FIGURE E2.1
Block diagram of a distillation column.

The material balance can be summarized in the following table:

	Feed	Overhead	Bottom
Benzene	0.4 F	?	?
Toluene	0.6 F	?	?

Independent equations

A set of equations are independent if you cannot derive one by adding and subtracting combinations of the others.

Is this set of equations independent?

$$x + 2y + z = 1$$
$$2x + y - z = 2$$
$$y + 2z = 5$$

Solution

Yes, the above three equations are independent because we cannot derive one by adding and subtracting combinations of the others.

Is this set of equations independent?

$$x + 2y + z = 1$$
$$2y + 4z = 10$$
$$y + 2z = 5$$

Solution

The above set is not independent because we can derive equation 2 by multiplying equation 3 by a value of 2.

Is this set of equations independent?

$$x + 2y + z = 1$$
$$2x + y - z = 2$$
$$3x + 3y = 3$$

Solution

The above set is not independent because we can derive equation 3 by adding equations 1 and 2.

Example 2.2   Separation process

Consider the process flow diagram in Figure E2.2 and perform a DFA.

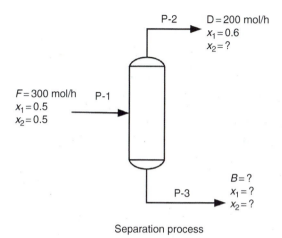

FIGURE E2.2
Flow sheet of a separation process.

Solution
a. The process flow sheet is shown in Figure E2.2.

Degree of freedom analysis
b. Number of unknowns (B, x_1): 2
c. Number of independent equations (two components 1, 2): 2
d. Number of relations: 0
e. $DF = 2 - 2 - 0 = 0$

Example 2.3 Binary component separation process
Consider the process flow diagram in Figure E2.3 and perform a DFA.

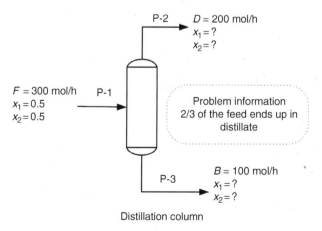

FIGURE E2.3
Flow sheet of a separation process.

Solution

    a. The process flow sheet is shown in Figure E2.3.

Degree of freedom analysis

    b. Number of unknowns (x_1 in D and x_1 in B): 2

    c. Number of independent equations (components 1, 2): 2

    d. Auxiliary relations: 0

    e. $DF = 2 - 2 = 0$

Discussion

The subsidiary relation is not considered because it is already allocated on the chart.

Example 2.4 Three component separation process

Consider the process flow diagram shown in Figure E2.4 and perform a DFA.

Solution

    a. The process flow sheet is shown in Figure E2.4.

Degree of freedom analysis

    b. Number of unknowns (F, n_{2F}, n_{1B}, n_{2B}): 4

    c. Number of independent equations (components 1, 2, 3): 3

    d. Auxiliary relations: 1

    e. $DF = 4 - 3 - 1 = 0$

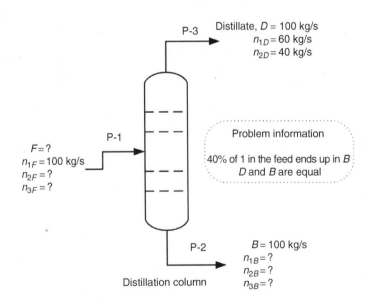

FIGURE E2.4

Flow sheet of a separation process.

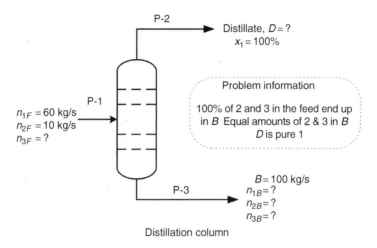

FIGURE E2.5
Flow sheet of a separation process.

Discussion
The first relationship is not useful because it is already known from the balance of n_{1f}, so it is not considered as a useful relation. The second relationship is useful, so it is considered as one relation.

Example 2.5   Tertiary component separation process
Consider the process flow diagram shown in Figure E2.5 and perform a DFA.

Solution
   a. The process flow sheet is shown in Figure E2.5.

Degree of freedom analysis
   b. Number of unknowns (n_{3F}, n_{1B}, n_{2B}, D): 4
   c. Number of independent equations (components 1, 2, 3): 3
   d. Auxiliary relations: 1
   e. $DF = 4 - 3 - 1 = 0$

Discussion
Three relationships are given but only one subsidiary relationship is useful. The first and the second relationships have one useful piece of information, so it is considered as one relationship. The third relationship is not useful because it is already specified in the chart as $100\% x_1$.

Example 2.6   Distillation column
Consider the process flow diagram in Figure E2.6 and perform a DFA.

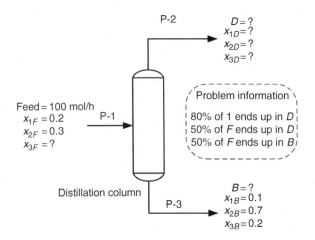

FIGURE E2.6
Flow sheet of a separation process.

Solution

a. The process flow sheet is shown in Figure E2.6.

Degree of freedom analysis

b. Number of unknowns (D, x_{1D}, x_{2D}, B): 4

c. Number of independent equations equals number of components: 3

d. Auxiliary relations (the first and second): 2

e. $DF = 4 - 3 - 2 = -1$

Discussion

This problem is overspecified since the number of pieces of information given is more than the number of unknowns. The first and second relationships are useful; by contrast, the third relationship is redundant, so it is not considered as a useful relationship.

Example 2.7   Binary component distillation column

An ethanol–methanol stream fed at the rate of 1000 kg/h is to be separated in a distillation column. The feed has 40% ethanol and the distillate has 90% methanol. The flow rate of the bottom product is 400 kg/h. Perform DFA.

Solution

a. The process flow sheet is shown in Figure E2.7.

Assumptions: Continuous process, steady state → accumulation = 0.
No chemical reaction → generation = consumption = 0.

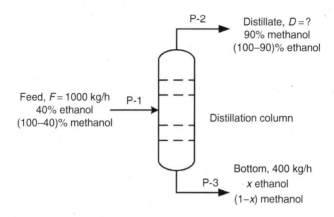

FIGURE E2.7
Flow sheet of a separation process.

Degree of freedom analysis
   b. Number of unknowns (D, x): 2
   c. Number of equations equals number of components: 2
   d. Number of relations: 0
   e. $DF = $ number of unknowns $-$ number of equations $= 2 - 2 = 0$

Example 2.8   Drying of solid material
Two hundred kilograms of wet leather is to be dried by heating and tumbling. The wet leather enters the drier with 1.5 g H_2O per gram BDL. The leather is to be dried to 20% moisture. (Note: BDL is bone dry leather— the leather component.) Perform a DFA.

Solution
   a. The process flow sheet is shown in Figure E2.8.

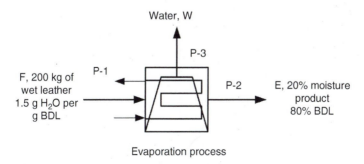

Evaporation process

FIGURE E2.8
Flow sheet of an evaporation process.

Degree of freedom analysis
    b. Number of unknowns (W, E): 2
    c. Number of independent equations equals number of components (water, BDL): 2
    d. Number of relations: 0
    e. $DF = 2 - 2 - 0 = 0$

Example 2.9   Binary component separation process

An ethanol–methanol stream fed at the rate of 1000 kg/h is to be separated in a distillation column. The feed has 40% ethanol and the distillate has 90% methanol. Eighty percent of the methanol is to be recovered as distillate. Perform a DFA.

Solution
    a. The process flow sheet is shown in Figure E2.9.

Degree of freedom analysis
    b. Number of unknowns (D, B, x): 3
    c. Number of independent equations equals number of components (ethanol, methanol): 2
    d. Number of relations: 1
    e. $DF = 3 - 2 - 1 = 0$

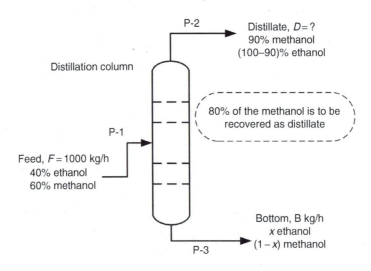

FIGURE E2.9
Flow sheet of a separation process.

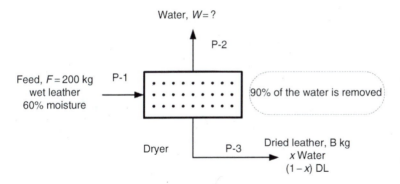

FIGURE E2.10
Flow sheet of a drying process.

Example 2.10   Drying process

Two hundred kilograms of wet leather is to be dried by heating and tumbling. The wet leather enters the drier with 60% moisture and 90% of the water is removed. Perform a DFA.

Solution

    a. The process flow sheet is shown in Figure E2.10.

Degree of freedom analysis

    b. Number of unknowns (W, B, x): 3

    c. Number of independent equations (water, dry leather [DL]): 2

    d. Number of relations: 1

    e. $DF = 3 - 2 - 1 = 0$

Example 2.11   Multicomponent separation column

An ethanol–methanol–propanol stream is fed at the rate of 1200 lb_m/h to a distillation column. The feed is 20% methanol. Ninety percent of the methanol is to be recovered in the distillate along with 60% of the ethanol. All of the 400 lb_m/h of propanol fed to the process must be sent to the bottom. Perform a DFA.

Solution

    a. The process flow sheet is shown in Figure E2.11.

Degree of freedom analysis

    b. Number of unknowns (D, Dm, De, B, Bm, Be): 6

    c. Number of independent equations (methanol, ethanol, propanol): 3

    d. Number of relations: 3

    e. $DF = 6 - 3 - 3 = 0$

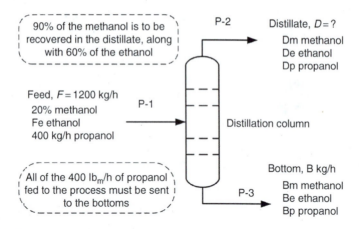

FIGURE E2.11
Flow sheet of a distillation process.

Example 2.12   Distillation column

An ethanol–methanol stream fed at the rate of 1000 kg/h is to be separated in a distillation column. The feed has 40 wt% ethanol and the distillate has 80% methanol by weight. Methanol (80 wt%) is to be recovered as distillate. Perform a DFA.

Solution

a. The process flow sheet is shown in Figure E2.12.

Remember, for a nonreactive system with N species, there are N independent equations. It is possible to formulate $N+1$ material balance equations, but only N of them will be independent.

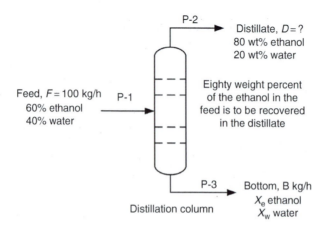

FIGURE E2.12
Flow sheet of a distillation column.

Degree of freedom analysis

    b. Number of unknowns (D, B, Xe): 3

    c. Number of independent equations (two components): 2

    d. Number of useful relations: 1

    e. $DF = 3 - 2 - 1 = 0$

Example 2.13   Mixing process

A stream of aqueous hydrochloric acid (HCl), 57.3 wt%, is mixed with water to produce a stream of 16.5% acid (Figure E2.13) and perform a DFA.

Solution

    a. The process flow sheet is shown in Figure E2.13.

Degree of freedom analysis

    b. Number of unknowns ($M1$, $M2$, $M3$): 3

    c. Number of independent equations (Water, HCl): 2

    d. Number of useful relations: 0

    e. $DF = 3 - 2 - 0 = 1$

Example 2.14   Absorber column

An absorber is used to remove acetone from a nitrogen carrier gas. The feed, with acetone weight fraction 0.213, enters at a rate of 200 kg/h. The absorbing liquid is water, which enters at a rate of 1000 kg/h. The exit gas stream is 0.8 wt% acetone and 2.9% water vapor. Perform a DFA.

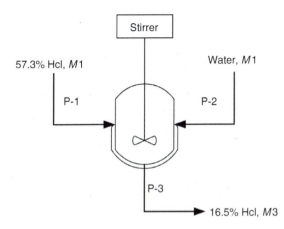

FIGURE E2.13
Flow sheet of a mixing process.

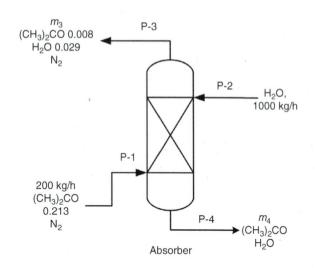

FIGURE E2.14
Flow sheet of an absorber column.

Solution

   a. The process flow diagram is shown in Figure E2.14.

Degree of freedom analysis

   b. Number of unknowns (m_3, m_4): 3

   c. Number of independent equations: 3

   d. Number of auxiliary relations: 0

   e. $DF = 3 - 3 - 0 = 0$

Example 2.15   Hydrocarbon mixtures

A hydrocarbon feed consisting of a mixture of 20 wt% propane, 30 wt% isobutane, 20 wt% isopentane, and 30 wt% n-pentane is fractionated at a rate of 1000 kmol/h into a distillate that contains all the propane and 78% of the isopentane in the feed. The mole fraction of isobutane in the distillate is 0.378. The bottom stream contains all the n-pentane fed to the unit.

Solution

   a. The process flow diagram is shown in Figure E2.15.

Degree of freedom analysis

   b. Number of unknowns (n_2, n_3, x_4, x_5, y_3): 5

   c. Number of independent equations: 4

   d. Number of auxiliary relations: 1

   e. $DF = 5 - 4 - 1 = 0$

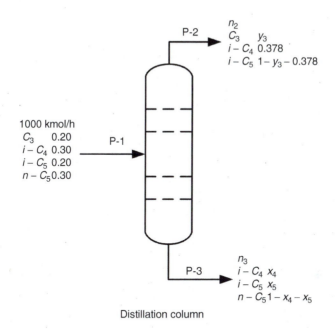

P-2

n_2
C_3 y_3
$i - C_4$ 0.378
$i - C_5$ $1 - y_3 - 0.378$

1000 kmol/h
C_3 0.20
$i - C_4$ 0.30
$i - C_5$ 0.20
$n - C_5$ 0.30

P-1

P-3

n_3
$i - C_4$ x_4
$i - C_5$ x_5
$n - C_5$ $1 - x_4 - x_5$

Distillation column

FIGURE E2.15
Flow sheet of an absorber column.

Example 2.16 Case Study

Triethanolamine (TEA) is used as a pH stabilizer as it is a weak base. It also has thickening and detergent properties. Over 1.2 billion pounds of TEA are produced each year in the United States. TEA is found in a number of cosmetic products, including shampoo, shaving cream, and skin moisturizers. TEA synthesis is performed in three reactions:

Reaction 1: $A + B \rightarrow R$
Reaction 2: $R + B \rightarrow S$
Reaction 3: $S + B \rightarrow T$

where
   A is ammonia
   B is ethylene oxide
   R is monoethanolamine
   S is diethanolamine
   T is TEA

You want to design a continuous system to manufacture TEA using the following information.

Your process will produce 100 lb-mol/h TEA. You have pure feeds of fresh ammonia and ethylene oxide available. Ammonia is the limiting reagent. The fresh ethylene oxide and fresh ammonia ratio is 5:1. Assume 100% conversion of the ammonia. Conversion of ethylene oxide is 50%. Ratio of R and S leaving the reactor is 1:3. Pure TEA is separated from the reactor effluent in a distillation column. Then, unreacted ethylene oxide is recovered from the intermediates R and S in a second distillation column. R and S are recycled, but ethylene oxide is not.

1. Draw a block diagram for this process. Name all pieces of equipment, number all streams, and indicate the components of the streams using the variables A, B, R, S, and T.

2. Perform a DFA to determine whether the entire process is specified and find the flow rates and compositions of every stream in the process. Explicitly state which specifications are flows, stream composition specifications, and system performance specifications? If the process is overspecified suggest a specification to relax it. If it is underspecified, suggest an additional independent specification. Do not solve for flows or compositions.

3. For each of the independent flow, stream composition specifications, and system performance specifications, write an equation relating the appropriate variables to the specified value. You do not have to write the material balance equation.

Solution

1. Block flow diagram of the process is shown in Figure E2.16.

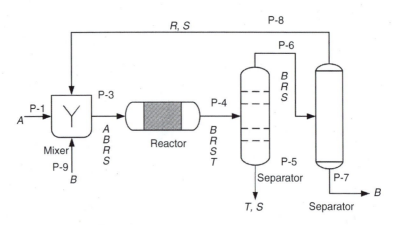

FIGURE E2.16
Flow sheet of a multiunit process flow diagram.

2. DFA of the entire process.

   Variable: Stream variables: 17, system variables (three reactions): 3, total variables $= 20$.

   Specified variables: Specified flows: 1 (100 lb-mol/h TEA)

   Specified stream compositions: 2 (5:1 B: A, 1:3 R: S)

   System performance: 1 (50% B conversion)

   Material balances: 4 (mixer) $+ 5$ (reactor) $+ 4$ (separator 1) $+ 3$ (separator 2) $= 16$

   Total specified variables $= 20$

3. Specified variables excluding material balance equations

   Flows: $\dot{n}_{TS} = 100$ lb-mol/h

   Stream component: $\dot{n}_{B2} = 5\ \dot{n}_{A1}$

   Ratio: $\frac{\dot{n}_{R4}}{\dot{n}_{S4}} = \frac{1}{3}$, system performance (conversion): $\frac{\dot{n}_{R3} - \dot{n}_{R4}}{\dot{n}_{R3}} = 0.5$

2.4 Summary of Degree of Freedom Analysis

There are two common situations where you will find fewer independent equations than species, and they are

1. Balance around a divider (splitter, Figure 2.25). Single input and two or more outputs with the same composition result in only one independent equation. Splitters are used for purge streams (reactor systems with recycle), total condensers at the top of distillation columns.

2. If two species are in the same ratio to each other wherever they appear in a process and this ratio is incorporated in the flowchart labeling, balances on those species will not be independent

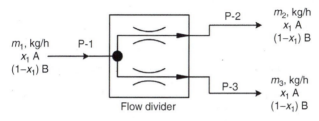

Flow divider

Inlet and exit composition are same but
mass flow in is different than mass flow out

FIGURE 2.25
Divider process flow sheet.

equations. This situation occurs frequently when air is present in a nonreactive process (21 mol% O_2; 79 mol% N_2), for example, vaporization of liquid carbon tetrachloride into an air stream.

2.5 Problems

2.5.1 Absorption of Acetone from Air

Air contains 3% acetone, and 2% water is fed to an absorber column. The mass flow rate of air is 1000 kg/h. Pure water is used as absorbent to absorb acetone from air. The air leaving the absorber should be free of acetone. The air leaving the absorber was found to contain 0.5% water. The bottom of the absorber is send to a distillation column to separate acetone from water. The bottom of the distillation column was found to contain 4% acetone and the balance is water. The vapor from the head of the absorber was condensed. The concentration of the condensate was 99% acetone and the balance water. Draw and label a process flowchart and perform a DFA.

2.5.2 Separation of Liquid Mixture

A liquid mixture containing 35.0 mol% toluene (T), 27.0 mol% xylene (X), and the remainder benzene (B) is fed to a distillation column. The bottom product contains 97.0 mol% X and no B, and 93.0% of the X in the feed is recovered in this stream. The overhead product is fed to a second column. The overhead product from the second column contains 5.0 mol% T and no X, and 96.0% of the benzene fed to the system is recovered in this stream.

   a. Draw and completely label a flow sheet of the process.
   b. Perform a DFA.

2.5.3 Absorber–Stripper Process

An absorber–stripper system is used to remove carbon dioxide and hydrogen sulfide from a feed consisting of 30% CO_2 and 10% H_2S in nitrogen. In the absorber a solvent selectively absorbs the hydrogen sulfide and carbon dioxide. The absorber overhead contains only 1% CO_2 and no H_2S. N_2 is insoluble in the solvent. The rich solvent stream leaving the absorber is flashed, and the overhead stream consists of 20% solvent, and contains 25% of the CO_2 and 15% of the H_2S in the raw feed to the absorber. The liquid stream leaving the flash unit is split into equal portions, one being returned to the absorber. The other portion, which contains 5% CO_2, is fed to the stripper. The liquid stream leaving the stripper consists of pure solvent and is returned to the absorber along with makeup solvent. The stripper over-head contains 30% solvent.

    a. Draw and completely label a flow sheet of the process.

    b. Perform a DFA.

2.5.4 Filtration Processes

From the following process description draw a flowchart that concisely summarizes each stage of the process (properly labeling all streams and species within those streams). An important, naturally occurring chemical, A, is to be removed from its ore (composed of A and various insoluble other junk). One hundred kilogram per hour of ore is fed to a dissolution tank where it is mixed with a stream of pure water, W, and a recycle stream. The tank is heated to 90°C so that all of the A (but none of the other junk, J) dissolves, forming a saturated solution. The material exits the tank and is sent to a filter (separator) where all of the junk and a small portion of the (still saturated) solution are removed. Finally, the remaining solution is fed to a crystallizer where it is cooled to 25°C in order to form some solid A's, which is shipped off to be packaged and sold (some filtrate also leaves with the solid). The remaining filtrate, now saturated at 25°C, is recycled to the dissolution tank. Upon analyzing the products from the filter and crystal-lizer, it is determined that the waste stream from the filter contains 40.0 kg/h of junk (J), 12 kg/h of A, and 4 kg/h of water (W) and that the final product stream from the crystallizer contains 42 kg/h A's, 6 kg/h A, and 6 kg/h W.

2.5.5 Evaporation Processes

Fresh feed containing 20% by weight KNO_3 (K) in H_2O (W) is combined with a recycle stream and fed to an evaporator. The concentrated solution leaving the evaporator, containing 50% KNO_3, is fed to a crystallizer. The crystals obtained from the crystallizer are 96% KNO_3 and 4% water. The supernatant liquid from the crystallizer constitutes the recycle stream and contains 0.6 kg KNO_3 per 1.0 kg of H_2O. Draw and label the process flowchart and perform a DFA.

Further Readings

1. Reklaitis, G.V. (1983) *Introduction to Material and Energy Balances*, John Wiley & Sons, New York.
2. Felder, R.M. and R.W. Rousseau (1999) *Elementary Principles of Chemical Processes*, 3rd edn, John Wiley, New York.
3. Himmelblau, D.M. (1974) *Basic Principles and Calculations in Chemical Engineering*, 3rd edn, Prentice-Hall, New Jersey.

4. Whirwell, J.C. and R.K. Toner (1969) *Conservation of Mass and Energy*, Blaisdell, Waltham, Massachusetts.
5. Cordier, J.-L., B.M. Butsch, B. Birou, and U. von Stockar (1987) The relationship between elemental composition and heat of combustion of microbial biomass. *Appl. Microbiol. Biotechnol. 25,* 305–312.
6. Atkinson, B. and F. Mavituna (1991) *Biochemical Engineering and Biotechnology Handbook*, 2nd edn, Macmillan, Basingstoke.
7. Scott Fogler, H. (1999) *Elements of Chemical Reaction Engineering*, 3rd edn, Prentice Hall Inc., New Jersey.

3

Material Balance in Single-Unit Processes

At the End of This Chapter You Should Be Able to

1. Recognize each term in the general balance equation.
2. Recognize process unit basic functions.
3. Explain the difference between an open and closed system, and batch, semibatch, and continuous processes.
4. Draw and label a process flowchart.
5. Choose a suitable basis for calculation.
6. Define a system and draw its boundaries.
7. Solve steady-state material balance problems in a single process unit with no chemical reactions (i.e., accumulation = generation = consumption = 0.0).

A material balance is simply an accounting of material. The term "balance" implies that, under steady-state operation and in the absence of generation/consumption processes, material flow rates entering and leaving the process must be equal, even if the process undergoes heating, mixing, drying, fermentation, or any other operation within the system (except nuclear reaction). Usually it is not feasible to measure the masses and compositions of all streams entering and leaving a system. Unknown quantities can be calculated using mass balance principles. Mass balance problems have a standard theme: Given the mass flow rates of input and output streams, calculate the mass flow rates of other streams. Because mass in biological systems is conserved at all times, the law of conservation of mass provides the theoretical framework for material balances.

Mass balances provide a powerful tool in chemical engineering analysis. Many complex situations are simplified by looking at the movement of mass and equating what goes into and what comes out of the system. Questions such as "what is the concentration of carbon dioxide in the fermenter off-gas?," "what fraction of the substrate consumed is not converted into products?," "how much reactant is needed to produce x grams of a product?," and "how much oxygen must be provided for this fermentation to proceed?" can be answered using mass balances. This chapter explains how the law of conservation of mass is applied to atoms, molecular species, and total mass, and describes formal techniques for solving material balance problems without reaction. For a given system, a material balance can be written in terms of the following conserved quantities:

1. Total mass (or moles).
2. Mass (or moles) of a chemical compound.
3. Mass (or moles) of an atomic species.

3.1 General Material Balance Equation

To apply material balance, you need to define the system and the quantities of interest (see Figure 3.1). A system is a region of space defined by a real or imaginary closed envelop (envelop = system boundary), and may be a single process unit, collection of process units, or an entire process.

$$\text{Accumulation} = \text{in} - \text{out} + \text{generation} - \text{consumption}$$

The general material balance equation:

$$\left\{ \begin{array}{c} \text{Accumulation} \\ \text{within the} \\ \text{system (buildup)} \end{array} \right\} = \left\{ \begin{array}{c} \text{input through} \\ \text{system} \\ \text{boundary} \end{array} \right\} - \left\{ \begin{array}{c} \text{output through} \\ \text{system} \\ \text{boundary} \end{array} \right\}$$

$$+ \left\{ \begin{array}{c} \text{generation} \\ \text{within the} \\ \text{system} \end{array} \right\} - \left\{ \begin{array}{c} \text{consumption} \\ \text{within the} \\ \text{system} \end{array} \right\}$$

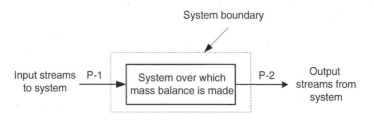

FIGURE 3.1
Process system.

3.1.1 Material Balance Simplifications

The following rules may be used to simplify the general material balance equation:

$$\text{Accumulation} = (\text{in} - \text{out}) + (\text{generation} - \text{consumption})$$

- If the system is at steady state, set accumulation $= 0$.
- If the balanced quantity is total mass, set generation $= 0$ and consumption $= 0$ (law of conservation of total mass).
- If the balanced substance is a nonreactive species (neither a reactant nor a product), set generation $= 0$ and consumption $= 0$.

3.2 Flowcharts

A flowchart is a convenient way of organizing process information for subsequent calculations. To obtain maximum benefit from the flowchart in material balance calculations, one must

1. Write the values and units of all known stream variables at the locations of the streams on the chart.
2. Assign algebraic symbols to unknown stream variables and write these variable names and their associated units on the chart.

3.2.1 Note on Notation

The use of consistent notation is generally advantageous. The following notation is used. For example, m (mass), $\dot{m}$ (mass flow rate), n (moles), $\dot{n}$ (mole flow rate), V (volume), $\dot{V}$ (volumetric flow rate), x (component fractions [mass or mole] in liquid streams), and y (component fractions in gas streams).

3.3 Problems Involving Material Balances on a Single Unit

Procedures will be outlined for solving single-unit processes where there are no reactions (consumption $=$ generation $= 0$), and when processes are continuous and steady state (accumulation $= 0$). These procedures will form the foundation for more complex problems involving multiple units and processes with reactions described in later chapters. For a stream to be fully specified, the flow rate and the composition of each component should be known. If any of these items are not given, then it will be considered as unknown.

Example 3.1   Distillation

Consider the process flow sheet in Figure E3.1; write a set of material balance equations.

Solution

The process-labeled flowchart is shown in Figure E3.1.

System: Distillation

We can write four equations in total, but only three are independent, because the fourth one is the sum of the other three. So, in doing the math, we have our choice on which of the three equations we want to use. Component balance ($\dot{m}$: mass flow rate):

$$A: \quad \dot{m}_{1,A} = \dot{m}_{2,A} + \dot{m}_{3,A}$$
$$B: \quad \dot{m}_{1,B} = \dot{m}_{2,B} + \dot{m}_{3,B}$$
$$C: \quad \dot{m}_{1,C} = \dot{m}_{2,C} + \dot{m}_{3,C}$$
$$\dot{m}_{3,A} = 0 \ (\text{no A exists in stream 3})$$

Total mass balance: $\dot{m}_{1,\text{Total}} = \dot{m}_{2,\text{Total}} + \dot{m}_{3,\text{Total}}$

Example 3.2   Ethanol–water separation process

A mixture containing 10% ethanol (E) and 90% H_2O (W) by weight is fed into a distillation column at the rate of 1000 kg/h. The distillate contains 60% ethanol and the distillate is produced at a rate of one tenth of that of the

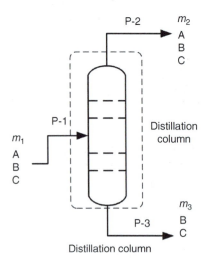

FIGURE E3.1

Process flow sheet of a distillation column.

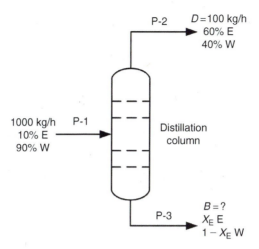

FIGURE E3.2
Flowchart of ethanol–water separation process.

feed. Draw and label a flowchart of the process. Calculate all unknown stream flow rates and composition.

Solution

The process-labeled flowchart is shown in Figure E3.2.
Basis: 1000 kg/h of feed
Assumptions: continuous process, steady state, no reactions

$$DF = \text{number of unknowns} - \text{number of independent equations}$$
$$- \text{number of relations}$$
$$= 2 - 2 - 0 = 0 \text{ (solvable)}$$

D is one-tenth of the feed $= 0.1 \, (1000) = 100 \, \text{kg/h}$
Total mass balance: $1000 \, \text{kg/h} = B + 100 \, \text{kg/h}$, $B = 900 \, \text{kg/h}$
Component balance (ethanol): $100 \, \text{kg/h} = 60 \, \text{kg/h} + (x_E)(900 \, \text{kg/h})$
$$\Rightarrow x_E = 0.044$$
Check water: $900 \, \text{kg} = 40 \, \text{kg} + (0.956)(900 \, \text{kg}) = 900 \, \text{kg}$

Example 3.3 Separation process

Calculate all unknown stream variables shown in Figure E3.3.

Solution

The process-labeled flowchart is shown in Figure E3.3.
Basis: 100 kg/min of the feed stream

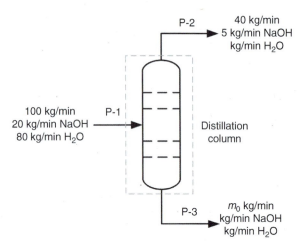

FIGURE E3.3
Flow sheet of a separation process.

The degree of freedom (DF) = 2 (unknown) − 2 (independent equations) = 0 (solvable).

Total mass balance: 100 kg/min = 40 kg/min + $\dot{m}_0$

Component mass balance (NaOH) = 20 kg/min = 5 kg/min + $\dot{m}_1$

Answer: $\dot{m}_0$ = 60 kg/min, $\dot{m}_1$ = 15 kg/min NaOH in the bottom stream

$\dot{m}_2$ = 60 − 15 = 45 kg/min H_2O in the bottom product stream

$\dot{m}_3$ = 40 − 5 = 35 kg/min H_2O in the top stream

3.4 Material Balance Fundamentals

Material balances (mass balances) are based on the fundamental law of conservation of mass (not volume). In particular, chemical engineers are concerned with writing mass balances around chemical processes. Writing a mass balance is similar in principle to accounting. In accounting, accountants do balances of what happens to a company's cash flow. Chemical engineers write mass balances to account for what happens to each of the chemicals that is used in a chemical process. Thus far, we have learned about the process variables that we need to describe the chemicals in a process stream. Now, we must learn how to specify a process stream, specify a process unit, do a mass balance on a single process unit, and do a mass balance on a sequence of process units.

3.4.1 Classification of Processes

3.4.1.1 *Based on How the Process Varies with Time*

3.4.1.1.1 *Steady-State Processes*

A steady state is one that does not change with time. Every time we take a snapshot of the process, all the variables have the same values as they did when measured the first time.

3.4.1.1.2 *Unsteady-State (Transient) Processes*

An unsteady-state process is a process that changes with time. Every time we take a snapshot, many of the variables have different values than in the first time.

3.4.1.2 *Based on How the Process Was Designed to Operate*

3.4.1.2.1 *Continuous Processes*

A continuous process is a process that has the feed streams and product streams moving chemicals into and out of the process all the time. At every instant, the process is fed and the product is produced. Examples include oil refineries and natural gas processing and polyethylene plants.

3.4.1.2.2 *Batch Processes*

A batch process is a process where the feed streams are fed to the process to get it started. The feed material is then processed through various process steps and the finished products are created during one or more of the steps. The process is fed and products result only at specific times. Examples include making a batch of a product, like soup, polyamide synthesis processes, polystyrene synthesis processes, or manufacturing a specialty chemical.

3.4.1.2.3 *Semibatch Processes*

A semibatch process is one that has some characteristics of continuous and batch processes. Some chemicals in the process are handled batch wise. Some chemicals are processed continuously.

3.4.2 Types of Balances

1. Differential balance is a balance taken at a specific point in time. It is generally applied to a continuous process. If the process is at steady state, a differential balance applied at any time gives the same result. We will apply differential balances to steady-state continuous processes. Each term in a differential balance represents a process stream and the mass flow rate of the chemicals in that particular stream. Differential balances are written in terms of

rates of change of the specified quantity with respect to time. For example, an employee is paid at a rate of $10.0/h. Accumulation is a differential term. This type of balance is typically applied to continuous steady-state processes. This type will be used extensively in this book.

2. Integral balance is a balance taken at two specific points in time. It describes what has happened over the period between the two time points. An integral balance is generally applied to the beginning and end of a batch process. It accounts for what happens to the batch of chemicals. We will apply integral balances to batch processes. Each term in an integral balance represents a process stream and the mass of chemicals in that stream. Integral balances are written in terms of the amounts of a specified quantity over a period. This type of balance is typically applied to batch processes. For example, a water storage tank at the start contains 5 L of water; after 30 min of water flowing to the tank, it is found to contain 50 L of water. In this case, the accumulated water after 30 min is 45 L of water.

$$\text{Accumulation} = (\text{final output} - \text{initial input}) = 50 - 5 = 45 \text{ L}$$

3.4.3 Stream Specifications

Streams are specified as suggested below:

1. If the stream composition is unknown (or if some of the component masses are known), represent the component masses directly and use a lower case letter for each chemical. For example, if stream F contains chemicals a, b, and c, label the flow rates as

$$F, a_F, b_F$$

and

$$c_F = F - a_F - b_F$$

2. If the stream composition is known from fractional compositions, represent the component masses directly and label them.

3. If the stream composition is partially known with fractional compositions and the total is known, represent the component masses indirectly and use lowercase x, y, and z for each fractional composition.

All material balance calculations are variations of a single theme: Given values of some input and output stream variables, derive and solve equations for the others. Solving the equations is a matter of simple algebra.

Example 3.4 Stream specification

Stream F contains 100 kg of O_2 and 700 kg of CH_4. Label the stream.

Solution

Note that the component masses must be added and it should be equal to the total (i.e., F). The total mass in F is 800 kg. Thus,

$$\text{Stream } F, F = 800 \text{ kg}$$

$$m_{O_2} = 100 \text{ kg}$$
$$m_{CH_4} = 700 \text{ kg}$$

Example 3.5 Labeling a stream

One thousand kilograms of a mixture of O_2, N_2, and CH_4 is fed to a process. The stream has 20% O_2 by mass. Label the stream (note: the mass of component i in the stream is equal to $F \times x_i$).

Solution

The composition is partially known. Note that the fractional compositions must add to 1; thus, we can write two alternatives. Using fractional composition

$$\text{Stream } F, F = 1000 \text{ kg}$$

$$x_{O_2} = 0.2$$
$$x_{N_2} = ?$$
$$x_{CH_4} = 1.0 - 0.2 - x_{N_2} = 0.8 - x_{N_2}$$

or using component masses

$$\text{Stream } F, F = 1000 \text{ kg}$$

$$m_{O_2} = 200 \text{ kg}$$
$$m_{N_2} = ?$$
$$m_{CH_4} = 1000 - 200 - m_{N_2} = 800 - m_{N_2}$$

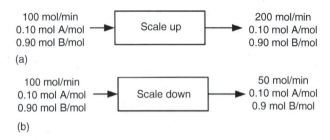

FIGURE 3.2
Scaling a process by a factor of 2. (a) Scaling up and (b) scaling down.

3.5 Scaling

Scaling—changing values of all stream amounts or flow rates of a certain process by a proportional amount while leaving the stream compositions unchanged.

Scaling up—the final stream quantities are larger than the original quantities.

Scaling down—the final stream quantities are smaller than the original quantities.

Figure 3.2 shows scaling up a feed stream by a factor of 2. The molar flow rates of *A* and *B* are doubled in the exit streams. By contrast, the composition will not be affected by scaling up or down.

3.6 Basis for Calculation

3.6.1 Concept

The amount or flow rate of one of the process streams can be used as a basis for calculation. It is recommended to consider that

1. If a stream amount or flow rate is given in the problem statement, use this as the basis for calculation.
2. If no stream amounts or flow rates are known, assume one, preferably a stream of known composition.
3. If mass fractions are known, choose the total mass or mass flow rate of that stream (e.g., 100 kg or 100 kg/h) as the basis.
4. If mole fractions are known, choose the total number of moles or the molar flow rate.

3.6.2 Method for Solving Material Balance Problems

1. Read the problem carefully and express what the problem statement asks you to determine. Analyze the information given in the problem.
2. Draw and fully label a process flowchart with all known and unknown process variables including their units.
3. Choose a basis for calculation.
4. Select a system and draw its boundaries.
5. State your assumptions (i.e., steady state, ideal gas, etc.).
6. Determine the number of unknowns and the number of equations that can be written to relate to them. That is, does the number of equations equal the number of unknowns?
7. Solve the equations.
8. Calculate the quantities requested in the problem statement if not already calculated. Check your solution and does it make sense?

Example 3.6 General material balance equation

1. Write the material balance equation in words, and briefly define each term, using the word "system" in your definitions.
2. List the four kinds of process units found on a block flow diagram, and briefly explain the function of each (Figure E3.6).
3. Some bacteria carry out a reaction that removes nitrates from water using methanol. The unbalanced reaction is

$$HNO_3 + CH_3OH \rightarrow C_3H_7NO_2 + CO_2 + H_2O$$

Write as a balanced chemical reaction equation.

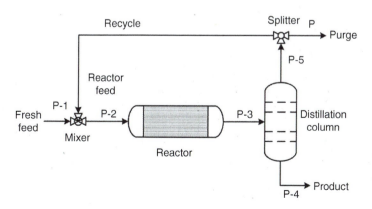

FIGURE E3.6
Process flow diagram that includes mixer, reactor, separator, and splitter.

Solution

1. The general material balance equation:

   Accumulation = (in − out) + (generation − consumption)

   Accumulation: Change in quantity of material inside system

   In: Material that enters system by crossing system boundary

   Out: Material that exits system by crossing system boundary

   Generation, consumption: Material that is generated or consumed by chemical reaction within the system

2. List the four kinds of process units found on a block flow diagram, and briefly explain the function of each.

   Mixer: Brings together several input streams, makes one output stream.

   Splitter: Takes an input stream and splits into two or more output streams, with each stream of identical composition.

   Reactor: Provides conditions to allow chemical reactions; reactants in input stream are converted to products by reaction and then leave through the output stream.

   Separator: Takes an input stream and separates it into two or more output streams, with each output stream of different composition.

3. Some bacteria carry out a reaction that removes nitrates from water using methanol. The unbalanced reaction is

$$HNO_3 + CH_3OH \rightarrow C_3H_7NO_2 + CO_2 + H_2O$$

Write a balanced chemical reaction equation.

If compounds are numbered 1 through 5 reading left to right, the element balance equations are

$$N: \nu_1 + \nu_3 = 0$$
$$C: \nu_2 + 3\nu_3 + \nu_4 = 0$$
$$H: \nu_1 + 4\nu_2 + 7\nu_3 + 2\nu_5 = 0$$
$$O: 3\nu_1 + \nu_2 + 2\nu_3 + 2\nu_4 + \nu_5 = 0$$

Now we pick ν_3 as a basis (other choices are also possible) and proceed to solve the set of equations to find $\nu_1 = -1$, $\nu_2 = -10/3$, $\nu_4 = 1/3$, and $\nu_5 = -11/3$.

The balanced chemical equation is

$$HNO_3 + \frac{10}{3}CH_3OH \rightarrow C_3H_7NO_2 + \frac{1}{3}CO_2 + \frac{11}{3}H_2O$$

Example 3.7   Mass, moles, mass fraction, and mole fraction

1. 502 lb of polystyrene (molar mass = 30,200 g/molar mass) is dissolved in 4,060 lb of styrene (C_8H_8, molar mass = 104 g/molar mass). Calculate the mass percent and mole percent of polystyrene in the mixture.

2. You have 180 g of a 12 wt% solution of glucose ($C_6H_{12}O_6$) in water. Calculate the molar mass of glucose and of water.

3. 2.7 lb of CO_2 is held in a vessel at 67°F and 1080 mmHg. Calculate the volume (cm^3) of the vessel using the ideal gas law.

Solution

1. The mass percent and mole percent of polystyrene in the mixture is calculated as

$$\text{Polystyrene mass percent} = \frac{502 \text{ lb}}{(502 \text{ lb} + 4060 \text{ lb})} \times 100\% = 11 \text{ wt\%}$$

There are $\dfrac{502 \text{ lb}}{30200 \text{ lb/lb-mol}} = 0.0166$ lb-mol polystyrene and $\dfrac{4060 \text{ lb}}{104 \text{ lb/lb-mol}} = 39.04$ lb-mol styrene.

$$\text{Polystyrene mole percent} = \frac{0.0166 \text{ lb-mol}}{0.0166 \text{ lb-mol} + 39.04 \text{ lb-mol}} \times 100\%$$

$$= 0.0425 \text{ lb-mol\%}$$

2. The molar mass of glucose and of water is calculated as

$$180 \text{ g solution} \times \frac{0.12 \text{ g glucose}}{\text{g solution}} \times \frac{1 \text{ molar mass glucose}}{180 \text{ g glucose}}$$

$$= 0.12 \text{ molar mass glucose}$$

$$180 \text{ g solution} \times \frac{0.88 \text{ g water}}{\text{g solution}} \times 1 \text{ molar mass water}/18 \text{ g water}$$

$$= 8.8 \text{ molar mass water}$$

3. The volume (cm^3) of the vessel using the ideal gas law is calculated as

$$n = 2.7 \text{ lb } CO_2 \times \frac{453.59 \text{ g}}{\text{lb}} \times \frac{1 \text{ molar mass}}{44 \text{ g}} = 27.8 \text{ molar mass } CO_2$$

$$T = (67°F - 32) \times (5/9) = 19.4°C + 273.15 = 292.6 \text{ K}$$

$$P = 1080 \text{ mmHg} \times \left(\frac{1 \text{ atm}}{760 \text{ mmHg}}\right) = 1.42 \text{ atm}$$

$$V = \frac{nRT}{P} = \frac{27.8 \text{ molar mass} \times 82.057 \text{ atm cm}^3/\text{molar mass K} \times 292.6 \text{ K}}{1.42 \text{ atm}}$$

$$= 470,000 \text{ cm}^3$$

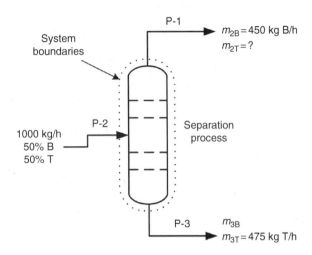

FIGURE E3.8
Flow sheet of binary liquid separation of benzene and toluene.

Example 3.8 Separation of a mixture of benzene and toluene

A 1000 kg/h mixture of benzene (B) and toluene (T) containing 50% benzene by mass is separated by distillation into two fractions. The mass flow rate of benzene in the top stream is 450 kg B/h and that of toluene in the bottom stream is 475 kg T/h. The operation is at steady state. Write balances for benzene and toluene to calculate the unknown component flow rate in the output streams.

Solution

The process-labeled flowchart is shown in Figure E3.8.
Basis: 1000 kg/h of feed
Assumptions: Steady state, no reaction
Material balance
In = out

$$B \text{ balance: } 0.5(1000) = 450\frac{\text{kg B}}{h} + m_{3B} \rightarrow m_{3B} = 50\frac{\text{kg B}}{h}$$

$$T \text{ balance: } 0.5(1000) = m_{2T} + 475\frac{\text{kg T}}{h} \rightarrow m_{2T} = 25\frac{\text{kg T}}{h}$$

Example 3.9 Methanol–water mixtures

Two methanol–water mixtures are contained in separate flasks. The first mixture contains 40.0 wt% methanol, and the second contains 70.0 wt% methanol. If 200 g of the first mixture is combined with 150 g of the second, what are the mass and composition of the product?

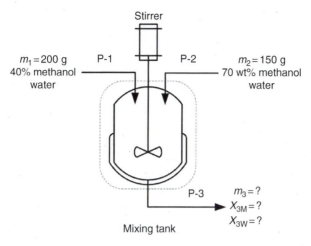

FIGURE E3.9
Flow sheet of methanol–water mixing process.

Solution
The process-labeled flowchart is shown in Figure E3.9.
Basis: 200 g of the 40 wt% methanol (i.e., stream 1)
Assumptions: Steady state, no reaction

Material balance

$$\text{Total balance: } 200 + 150 = m_3 \Rightarrow m_3 = 350$$

$$\text{M balance: } 0.4(200) + 0.7(150) = x_{3M}(350) \Rightarrow x_{3M} = 0.53$$

Example 3.10   Synthesis of strawberry jam
Strawberries contain about 15 wt% solids and 85 wt% water. To make strawberry jam, crushed strawberries and sugar are mixed in a 45:55 mass ratio and the mixture is heated to evaporate water until the residue contains one-third water by mass. Calculate how many pounds of strawberries are needed to make 1 kg of jam and how many pounds of water have evaporated.

Solution
The process-labeled flowchart is shown in Figure E3.10.

$$\frac{\text{Strawberries}}{\text{sugar}} = \frac{m_1}{m_2} = \frac{45}{55}$$

Basis: 1 kg of jam produced; $m_3 = 1$ kg
Assumption: Steady state, no reaction

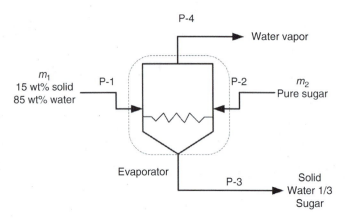

FIGURE E3.10
Flow sheet of strawberry jam evaporator.

Material balance

Solid balance: $0.15m_1 = \frac{2}{3}(1) \rightarrow m_1 = \frac{2}{3}\frac{1}{0.15}B \Rightarrow m_1 = 4.44$ kg

Using the reaction: $\frac{m}{m_2} = \frac{45}{55} \Rightarrow m_2 = m_1 \times \frac{55}{45} = 5.41$ kg

Water balance: $0.85(m) = \frac{1}{3}(m_3) + m_4$

4.44 kg $= \frac{1}{3}(5.4$ kg$) + m_4 \Rightarrow m_4 = 1.974$ kg

Example 3.11 Roasting of cement raw materials

Cement is produced by roasting the raw materials in a rotating kiln kept at high temperature. One ton per minute of raw material enters the kiln producing 0.7 ton of cement. It is known that gaseous by-products are produced during roasting. From the data, determine their emission rates.

Solution

The process-labeled flowchart is shown in Figure E3.11.

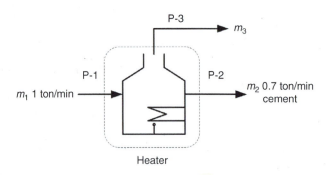

FIGURE E3.11
Flow sheet of cement furnace.

Basis: 1 ton feed
Assumption: Steady state, no reaction
Total mass balance

$$\dot{m}_1 = \dot{m}_2 + \dot{m}_3$$

$$1\frac{\text{ton}}{\text{min}} = 0.7\frac{\text{ton}}{\text{min}} + m_3 \Rightarrow m_3 = 0.3\frac{\text{ton}}{\text{min}}$$

Example 3.12   Partial vaporization

A liquid mixture of benzene and toluene contains 55.0% benzene by mass. The mixture is to be partially evaporated to yield a vapor containing 85.0% benzene and a residual liquid containing 10.6% benzene by mass:

(a) Suppose that the process is to be carried out continuously and at steady state, with a feed rate of 100.0 kg/h of the 55% mixture. Let V (kg/h) and L (kg/h) be the mass flow rates of the vapor and liquid product streams, respectively. Draw and label a process flowchart, then write and solve balances on total mass and on component balance of benzene. For each balance, state which terms of the general balance equation you discarded and why you discarded them.

Solution

The process-labeled flowchart is shown in Figure E3.12.

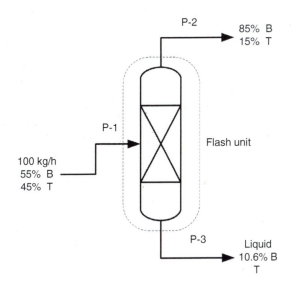

FIGURE E3.12
Flow sheet of flash unit.

Basis: 100 kg of feed
Assumption: Steady state, no reaction

Material balance
Total balance $100 = V + L$
B balance: $0.55(100) = 0.85V + 0.106\ L$
$55 = 0.85V + 0.106(100 - V) \rightarrow \dot{V} = 50.67$ kg/h
$55 = 0.85V + 10.6 - 0.106V \rightarrow \dot{L} = 40.33$ kg/h

Example 3.13 Drying using condensation process
A gas stream containing 40% O_2, 40% H_2, and 20 mol% H_2O is to be dried by cooling the steam and condensing out the water. If 100 mol/h of a gas stream is to be processed, what is the rate at which the water will be condensed out and what is the composition of dry gas?

Solution
The process-labeled flowchart is shown in Figure E3.13.
Basis: 100 mol/h wet gas
System: dryer

Degree of freedom analysis
Number of unknowns $= 3$ (D, C, X_2)
Number of independent material balance equations $= 3$ (O_2, H_2, H_2O)
$DF = 3 - 3 = 0$ (the problem is solvable)
Total material balance: $100 = D + C$
Component balance (H_2O): $0.2(100) = C$
Component balance (O_2): $0.4(100) = x_2\ D$
Solving the three material balance equations leads to the following answers:
$C = 20$ mol/h, $D = 80$ mol/h, and $x_2 = 0.5$

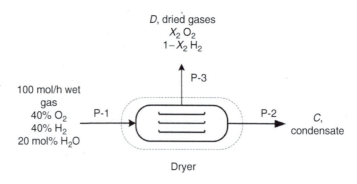

Dryer

FIGURE E3.13
Flow sheet of condenser dryer.

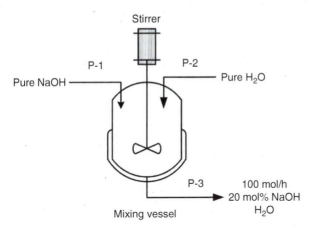

FIGURE E3.14
Mixing process flow sheet.

Example 3.14   Mixing of binary liquids
An aqueous solution at 20 mol% NaOH is to be prepared on a continuous basis by mixing pure NaOH and water. What is the addition rate of each stream required to prepare 100 mol/h solutions?

Solution
The process-labeled flowchart is shown in Figure E3.14.
System: mixer
Basis: 100 mol/h of product stream (stream no. 3)

Material balance
Total balance: $\dot{n}_1 + \dot{n}_2 = 100$ mol/h
Component balance (NaOH): $\dot{n}_1 = 0.2(100) = 20$ mol/h
From total balance equation: 20 mol/h $+ \dot{n}_2 = 100$ mol/h $\Rightarrow \dot{n}_2 = 80$ mol/h

Example 3.15   Extraction
An oil-free vegetable protein meal can be obtained from cottonseed by using hexane to extract the seed oil from cleaned seed. Given raw cottonseed consisting of 14 wt% cellulose material, 37 wt% meal, and 49 wt% oil, calculate the composition of oil extract obtained when 3 lb_m hexane is used per 1 lb_m mass raw seeds.

Solution
The process-labeled flowchart is shown in Figure E3.15.

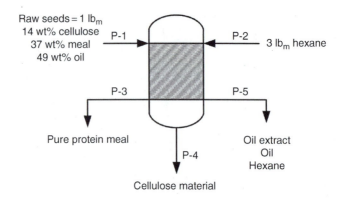

FIGURE E3.15
Extraction process flow sheet.

Basis: 1 lb_m of raw cottonseeds

Material balance
Component balance (meal balance): $0.37(1 \ lb_m) = m_3$
The total protein meal per 1 lb_m of raw cottonseeds $= m_3 = 0.37 \ lb_m$
Component balance (cellulose): $0.14(1 \ lb_m) = m_4$
Component balance (oil balance): $0.49 \ (1 \ lb_m) = x_{oil} \ m_5$
Component balance (hexane balance): $3 \ lb_m = (1-x_{oil}) \ m_5$
Total mass balance: $1 \ lb_m + 2 \ lb_m = 0.37 \ lb_m + 0.14 \ lb_m + m_5$
Mass of the extracted oil $= m_5 = 2.49 \ lb_m$
Substitute m_5 in the oil balance equation

$$0.49(1 lb_m) = x_{oil} \ 2.49 \Rightarrow x_{oil} = 0.197$$

Example 3.16   Multicomponent distillation column
In a distillation column, an equimolar mixture of ethanol, propanol, and butanol is separated into an overhead stream containing 2/3% ethanol and no butanol, and a bottom stream containing no ethanol. Calculate the compositions of the overhead and bottom streams for a feed rate of 1000 mol/h.

Solution
The process-labeled flowchart is shown in Figure E3.16.
Basis: 1000 mol/h of feed

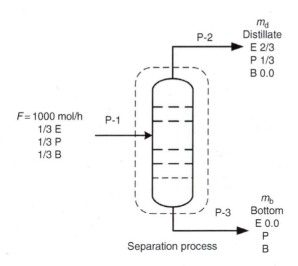

FIGURE E3.16
Flow sheet of tertiary component separation.

Degree of freedom analysis
Number of unknowns $= 3$
Number of independent equations $= 3$
Number of relations $= 0$
$DF = 0$

Material balance
Total balance: $1000 \text{ mol/h} = \dot{m}_d + \dot{m}_b$
Component balance (E): $(1/3)(1000 \text{ mol/h}) = (2/3)\dot{m}_d + 0.0 \Rightarrow \dot{m}_d = 500 \text{ mol/h}$
From the total material balance:
$1000 \text{ mol/h} = 500 \text{ mol/h} + \dot{m}_b \Rightarrow \dot{m}_b = 500 \text{ mol/h}$
Component balance (P):
$(1/3)(1000 \text{ mol/h}) = (1/3)\dot{m}_d + x_p\dot{m}_b$
Substitute m_b, m_d in the P component balance
$(1/3)(1000 \text{ mol/h}) = (2/3)(500 \text{ mol/h}) + x_b(500 \text{ mol/h})$
$(1/3)(2) = (1/3) + x_b \Rightarrow x_b = 1/3$

Example 3.17   Benzene–toluene separation process
The feed to a distillation column contains 36% benzene (B) by weight, the remainder being toluene (T). The overhead distillate is to contain 52%

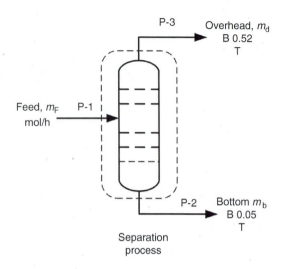

FIGURE E3.17
Binary distillation column flow sheet.

benzene by weight, while the bottom distillate is to contain 5% benzene by weight. Calculate

(a) Percentage of the benzene feed that is contained in the distillate.
(b) Percentage of the total feed that leaves as distillate.

Solution
The process-labeled flowchart is shown in Figure E3.17.

Degree of freedom analysis
Number of unknowns $= 3$
Number of independent equations $= 2$
Number of relations $= 0$
$DF = 1$

The DF is 1; hence, the problem cannot be solved. Since none of the stream flow rates are specified, a basis should be assumed.
    Basis: 100 kg/h of feed to the column

$$\text{Total balance: } 100 \text{ kg/h} = \dot{m}_d + \dot{m}_b \tag{3.1}$$

$$\text{Component balance } (B): 0.36(100 \text{ kg/h}) = 0.52\dot{m}_d + 0.05\dot{m}_b \tag{3.2}$$

Substitute m_b from Equation 3.1 into Equation 3.2

$$0.36(100 \text{ kg/h}) = 0.52 \, \dot{m}_d + 0.05(100 - \dot{m}_d)$$

$$36 \text{ kg/h} = 0.52 \, \dot{m}_d + 5 - 0.05 \, \dot{m}_d \Rightarrow \dot{m}_d = 66 \text{ kg/h}$$

(a) Percentage of the benzene feed that is contained in the distillate is

$$\frac{0.52(66)}{0.36(100)} \times 100\% = 95\%$$

(b) Percentage of the total feed that leaves as distillate is

$$\frac{66}{100} \times 100\% = 66\%$$

Example 3.18 Growth rate of certain organisms

An experiment on the growth rate of certain organisms requires an environment of humid air enriched in oxygen. Three input streams are fed into an evaporation chamber to produce an output stream with the desired composition.

A: Liquid water, fed at a rate of 20.0 cm^3/min

B: Air (21 mol% O_2, the balance N_2)

C: Pure oxygen, with a molar flow rate one-fifth of the molar flow rate of stream B

The output gas is analyzed and is found to contain 1.5 mol% water. Draw and label a flowchart of the process, and calculate all unknown stream variables.

Solution

The process-labeled flowchart is shown in Figure E3.18.

Molar flow rate of pure $O_2 = \dfrac{1}{3}(\dot{n}_B)$

Basis: $20 \dfrac{cm^3}{min}$ of liquid water

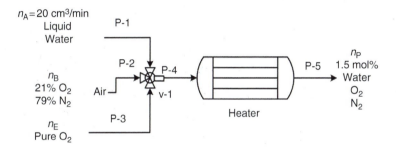

FIGURE E3.18
Flow sheet of evaporation chamber.

Mass of water $= \dot{V}\rho = \left(20\dfrac{cm^3}{min}\right)\left(\dfrac{1g}{cm^3}\right) = 20\,\dfrac{g}{min}$

Moles of water $= 20\,\dfrac{g}{min} \div \dfrac{1}{M_w} = 20\,\dfrac{1}{min}\dfrac{g}{18\ g/mol} = 1.11\,\dfrac{mol}{min}$

Assumption: Steady state, no reaction

Material balance

Water balance: $\dot{n}_A = 0.015\dot{n}_p$

$1.11\,\dfrac{mol}{min} = 0.015\,\dot{n}_p \rightarrow \dot{n}_p = 74.0\ mol$

Total mole balance: $\dot{n}_A + \dot{n}_B + \dot{n}_C = \dot{n}_P \Rightarrow 1.11\,\dfrac{mol}{min} + \dot{n}_B + \dfrac{1}{3}\dot{n}_B = 47.0 \Rightarrow$

$\dot{n}_B = 60.8$

O_2 balance: $0.79\,\dot{n}_B = n_{PN_2} \times \dot{n}_P$

$0.79(60.8) = (n_{N_2})_P \times 74.1 \Rightarrow (n_{N_2})_P = 0.335$

Check O_2: $(0.21)(60.8) + \dfrac{1}{5}(60.8) = 0.335(74.1) \Rightarrow 25 = 25$

Example 3.19   Ethanol–methanol separation process

A 100 kg/h ethanol–methanol stream is to be separated in a distillation column. The feed has 40% ethanol and the distillate has 90% methanol. The flow rate of the bottom stream product is 400 kg/h. Determine the percentage of methanol in the bottom stream.

Solution

The process-labeled flowchart is shown in Figure E3.19.

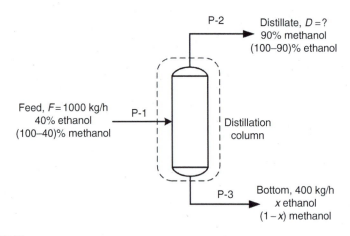

FIGURE E3.19

Flow sheet of binary separation of ethanol–methanol mixture.

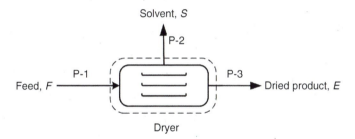

FIGURE E3.20
Flow sheet of wet leather drying process.

Basis: 1000 kg/h of feed
Assumptions: Continuous process, steady state $\rightarrow$ accumulation $= 0$
No chemical reaction $\rightarrow$ generation $=$ consumption $= 0$

Degree of freedom analysis
Number of unknowns $= 2$ (D, x)
Number of equations $= 2$, which equals number of components
DF $=$ number of unknowns $-$ number of equations $= 2 - 2 = 0$
Total mass balance: 1000 kg/h $= D + 400$ kg/h $D \rightarrow 600$ kg/h
Component balance for methanol: $0.6(1000 \text{ kg/h}) = 0.9 \ (600 \text{ kg/h}) + (1 - x)$
400, $x = 0.85$

Example 3.20   Drying of wet leather
Two hundred kilograms of wet leather is to be dried by heating and
tumbling. The wet leather enters the drier with 1.5 g H_2O per gram bone
dry leather (BDL). The leather is to be dried to 20% moisture. Determine the
mass of water removed.

Solution
The process-labeled flowchart is shown in Figure E3.20.
Total balance: $200 = S + E$
Component balance (BDL): BDL $= 0.8E$
Feed content: $200 = W + \text{BDL}$, $W/\text{BDL} = 1.5$
Results: BDL $= 80$, $E = 100$, $S = 100$, $W = 100$ kg

**Example 3.21   Separation of propane, isobutane, isopentane,
and *n*-pentane**
A hydrocarbon feed consisting of a mixture of propane, isobutane, isopentane, and *n*-pentane is fractionated at a rate of 1000 kmol/h into a distillate
that contains all the propane and 78% of the isopentane in the feed. The
mole fraction of isobutane in the distillate is 0.378. The bottom stream

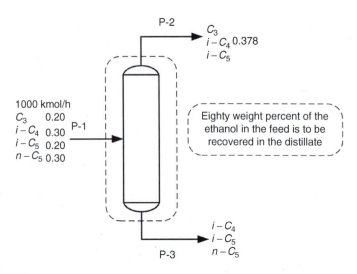

FIGURE E3.21a
Flow sheet of multicomponent separation process.

contains all the *n*-pentane fed to the unit. Determine the flow rates and composition of the bottom stream.

Solution
The process-labeled flowchart is shown in Figure E3.21a.

DF Analysis	Distillation
Number of unknowns	$5\ (n_2, n_3, x_2, x_3, y_3)$
Number of independent equations	4
Number of relations	1
DF	0

Basis: 1000 kmol/h
Total mass balance: $1000 = \dot{n}_2 + \dot{n}_3$
Component mass balance:
C_3: $0.2(1000) = x_2 \dot{n}_2$
$i - C_4$: $0.3(1000) = 0.378\dot{n}_2 + x_3\dot{n}_3$
$i - C_5$: $0.2(1000) = (1 - x_2 - 0.375)\dot{n}_2 + y_3\dot{n}_3$
$n - C_5$: $0.3(1000) = (1 - x_3 - y_3)\dot{n}_3$
Relation: $0.2(1000) \times 0.78 = (1 - x_2 - 0.378)\dot{n}_2$

The generated material balance equations are solved using the E–Z solve software (Figure E3.21b).

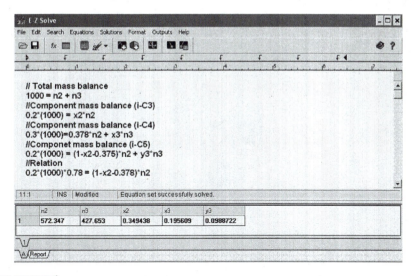

FIGURE E3.21b
Solving the set of equations using *E–Z* solve.

3.6.3 Material Balance on Bioprocesses

Example 3.22 Biochemical processes: Fermentation

A continuous rotary vacuum filter is used to filtrate 120 kg of fermentation slurry that contains 6% *Streptomyces kanamyceticus* cell solids. To improve filtration rates, particles of diatomaceous earth filter aid are added at a rate of 10 kg/h. The concentration of kanamycin in the slurry is 0.05% by weight. Liquid filtrate is collected at a rate of 112 kg/h; the concentration of kanamycin in the filtrate is 0.045% by weight. Filter cake containing cells and filter aid is continuously removed from the filter cloth.

(a) Draw and fully label the process flowchart.

(b) Calculate all unknown streams flow rate and compositions.

Solution
The process-labeled flowchart is shown in Figure E3.22.
  Basis: 120 kg/h of fermentation slurry

DF Analysis	Filter
Number of unknowns	4
Number of independent equations	4
Number of relations	0
DF	0

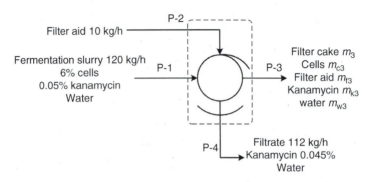

FIGURE E3.22
Flow sheet of rotary vacuum filter.

Material balance
Total balance: $120 \text{ kg/h} + 10 \text{ kg/h} = m_3 + 112 \text{ kg/h}$
Component balance: Filter aid: $10 \text{ kg/h} = m_{f3}$
Kanamycin: $0.0005(120 \text{ kg/h}) = m_{k_3} + 0.00045(112 \text{ kg/h})$
Water: $(1 - 0.06 - 0.0005)(120 \text{ kg/h}) = m_{w_3} + (1 - 0.00045)(112 \text{ kg/h})$.

Example 3.23 Bioprocesses: Batch mixing

Corn-steep liquor contains 2.5% invert sugars and 50% water; the rest can be considered solids. Beet molasses containing 50% sucrose, 1% invert sugars, 18% water, and the remainder made up of solids is mixed with corn-steep liquor in a mixing tank. Water is added to produce a diluted sugar mixture containing 2 wt% invert sugars. Corn-steep constituting 125 and 45 kg molasses are fed into the tank. What is the concentration of sucrose in the final mixture?

Solution
The process-labeled flowchart is shown in Figure E3.23.

Solution

DF Analysis	Mixer
Number of unknowns	4
Number of independent equations	4
Number of relations	0
DF	0

Component balance (invert sugar): $P = 178.75 \text{ kg}$
Component balance (solid): $0.475(125 \text{ kg}) + 0.31(45) = S_P \Rightarrow S_P = 73.325 \text{ kg}$

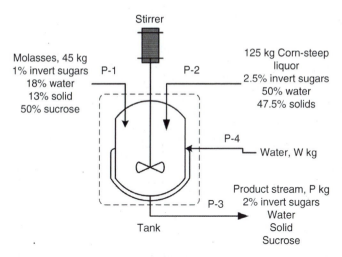

FIGURE E3.23
Flow sheet for mixing process.

Component balance (sucrose): $0.5(45) = $ mass of sucrose in $P = Sc_P \Rightarrow$
$Sc_P = 22.5$ kg
Component balance (water): $0.5(125 \text{ kg}) + 0.18(45 \text{ kg}) + W = 79.35 \text{ kg} \Rightarrow$
$W = 8.75$ kg

3.7 Problems

3.7.1 Separation of Ethanol–Methanol Process Stream

One thousand kilogram per hour of an ethanol–methanol stream is to be separated in a distillation column. The feed has 40% ethanol and the distillate has 90% methanol. Eighty percent of the methanol is to be recovered as distillate. Determine the percent methanol in the bottoms product.

(Answer: $B = 466.557$, $C = 533.33$, % methanol in the bottoms product $= 86.7\%$)

3.7.2 Wet Leather Drying Process

Two hundred kilograms of wet leather is to be dried by heating and tumbling. The wet leather enters the drier with 60% moisture and 90% of the water is removed. Determine the mass of the dried leather.

(Answer: $B = 92$ kg/h, $W = 108$ kg/h, and $x = 0.13$)

3.7.3 Separation of Ethanol–Methanol–Propanol Mixture

A 1200 kg/h ethanol–methanol–propanol stream is fed to a distillation column. The feed is 20% methanol. Ninety percent of the methanol is to be recovered in the distillate along with 60% of the ethanol. All of the 400 kg/h of propanol fed to the process must be sent to the bottom of the distillate. Calculate all unknown stream flow rates and composition.

(Answer: $B = 648$ kg/h, $D = 552$ kg/h, bottom $(E = 224$, $M = 184$, $P = 400$ kg/h)

3.7.4 Ethanol–Water Separation

A 1000 kg/h ethanol–methanol stream is to be separated in a distillation column. The feed has 40 wt% ethanol and the distillate has 80% methanol by weight. Eighty percent (weight) of the methanol is to be recovered as distillate. Determine the composition of methanol (in wt%) in the bottom stream.

(Answer: $D = 60$ kg/h, $B = 40$ kg/h, and $X_w = 0.7$)

3.7.5 Mixing of Hydrochloric Acid with Water

A stream of aqueous hydrochloric acid that is 57.3 wt% HCl is mixed with water to produce a stream of 16.5% acid. What ratio of water to concentrated acid should be used?

(Answer: $M_2/M_1 = (0.573 - 0.165)/0.165 = 2.473$)

3.7.6 Removal of Acetone from Nitrogen Using an Absorber

An absorber is used to remove acetone from a nitrogen carrier gas. The feed, with acetone weight fraction of 0.213, enters at a rate of 200 kg/h. The absorbing liquid is water, which enters at a rate of 1000 kg/h. The exit gas stream is 0.8 wt% acetone and 2.9% water vapor. Determine the mass flow rates and composition of the bottom stream.

(Answer: $\dot{m}_3 = 164$ kg/h, $\dot{m}_4 = 1036$ kg/h, $x = 0.04$, and $1 - x = 0.96$)

3.7.7 Separation of Benzene/Toluene Mixture

A 1000 kg/h mixture containing equal parts by mass of benzene and toluene is distilled. The flow rate of the overhead product stream is 488 kg/h, and the bottom stream contains 7.11 wt % benzene. Draw and label a flowchart of the process. Calculate the mass and mole fractions of benzene and the molar flow rates of benzene and toluene (mol/h) in the overhead product stream.

3.7.8 Dilution of Methanol Mixture

A stream containing 25 wt% methanol in water is to be diluted with a second stream containing 10% methanol to form a product containing 17% methanol.

(a) Choose a convenient basis for calculation, draw and label a flowchart of this process, and calculate the ratio (kg 17% solution/kg 25% solution).

(b) What feed rate of the 10% solution would be required to produce 1250 kg/h of the product?

3.7.9   Humidification Chamber

Liquid water and air flow into a humidification chamber in which the water evaporates completely. The entering air contains 1.00 mol% H_2O (ν) and 20.8% O_2; the balance consists of N_2, and the humidified air contains 10.0 mol% H_2O. Calculate the volumetric flow rate (ft^3/min) of liquid required to humidify 200 lb-mol/min of the entering air.

3.7.10   Absorption of Water from a Gas Mixture

A gas containing equal parts (on a molar basis) of H_2, N_2, and H_2O is passed through a column of calcium chloride pellets, which absorb 97% of the water but none of the gases. The column packing was initially dry and had a mass of 2.00 kg. Following 6 h of continuous operation, the pellets are reweighed and are found to have a mass of 2.21 kg. Calculate the molar flow rate (mol/h) of the feed gas and the mole fraction of water vapor in the product gas.

3.7.11   Drying of Wet Sugar

Wet sugar that contains 20% water is sent through a dryer where 75% of the water is removed. Taking 100 kg feed as the basis calculate the mass fraction of dry sugar in the wet sugar that leaves the dryer. Calculate the ratio of H_2O removed/kilogram wet sugar leaving the dryer. If 1000 t/day of wet sugar are fed to the dryer, how much additional water must be removed from the outlet sugar to dry it completely, and how much revenue can be expected if dry sugar sells for $0.25/lb?

Further Readings

1. Himmelblau, D.M. (1996) *Basic Principles and Calculations in Chemical Engineering*, 6th edn, Prentice-Hall, New Jersey.
2. Reklaitis, G.V. (1983) *Introduction to Material and Energy Balances*, John Wiley & Sons, New York.
3. Felder, R.M. and R.W. Rousseau (1999) *Elementary Principles of Chemical Processes*, 3rd edn, John Wiley, New York.
4. Atkinson, B. and F. Mavituna (1991) *Biochemical Engineering and Biotechnology Handbook*, 2nd edn, Macmillan, Basingstoke.
5. Doran, P. (1995) *Bioprocess Engineering Principles*, Academic Press, London.
6. Bailey, J.E. and D.F. Ollis (1986) *Biochemical Engineering Fundamentals*, 2nd edn, McGraw-Hill, New Jersey.
7. Shuler, M.L. and F. Kargi (2002) *Bioprocess Engineering—Basic Concepts*, 2nd edn, Prentice-Hall International Series, New Jersey.

4

Multiple-Unit Process Calculations

At the End of This Chapter You Should Be Able to

1. Write material balances for a complex process involving more than one unit.

2. Draw a process flow diagram for problems involving recycle, bypass, and purge.

3. Explain the purpose of a recycle stream, a bypass stream, and a purge stream.

4. Apply the degree of freedom analysis (DFA) to solve steady-state problems.

5. Solve problems involving several connected units by applying the DFA.

Until now, the processes we have studied involved only one process unit, and this unit constituted the system. However, industrial processes rarely involve only one process unit. Before we can start to analyze multiple-unit processes, we first need to define what we mean by "the system." A system is any portion of a process that can be enclosed within a boundary. The system may be the entire process, a single process unit, a combination of process units, or a point where two or more process streams come together or one stream splits into branches. The inputs and outputs to a system are the process streams that intersect the system boundary.

4.1 Multiple-Unit Process

A multiple-unit process is a process that contains more than one unit such as the following process (Figure 4.1), which consists of one mixing point, two unit operations, and one splitter.

In the above process, five systems do exist: overall process system (S1), mixing point (S2), unit one (S3), splitter system (S4), and unit two (S5).

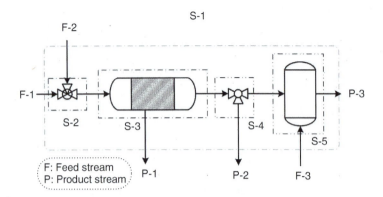

FIGURE 4.1
Multiple-unit process.

The procedure for material balance calculations on multiple-unit processes is the same as that for single-unit processes. However, now you need to isolate and write balances on several subsystems of the process in order to determine all the process variables. For each system:

(a) Perform a DFA.

(b) Write balance equations (total and component balances).

Start solving the material balance set of equations for the system that has a degree of freedom (DOF) equal to 0, that is, $n_{df} = 0$.

Example 4.1 Degree of freedom analysis for multiple units

Perform a DFA for the following process flow diagram shown in Figure E4.1.

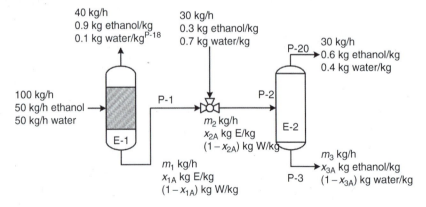

FIGURE E4.1
Flow sheet of a two-unit process.

Solution

The system with 0 DOF is the first to start with.
  The DFA is shown in the following table.

| | Systems | | | |
DFA	Unit 1	Mixing Point	Unit 2	Process
Number of unknowns	2	4	4	2
Number of independent equations	2	2	2	2
Number of relations	—	—	—	—
DF	0	2	2	0

4.2 Recycle, Bypass, Purge, and Makeup

Suppose we have the following chemical reaction taking place in a reactor:

$$A + B \rightarrow C + D$$

Since it is rare for any chemical reaction to proceed to completion, some of A will remain in the product stream (Figure 4.2). This is not an ideal situation as some unreacted A is left in the product stream (wasted) and the final product is not very pure in B. How can we improve this situation?

4.2.1 Recycle

Recycle is common in chemical processes. There are several reasons for using recycle in a chemical process: Recovering and reusing unconsumed reactants, recovery of catalyst, dilution of a process stream, control of a process variable, circulation of a working fluid, and for nonaccumulation of mass since input = output (overall balance). The process flow sheet with recycle stream is shown in Figure 4.3.

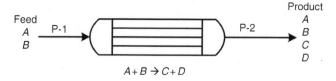

FIGURE 4.2
Flow sheet of a reaction process.

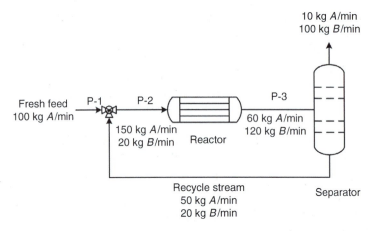

FIGURE 4.3
Process system with recycle.

4.2.2 Bypass

Bypass is a fraction of the feed to a process unit that is diverted around the unit and combined with the output stream (Figure 4.4), thus varying the composition and properties of the product.

Chemical processes involving bypass streams are treated in exactly the same manner as processes containing recycle streams: The flow sheet is drawn and labeled and balances around the process unit or the stream mixing point following the process unit are used to determine the unknown process variables. The methods for solving recycle and bypass problems are basically the same. In the steady state, there is no buildup or depletion of material within the system or recycle stream of a properly designed and operated process. When you are solving, you can write balances (total material or component) around the entire process structure, the mixing point, the splitter, and the process unit (see Figure 4.5). Only three of these will be independent. If you pick the right balances, you may be able to organize the problem for a sequential solution. In particular, when you

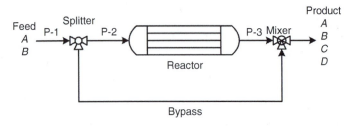

FIGURE 4.4
Flow sheet of unit process with bypass.

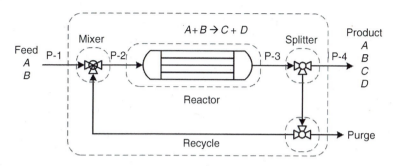

FIGURE 4.5
Flow sheet with recycle and purge.

write the balance around the entire process system, terms describing the recycle or bypass stream do not appear; only the fresh feed and the product are required.

4.2.3 Purge

A purge stream is a small stream bled off from a recycle loop to prevent buildup of inerts or impurities in the system. Often, the purge flow is so much smaller than the recycle flow that it can be neglected in the steady-state overall material balance of the process. The process flowchart with recycle and purge streams is shown in Figure 4.5.

4.2.4 Makeup

A makeup stream is required to replace losses to leaks, carryover, etc. within the recycle loop. Proper sizing and location of makeup and purge streams can prevent many problems with process plants. A process flow sheet with recycle and makeup stream is shown in Figure 4.6.

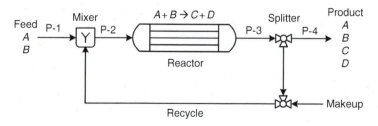

FIGURE 4.6
Flow sheet with recycle and makeup streams.

Example 4.2   Two distillation columns in series

Two columns in sequence are used to separate the components of a feed consisting of 30% benzene (B), 55% toluene (T), and 15% xylene (X). The analysis of the overhead stream from the first column is 94.4% B, 4.54% T, and 1.06% X. The second column is designed to recover 92% of the toluene in the original feed in the overhead stream at a composition of 0.946. The bottoms are intended to contain 92.6% of xylene at a composition of 0.776. Compute the compositions of all streams.

Solution

The process flowchart is shown in Figure E4.2.

Degree of freedom analysis

DFA	Column (D-1)	Column (D-2)	Overall Process
Number of unknowns	5	7	6
Number of independent equations	3	3	3
Number of relations	—	—	2
DF	2	4	1

Basis = 100 kg/h
System: overall
Total mass balance: $100 = m_2 + m_4 + m_5$

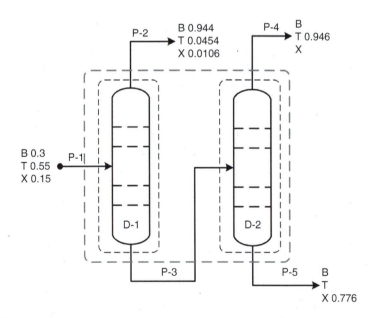

FIGURE E4.2
Two distillation columns in series.

Component mass balance:

$$\text{B: } (0.3)(100) = (0.944)m_2 + x_4 m_4 + x_5 m_5$$

$$\text{T: } (0.55)(100) = (0.0454)m_2 + (0.946)m_4 + (1 - 0.776 - x_5)m_5$$

$$\text{X: } (0.15)(100) = (0.0106)m_2 + (1 - 0.946 - x_4)m_4 + (0.776)m_5$$

Relation:

$$(0.92)(0.55)(100) = 0.946 m_4 \Rightarrow m_4 = 53.4884 \text{ kg/h}$$

$$(0.926)(0.15)(100) = 0.776 m_5 \Rightarrow m_5 = 17.90 \text{ kg/h}$$

Answer: $m_2 = 28.6121$ kg/h, $x_4 = 0.039918$, $x_5 = 0.05075$

Example 4.3 Purifier column

A separator is designed to remove exactly two-thirds of dimethyl forma-mide (DMF) fed to it. However, we are required to reduce the DMF content of the raw feed from 55 mol% (the balance is nitrogen) to 10%. To achieve this objective, a recycle loop is used. Compute the fraction of the purifier exit stream that must be recycled.

Solution

The process flow sheet is shown in Figure E4.3.

DFA	Mixing	Purifier	Splitter	Overall
Number of unknowns	4	4	3	3
Number of independent equations	2	2	2	2
Number of relations	—	1	—	—
DF	2	1	1	1

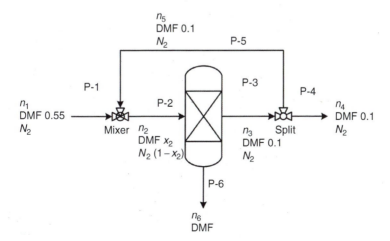

FIGURE E4.3
Flow sheet of purifier column.

Bases: 100 mol
System: overall process
Over all material balance:

$$100 = n_6 + n_4$$

DMF component balance:

$$0.55(100) = n_6 + 0.1n_4$$
$$n_4 = 50 \text{ mol}, n_6 = 50 \text{ mol}$$

System: purifier

$$\text{Total mass balance: } n_2 = n_6 + n_3 \tag{4.1}$$

$$\text{DMF balance: } x_2 n_2 = n_6 + n_3^*(0.1) \tag{4.2}$$

Relations: $n_6 = (2/3)n_2 \times x_2 \Rightarrow 50 = (2/3)n_2 \times x_2 \Rightarrow n_2 \times x_2 = 75$
From Equation 4.2: $75 = 50 + n_3 (0.1) \Rightarrow n_3 = 250$
System: splitter

$$n_3 = n_5 + n_4 \Rightarrow 250 = n_5 + 50 \Rightarrow n_5 = 200 \text{ mol}$$

Example 4.4   Absorber column

The raw feed to a sulfur removal system contains 15 mol% CO_2, 5% H_2S, and 1.41% COS. The balance is CH_4. The original absorber design placed a maximum flow limit of 820 kmol/h and yielded a product stream with only 1% H_2S and 0.3% COS. The feed to the unit is 1000 kmol/h. The excess feed flow is bypassed. Perform degree of freedom analysis. Find the flow rates and compositions of all streams.

Solution

Basis: 1000 kmol/h of fresh feed
System: overall
Assumptions: steady state, no reaction

Degree of freedom analysis

DFA	Absorber	Mixer	Overall
Number of unknowns	8	6	4
Number of independent equations	4	4	4
Number of relations	0	0	0
DF	4	2	0

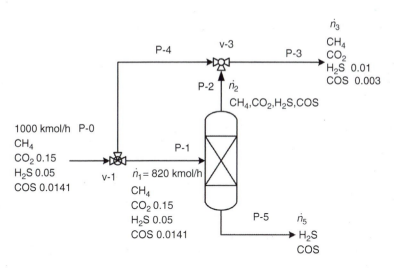

FIGURE E4.4
Flow sheet of absorption process.

The process flow sheet is shown in Figure E4.4.
Total material balance: $1000 = \dot{n}_3 + \dot{n}_5$
Component balance:

$$CO_2: 0.15(1000) = (x_{CO_2,3})\dot{n}_3 + 0$$

$$H_2S: 0.05(1000) = 0.01\dot{n}_3 + (x_{H_2S,5})\dot{n}_5$$

$$COS: 0.0141(1000) = 0.003\dot{n}_3 + x_{COS,5}\dot{n}_5$$

$$x_{COS,5} = 1 - x_{H_2S,5}$$

Solving the generated set of material balance equations using E–Z solve gives the following results: $n_3 = 948.227$ and $n_5 = 51.773$.

Example 4.5   Orange juice production process
Consider a process in which raw oranges are processed into orange juice. The oranges enter a crusher, in which all of the water contained within the oranges is released. The crushed oranges enter a strainer. The strainer is able to capture 90% of the solids; the remainder exit with the orange juice as pulp. The velocity of the orange juice stream was measured to be 30 m/s and the radius of the piping was 8 in. Assuming that the captured solid has negligible amount of water, calculate

(a) Mass flow rate of the orange juice product.
(b) Number of oranges per year that can be processed with this process if it is run 8 h a day and 360 days a year. Ignore changes due to unsteady state at startup.

Use the following data: mass of an orange, 0.4 kg; water content of an orange, 80%; density of the solids (being mostly sugars), it is about the density of glucose, which is 1.54 g/cm^3.

Solution

The process flow sheet is shown in Figure E4.5.

Degree of freedom analysis

DFA	Crusher	Strainer	Overall
Number of unknowns	3	5	4
Number of independent equations	2	2	2
Number of relations	—	2	2
DF	1	1	0

The detailed analysis of the process:

System: crusher

- There are three unknowns (m_1, m_2, and x_{S2}).
- We can write two independent mass balances.
- Thus, the crusher has DF $= 3 - 2 = 1$.

System: strainer

- There are five unknowns (m_2, x_{S2}, m_3, x_{S4}, m_4).
- We can write two independent mass balances on the overall system (one for each component).
- Relation 2 (90% solid captured in the strainer and the volumetric flow rate of product).
- DF $= 5 - 2 - 2 = 1$.

System: block

- There are four unknowns (m_1, m_3, m_4, x_{S4}).
- We can write two independent mass balance on the overall system (one for each component).

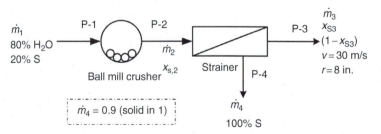

FIGURE E4.5
Flow sheet of orange juice process plant.

- Relation 2 (90% of the solid in the feed is released in the strainer, volumetric flow of stream 4).
- Thus, the DOF is $4 - 2 - 2 = 0$.

System: system as a whole:

- Sum of DOF for unit operations is $1 + 1 = 2$.
- Number of intermediate variables $= 2$ (m_2 and x_{S2}).
- Total DOF $= 2 - 2 = 0$.

Hence, the problem is specified and can be solved. Here, the most sensible choice is either to convert everything to the CGS system or to the m-kg-s system, since most values are already in metric. Here, the latter route is taken.

$$r_4 = 8 \text{ in.} \left| \frac{2.54 \text{ cm}}{\text{in.}} \right| \frac{1 \text{ m}}{100 \text{ cm}} = 0.2032 \text{ m}$$

$$\rho_s = 1.54 \frac{\text{g}}{\text{cm}^3} = 1540 \frac{\text{kg}}{\text{m}^3}$$

Volumetric flow rate of stream 4 $= \left(\pi \times 0.2032^2 \text{ m}^2 \right) \left| \frac{30 \text{ m}}{\text{s}} \right. = 3.8915 \frac{\text{m}^3}{\text{s}}$

$$\rho_4 = \frac{\dot{m}_4}{\text{Volumetric flow rate}} = \frac{\dot{m}_4}{3.8915}$$

Now that everything is in the same system, we can move on to the next step. The density of the mixture can be calculated from

$$\frac{1}{\rho_{mix}} = \sum \frac{1}{\rho_i} \rightarrow \frac{x_{S4}}{\rho_{S4}} + \frac{1 - x_{S4}}{\rho_{W4}} = \frac{1}{\rho_4}$$

$$\frac{x_{S4}}{1540} + \frac{1 - x_{S4}}{1000} = \frac{1}{\rho_4} = \frac{3.8915 \frac{\text{m}^3}{\text{s}}}{\dot{m}_4 \frac{\text{kg}}{\text{s}}}$$

$$0.2 \times \dot{m}_1 = (0.9 \times 0.2 \times \dot{m}_1) \times 1 + x_{S4} \times \dot{m}_4 \Rightarrow \dot{m}_1 = 0.9 \times 0.2 \times \dot{m}_1 + \dot{m}_4$$

Using *E–Z* solve leads to the following answers:

$$\dot{m}_1 = 4786 \frac{\text{kg}}{\text{s}}, \ \dot{m}_4 = 3925.07 \frac{\text{kg}}{\text{s}}, \ x_4 = 0.0244$$

$$\left| \frac{4786 \text{ kg}}{\$} \right| \frac{1 \text{ orange}}{0.4 \text{ kg}} \left| \frac{3600 \$}{\text{h}} \right| \frac{8 \text{ h}}{\text{day}} \left| \frac{360 \text{ days}}{\text{yr}} \right. = 1.24 \times 10^{11} \frac{\text{oranges}}{\text{year}}$$

$$\text{Yearly production} = 1.24 \times 10^{11} \frac{\text{oranges}}{\text{year}}$$

Example 4.6   Splitter and recycle

Hexane and pentane are being continuously split in a distillation column with a reflux ratio of 0.69 (reflux ratio $= R/D$). If the feed is 50% hexane, the distillate is 5% hexane, and the bottom stream is 96% hexane (all by weight), determine the distillate, bottom, and overhead flow rates for a feed of 100 kg/h. Hint: The compositions do not change at the splitter, but the total masses are different; therefore, there is only one independent material balance equation. This will always be true of a splitter. It just divides the flow rates, so there is only one independent balance that can be performed.

Solution

The process flow sheet is shown in Figure E4.6.
Basis: 100 kg/h of feed

The DFA is summarized in the following table.

DFA	Still	Splitter	Overall Process
Number of unknowns	3	3	2
Number of independent equations	2	1	2
Number of relations	0	1	0
DF	1	1	0

A zero here indicates that this is where we should start and the problem is uniquely defined.

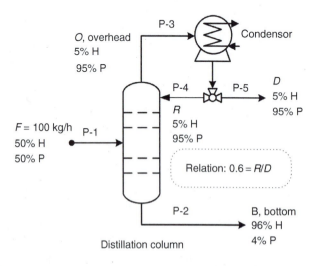

FIGURE E4.6
Flow sheet of distillation column.

System: overall

$$\text{Total: } 100 \text{ kg} = D + B$$

$$\text{H: } 50 \text{ kg} = 0.05 \times D + 0.96 \times B$$

Answer: $B = 49.5$ kg $D = 50.5$ kg (check with P balance)
System: splitter
Total balance: $O = R + D$
Relation: $R/D = 0.6$
Solving these two equations for the two unknowns

$$R = 0.6 \, D = 0.6(50.5) = 30.3 \text{ kg/h}$$
$$O = R + D$$
$$O = 30.3 + 50.5 = 80.8 \text{ kg/h (check with P balance)}$$

Scale: now convert to 100 kg mol basis

Feed: $n_H = (50 \text{ kg H}) / \left(\frac{86 \text{ kg H}}{\text{kg mol H}} \right) = 0.581$ kg mol H

Purge: $n_P = (50 \text{ kg P}) / \left(\frac{72 \text{ kg P}}{\text{kg mol P}} \right) = 0.694$ kg mol P

$$n_F = n_H + n_P = 1.276 \text{ kg mol}$$

$$O: (0.05)(80.8)/86 + (0.95)(80.8)/72 = 1.07 \text{ kg mol}$$

$$D: (0.05)(80.8)/86 + (0.95)(50.5)/72 = 0.70 \text{ kg mol}$$

$$B: (0.96)(49.5)/86 + (0.04)(49.5)/72 = 0.58 \text{ kg mol}$$

Scale factor

$$(100 \text{ kg mol F/h})/(1.276 \text{ kg mol F}) = 78.4/h$$

Results after scaling up

$$F = (1.26 \text{ kg mol})(78.4/h) = 100 \text{ kg mol/h}$$
$$O = (1.07 \text{ kg mol})(78.4/h) = 83.9 \text{ kg mol/h}$$
$$B = (0.58 \text{ kg mol})(78.4/h) = 45.5 \text{ kg mol/h}$$
$$D = (0.70 \text{ kg mol})(78.4/h) = 54.9 \text{ kg mol/h}$$

Example 4.7   Evaporator–crystallizer units

Fresh feed containing 20% by weight KNO_3 (K) in H_2O (W) is combined with a recycle stream and fed to an evaporator. The concentrated solution leaving the evaporator, containing 50% KNO_3, is fed to a crystallizer. The crystals obtained from the crystallizer are 96% KNO_3 and 4% water. The supernatant liquid from the crystallizer constitutes the recycle stream and contains 0.6 kg KNO_3 per 1.0 kg of H_2O. Calculate all stream values and compositions.

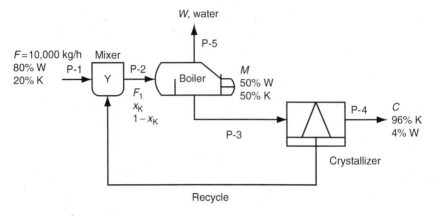

FIGURE E4.7
Flow sheet of evaporator–crystallizer process.

Solution
The labeled process flow sheet is shown in Figure E4.7.

Solution
Basis: 10,000 kg/h feed
Mass fractions of K in recycle:

$$\left(0.6\,\frac{\text{kg K}}{\text{h}}\middle/\frac{\text{kg H}_2\text{O}}{\text{h}}\right) = \left(0.6\,\frac{\text{kg K}}{\text{h}}\right)\middle/\left(1\,\frac{\text{kg H}_2\text{O}}{\text{h}} + 0.6\,\frac{\text{kg K}}{\text{h}}\right) = 0.375$$

Thus, $x_{K,R} = 0.375$ K and $x_{W,R} = 1 - x_{K,R} = 0.625$ W

Degree of freedom analysis

DFA	Evaporator	Crystallizer	Union	Overall Process
Number of unknowns	4	4	3	2
Number of independent equations	2	2	2	2
Number of relations	0	1	0	0
DF	2	1	1	0

System: overall process

$$\text{Overall balance: } 10{,}000\,\frac{\text{kg}}{\text{h}} = W + C$$

$$\text{Component balance (K): } 2000\,\frac{\text{kg}}{\text{h}} = (0.96)C$$

Solve this first since it is a tie component.

Results: $C = 2083 \frac{kg}{h}$, $W = 7917 \frac{kg}{h}$ (check your answer with W balance)

$$8000 \text{ kg} = 7917 \frac{kg}{h} + (0.04)\left(2083 \frac{kg}{h}\right) = 8000 \frac{kg}{h}$$

System: crystallizer

$$\text{Overall: } M = C + R = 2083 + R$$

$$K: 0.5M = (0.96)(2083) + (0.375)R$$

Answer: $$M = 9748 \frac{kg}{h}, R = 7665 \frac{kg}{h}$$

Now DOF for the evaporator and the union has been reduced to zero and either unit could be solved next:
System: union

$$\text{Overall: } 10{,}000 \frac{kg}{h} + 7665 \frac{kg}{h} = F_1 = 17{,}668 \frac{kg}{h}$$

$$K: 2000 \frac{kg}{h} + (0.375)\left(7665 \frac{kg}{h}\right) = x_K\left(17665 \frac{kg}{h}\right)$$

Results: $x_K = 0.276$

Example 4.8   Absorber–stripper process units

The absorber–stripper system shown below is used to remove carbon dioxide and hydrogen sulfide from a feed consisting of 30% CO_2 and 10% H_2S in nitrogen. In the absorber a solvent selectively absorbs the hydrogen sulfide and carbon dioxide. The absorber overhead contains only 1% CO_2 and no H_2S. N_2 is insoluble in the solvent. The rich solvent stream leaving the absorber is flashed, and the overhead stream consists of 20% solvent, and contains 25% of the CO_2 and 15% of the H_2S in the raw feed to the absorber. The liquid stream leaving the flash unit is split into equal portions, one being returned to the absorber. The other portion, which contains 5% CO_2, is fed to the stripper. The liquid stream leaving the stripper consists or pure solvent and is returned to the absorber along with makeup solvent. The stripper overhead contains 30% solvent. Calculate all flow rates and compositions of unknown streams?

Solution

The labeled process flow sheet is shown in Figure E4.8.

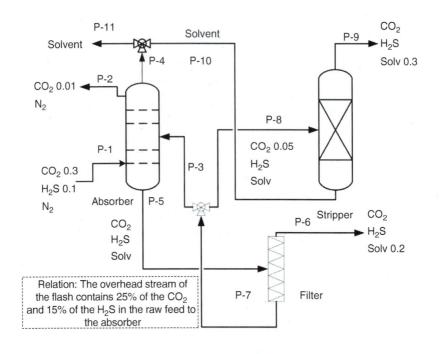

FIGURE E4.8
Flow sheet of multiple-unit process.

Degree of freedom analysis

DFA	Absorber	Flash	Stripper	Splitter	Mixer	Overall
Number of unknowns	8	7	5	4	3	7
Number of independent equations	4	3	3	1	1	4
Number of relations	0	2	0	1	0	2
DF	4	2	2	2	2	1

The lowest DOF value is for the overall process, taking this as the basis will reduce it to zero.
Basis: 100 mol/h of stream 1
System: overall process (i.e., entire block)

Material balance

$$\text{Total mole balance: } 100 + n_{11} = n_2 + n_6 + n_{10}$$

$$CO_2: 0.3(100) = 0.01n_2 + n_{10}x_{CO_2, 10} + n_6 x_{CO_2, 6}$$

$$H_2S: 0.1(100) = 0 + n_{10}(1 - 0.3 - x_{CO_2, 10}) + (1 - x_{CO_2, 6} - 0.2)n_6$$

$$\text{Solvent: } n_{11} = 0.2n_6 + 0.3n_{10}$$

$$\text{Relation 1: } 0.25(0.3100) = x_{CO_2,6}n_6$$

$$\text{Relation 2: } 0.15 \times (0.1 \times 100) = (1 - x_{CO_2,6} - 0.2)n_6$$

Answer: (using E–Z solve)

$n_2 = 60.61$, $n_6 = 11.25$, $n_{10} = 43.42$, $n_{11} = 15.276$, $x_{CO_2,10} = 0.504$, $x_{CO_2,6} = 0.67$

System: stripper

$$\text{Overall material balance: } n_8 = n_9 + 43.4$$

$$CO_2\text{: } 0.05 \times n_8 = 0.504 \times 43.4$$

$$\text{Solvent: } x_{s,8} \times n_8 = (0.3 \times 43.4) + n_9$$

Answer: $n_8 = 437.472$, $n_9 = 394.072$, $x_{s8} = 0.93$

System: splitter
Total mole balance: $n_7 = 437.472 + n_3$

$$\text{Relation: } n_8 = n_3$$

This leads to $n_7 = 875.8$ mol/h
System: flash
Total mole balance:

$$n_5 = n_6 + n_7$$

$$n_5 = 11.25 + 875.8$$

$$n_5 = 887.0 \text{ mol/h}$$

System: absorber
Total mole balance:

$$n_1 + n_3 + n_4 = n_2 + n_5$$

Total mole balance:

$$100 + 437.4 + n_4 = 60.6 + 887$$

$$n_4 = 410 \text{ mol/h}$$

Example 4.9   Toluene–xylene–benzene mixture

A liquid mixture containing 35.0 mol% toluene (T), 27.0 mol% xylene (X), and the remainder benzene (B) is fed to a distillation column. The bottom product contains 97.0 mol% X and no B, and 93.0% of the X in the feed is recovered in this stream. The overhead product is fed to a second column. The overhead product from the second column contains 5.0 mol% T and no X, and 96.0% of the benzene fed to the system is recovered in this stream.

(a) Draw and completely label a flow sheet of the process.
(b) Calculate (i) the composition of the bottoms stream from the second column and (ii) the percentage of toluene contained in the process feed that emerges in the bottom product from the second column.

Solution

(a) The labeled process flow sheet is shown in Figure E4.9.
(b) Since no stream flows are specified, assume a basis.

Basis: $\dot{n}_f = 100$ kmol/h
Given information:
93% of feed X is recovered in the bottom stream of the first column.

$$0.93(100)(0.27) = \dot{n}_1(0.97) \rightarrow \dot{n}_1 = 25.89 \text{ kmol/h}$$

96% of feed B is recovered in the overhead stream of the second column.

$$(0.96)(100)(0.38) = \dot{n}_3(0.95) \rightarrow \dot{n}_3 = 38.40 \text{ kmol/h}.$$

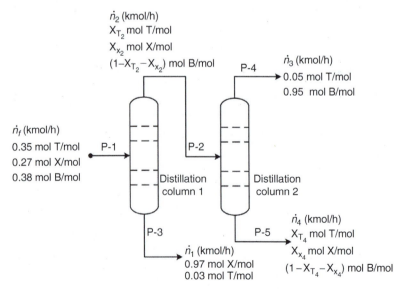

FIGURE E4.9
Binary separation process.

Balances on overall system (steady state, no reaction)

$$\text{Total moles: } 100 = \dot{n}_1 + \dot{n}_3 + \dot{n}_4$$
$$\text{Benzene: } 38 = 0.95\,\dot{n}_3 + x_{B_4}\dot{n}_4$$
$$\text{Xylene: } 27 = 0.97\,\dot{n}_1 + x_{X_4}\dot{n}_4$$

In the above set of equations the two variables, $\dot{n}_1$ and $\dot{n}_3$, are known and the remaining three variables are unknown, solve

$\dot{n}_4 = 35.71$ kmol/h, $x_{B_4} = 0.043$ mol B/mol, $x_{X_4} = 0.053$ mol X/mol, $\Rightarrow x_{T_4}$
$= 0.904$ mol T/mol

The bottom stream from the second column contains 4.3 mol% benzene, 5.3 mol% xylene, and 90.4 mol% toluene.

Fraction of toluene recovered in the bottom stream $= 1000\times$

$$\frac{(0.904)(35.71)}{35} = 92.3\%$$

Example 4.10   Benzene–toluene separation process

A feed stream to a distillation column consists of 50 mol% liquid benzene, and 50 mol% toluene is separated into two product streams by distillation. The feed stream flow rate is 100 mol/s. The vapor leaving the top of the column, which contains 97 mol% benzene, is completely condensed and split into two equal fractions: one is removed as the overhead product stream, and the other (the reflux) is recycled to the top of the column. The overhead product stream contains 89% of the benzene fed to the column. The liquid leaving the bottom of the column is fed to a partial reboiler in which 45% of it is vaporized. The vapor generated in the reboiler (the boilup) is recycled to become the rising vapor stream in the column, and the residual reboiler liquid is removed as the bottom product stream. The relation that governs the compositions of the streams leaving the reboiler is

$$\frac{y_B/(1-y_B)}{x_B(1-x_B)} = 2.25,$$

where y_B and x_B are the mole fractions of benzene in the vapor and liquid streams, respectively:

a. Draw and completely label a flowchart.
b. How many systems exist and what are they?
c. Perform a DFA.
d. Calculate the molar amounts of the overhead and bottom products and the mole fraction of benzene in the bottom product.
e. The percentage recovery of toluene in the bottom product stream (100 × mole toluene in bottoms/mole toluene in feed).

Solution

(a) The diagram below (Figure E4.10a) shows the schematic of the distillation column used to separate binary liquid mixtures of benzene and toluene. Inside the column, a liquid stream flows downward and the vapor stream rises. At each point in the column, some of the liquid vaporizes and some of the vapor condenses.

(b) There are four systems (overall process, column, condenser, and reboiler). Do the DOF analysis and identify a system with which the process analysis might be appropriately beginning with 1 and 0 DOF.

(c) DFA.

	System			
DFA	Condenser	Reboiler	Column	Overall Process
Number of unknowns	2	6	6	3
Number of independent equations	1	2	2	2
Number of relations	—	2	1	1
DF	1	2	3	0

(d) Calculate all flow rates and compositions of unknown streams.

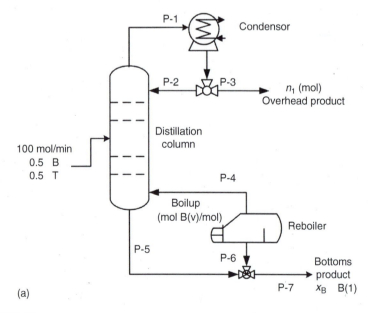

(a)

FIGURE E4.10a
Distillation column.

Solution

Starting with the system of 0 DOF
Basis: 100 mol/s of feed
System: overall process
Material balance for the selected system:
Total balance:

$$100 \text{ mol} = n_1 + n_3 \tag{4.3}$$

Component balance (B):

$$0.5(100 \text{ mol}) = 0.97n_1 + x_B n_3 \tag{4.4}$$

Auxiliary relation:

$$0.89(0.5 \times 100) = 0.97n_1 \tag{4.5}$$

We have three equations with three unknowns; the unknown variables can be found using E–Z solve or by manual calculation (Figure E4.10b).

Our second system will be the condenser.

System: condenser

Total material balance around the condenser:

$$n_2 = n_1 + n_1 = 2n_1$$

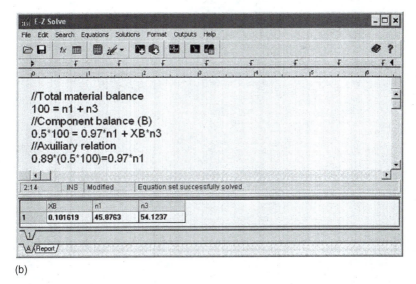

(b)

FIGURE E4.10b
Results using E–Z solve.

n_1 has already been calculated for the overall process system.

$$n_2 = 2n_1 = 2(45.88)$$
$$n_2 = 91.76 \text{ mol/s}$$

System: partial reboiler
Total material balance for the reboiler:

$$n_5 = 54.123 + n_4 \tag{4.6}$$

Component balance (B):

$$z_B n_5 = 0.1(54.1237) + y_B n_4 \tag{4.7}$$

Relation 1:

$$n_4 = 0.45n_5 \tag{4.8}$$

Relation 2:

$$\frac{y_B(1 - x_B)}{x_B(1 - y_B)} = 2.25 \tag{4.9}$$

We have four equations and four unknown variables (n_4, n_5, y_B, z_B), so the equations can be solved simultaneously using E–Z solve (Figure E4.10c).

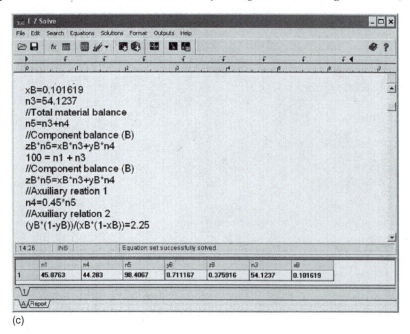

(c)

FIGURE E4.10c
Results obtained with E–Z solve software.

All unknown variables have been calculated and there is no need to do material balance around the column system.

(a) The percentage recovery of toluene $= \dfrac{0.101619 \times 54.1237}{50} \times$ 100% = 11%

Example 4.11 Acetone absorber

Air containing 3% acetone and 2% water is fed to an absorber column. The mass flow rate of air is 1000 kg/h. Pure water is used as absorbent to absorb acetone from air. The air leaving the absorber should be free of acetone. The air leaving the absorber was found to contain 0.5% water. The bottom of the absorber is sent to a distillation column to separate acetone from water. The bottom of the distillation column was found to contain 4% acetone and the balance is water. The vapor from the head of the absorber was condensed. The concentration of the condensate was 99% acetone and the balance water (Figure E4.11).

(1) Draw and label a process flowchart.
(2) Calculate the flow rate of all unknown streams.

Solution

Start the solution with a system of 0 DOF.

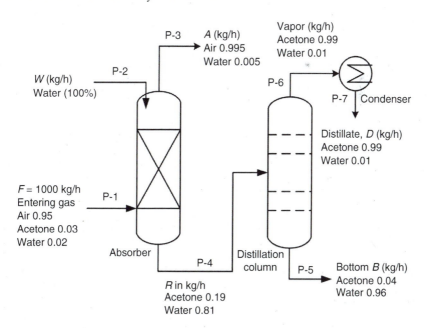

FIGURE E4.11
Absorption–separation process.

The DFA is shown in the following table.

	Systems		
DFA	**Absorber**	**Distillation**	**Overall Process**
Number of unknowns	3	3	4
Number of independent equations	3	2	3
Number of relations	—	—	—
DF	0	1	1

System: absorber
Component balance (air):

$$0.95(1000) = 0.995A$$

$$A = 954.774 \text{ kg/h}$$

Component balance (acetone):

$$0.03(1000) = 0.19R$$

$$R = 1578.95 \text{ kg/h}$$

Overall balance:

$$W + 100 = A + R$$

$$W + 100 = 954.77 + 1578.95$$

$$W = 2433.72 \text{ kg/h}$$

System: distillation column
Total balance:

$$R = V + B$$

$$1578.95 = V + B$$

Component balance (acetone):

$$0.19(1578.95) = 0.99V + 0.04B$$

The above equations can be solved using E–Z solve or manual calculations by substituting V from the total balance equation in the component balance equation and solving for B as follows:

$$1578.95 = V + B$$

$$1578.95 - B = V$$

$$0.19(1578.95) = 0.99(1578.95 - B) + 0.04B$$

$$0.19(1578.95) = 0.99 \times 1578.95 - 0.99B + 0.04B$$

$$0.95B = 1263.15$$

$$B = 1329.63 \text{ kg/h}$$

$$B = 1329.63 \text{ kg/h}, \quad V = 249.308 \text{ kg/h}$$

$$V = D = 249.308 \text{ kg/h}$$

Example 4.12   Case study

Consider a process shown in Figure E4.12a, in which freshly mined ore is to be cleaned so that later processing units are not contaminated with dirt. 3000 kg/h of dirty ore is dumped into a large washer, in which water is allowed to soak the ore on its way to a drain at the bottom of the unit. The amount of dirt remaining on the ore after this process is negligible, but water remains absorbed on the ore surface such that the net mass flow rate of the cleaned ore is 3100 kg/h. The dirty water is cleaned in a settler, which is able to remove 90% of the dirt in the stream without removing significant amount of water. The cleaned stream is then combined with a fresh water stream before re-entering the washer. The wet, clean ore enters a dryer, in which all of the water is removed. Dry ore is removed from the dryer at 2900 kg/h.

(a) Calculate the necessary mass flow rate of fresh water to achieve this removal at steady state.

(b) Suppose the solubility of dirt in water is 0.4 g dirt/cm³ H_2O. Assuming that the water leaving the washer is saturated with

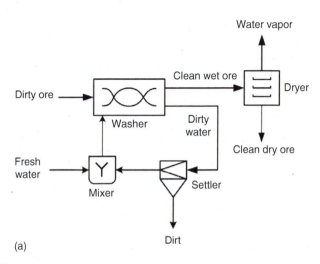

(a)

FIGURE E4.12a
Process block diagram of dirty ore cleaning.

dirt, calculate the mass fraction of dirt in the stream that enters the washer (after it has been mixed with the freshwater stream).

The design schematic for this process is as follows.

Solution

A schematic is given in the problem statement but it is incomplete (Figure E4.12a), since it does not contain any of the design specifications (i.e., flow rates and compositions). Therefore, it is highly recommended that you draw your own process flow sheet even when one is provided for you. Make sure you label all of the streams, and the unknown concentrations (see Figure E4.12b).

The summary of the DOF is shown in Table E4.12.

Detailed explanation of the DFA of each unit is shown below:

System: washer: six independent unknowns $(x_{O1}, \dot{m}_2, x_{D2}, \dot{m}_3, x_{D3}, x_{O4})$, three independent mass balances (ore, dirt, and water), and one solubility as relation. The washer has 2 DOF.

System: dryer: two independent unknowns $(x_{O4}, \dot{m}_5)$ and two independent equations. The dryer has 0 DOF.

System: settler: five independent unknowns $(\dot{m}_3, x_{D3}, \dot{m}_7, \dot{m}_8, x_{D8})$, two mass balances (dirt and water), the solubility of saturated dirt, and one additional information (90% removal of dirt), leaving us with 1 DOF.

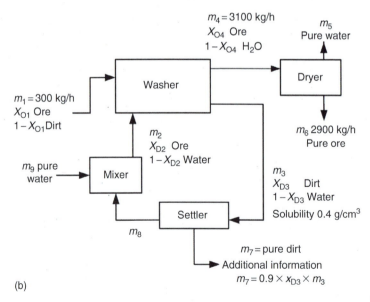

(b)

FIGURE E4.12b
Labeled block diagram for ore cleaning process.

TABLE E4.12

Degree of Freedom Analysis

	System				
DFA	Washer	Dryer	Settler	Mixing Point	Overall Process
Number of unknowns	6	2	5	5	3
Number of independent equations	3	2	2	2	3
Number of relations	1	—	2	—	—
DF	2	0	1	3	0

System: mixing point: we need to include this in order to calculate the total DOF for the process; otherwise we are not counting m_9 anywhere. Five unknowns ($\dot{m}_2$, x_{D2}, $\dot{m}_8$, x_{D8}, $\dot{m}_9$) and two mass balances leave us with 3 DOF.

Therefore, overall $= 3 + 2 + 1 - 6$ intermediate variables (not including x_{O4} since this is going to the dryer) $= 0$. The problem is well defined.

Solution

1. Start with a unit operation or some combination of unit with 0 DOF, calculate those variables, and then recalculate the DOF until everything is accounted for.

2. From our initial analysis, the dryer had 0 DOF so we can calculate the two unknowns x_{O4} and m_5. Now we can consider x_{O4} and m_5 known and redo the DFA on the unit operations.

Washer: we only have five unknowns now (x_{O1}, $\dot{m}_2$, x_{D2}, $\dot{m}_3$, x_{D3}), but still only three equations and the solubility, 1 DOF.

Settler: nothing has changed here since x_{O4} and m_5 are not connected to this operation.

Overall system: we have three unknowns (x_{O1}, $\dot{m}_7$, $\dot{m}_9$) since $\dot{m}_5$ is already determined, and we have three mass balances (ore, dirt, and water). Hence, we have 0 DOF for the overall system.

Now we can say we know x_{O1}, $\dot{m}_7$, and $\dot{m}_9$.

Settler again: since we know m_7 the settler now has 0 DOF and we can solve for $\dot{m}_3$, x_{D3}, $\dot{m}_8$, and x_{D8}.

Around the washer again: now we know m_8 and x_{D8}. How many balances can we write?

If we try to write a balance on the ore, we will find that the ore is already balanced because of the other balances we have done. If you try to write an ore balance, you will see you already know the values of all the unknowns in the equations. Hence, we cannot count this balance as an equation we can use. The washer, therefore, has two unknowns (m_2, x_{D2}) and two equations (the dirt and water balances), hence DOF is 0. Balances on the recombination point can also do this final step. Once we have m_2 and x_{D2} the system is completely determined. The only given information in inconsistent units is the solubility, which is given as 0.4 g dirt/cm^3 H$_2$O. However, we know the density of water (or can look it up), and we can convert this to kilogram dirt per kilogram water as follows: $\frac{0.4 \text{ g dirt}}{\text{cm}^3\text{H}_2\text{O}} \Big| \frac{1 \text{ cm}^3\text{H}_2\text{O}}{\text{g H}_2\text{O}} \Big| = \frac{0.4 \text{ g dirt}}{\text{g H}_2\text{O}} = 0.4 \frac{\text{kg dirt}}{\text{kg H}_2\text{O}}$.

Now this information is in the same units as the mass flow rates; we can proceed to the next step. First, do any two mass balances on the dryer. I choose total and ore balances. Remember that the third balance is not independent of the first and two.

$$\text{Overall balance: } \dot{m}_4 = \dot{m}_5 + \dot{m}_6$$

$$\text{Ore balance: } \dot{m}_4 x_{O4} = \dot{m}_5 x_{O5} + \dot{m}_6 x_{O6}$$

Substituting the known values:

$$\text{Overall: } 3100 = \dot{m}_5 + 2900$$

$$\text{Ore: } x_{O4} \times 3100 = 1 \times 2900$$

Solving gives: $\dot{m}_5 = 200 \frac{\text{kg}}{\text{h}}$, $x_{O4} = 0.935$

Now that we are finished with the dryer we move to the next step in our plan, which is the overall system balance:

$$\text{Water balance: } \dot{m}_9 = \dot{m}_5$$

$$\text{Ore balance: } x_{O1} \times \dot{m}_1 = \dot{m}_6$$

$$\text{Dirt balance: } (1 - x_{O1}) \times \dot{m}_1 = \dot{m}_7$$

$$\dot{m}_9 = 200 \frac{\text{kg}}{\text{h}}, \ \dot{m}_7 = 100 \frac{\text{kg}}{\text{h}}, \ x_{O1} = 0.967$$

Next we move to the settler as planned. This one is a bit trickier since the solutions are not immediately obvious but a system must be solved.

$$\text{Overall balance: } \dot{m}_3 = \dot{m}_7 + \dot{m}_8$$

$$\text{Dirt balance: } \dot{m}_3 \times x_{D3} = \dot{m}_7 \times x_{D7} + \dot{m}_8 \times x_{D8}$$

$$\text{Efficiency of removal: } \dot{m}_7 = 0.9 \times m_3 \times x_{D3}$$

Using the solubility is slightly tricky. You use it by noticing that the mass of dirt in stream 3 is proportional to the mass of water, and you can write that:

Mass dirt in stream 3 $= 0.4 \times$ mass water in stream 3
Solubility: $\dot{m}_3 \times x_{D3} = 0.4 \times \dot{m}_3 \times (1 - x_{D3})$

Plugging in known values, the following system of equations is obtained:

$$\dot{m}_3 = 100 + \dot{m}_8$$

$$\dot{m}_3 \times x_{D3} = 100 + \dot{m}_8 \times x_{D8}$$

$$\dot{m}_3 \times x_{D3} = 111.11$$

$$\dot{m}_3 \times x_{D3} = 0.4 \times \dot{m}_3 \times (1 - x_{D3})$$

Solving these equations for the four unknowns, the solutions are

$$\dot{m}_3 = 388.89 \; \frac{kg}{h}, \; \dot{m}_8 = 288.89 \; \frac{kg}{h}, \; x_{D3} = 0.286, \; x_{DS} = 0.0385$$

Finally, we can go to the mixing point, and say

$$\text{Overall: } \dot{m}_8 + \dot{m}_9 = \dot{m}_2$$

$$\text{Dirt: } \dot{m}_8 \times x_{D8} = \dot{m}_2 \times x_{D2}$$

From these the final unknowns are obtained: $\dot{m}_2 = 488.89 \; \frac{kg}{h}, \; x_{D2} = 0.0229$

Since the problem was to solve for m_2, we are now finished.

Check your work
These values should be checked by making a new flowchart with the numerical values, and ensuring that the balances on the washer are satisfied. This is left as an exercise for the reader.

4.3 Problems

4.3.1 Separations of Benzene, Toluene, Xylene Mixtures

Benzene (B), molecular weight $= 78$, toluene (T), molecular weight $= 92$, and xylene (X), molecular weight $= 106$, are to be separated using the two stage process. The feed to this process contains 50 wt% benzene, 30 wt% toluene, and 20 wt% xylene. The feed enters the process at a flow rate of 30,000 kg/h. The overhead stream from the first unit contains 95 wt% benzene, 3 wt% toluene, and 2 wt% xylene. The overhead stream from the second unit contains 3 wt% benzene, 95 wt% toluene, and 2 wt% xylene. The flow rate of the overhead from the first unit is 52% of the feed flow rate. The amount of benzene in the overhead from the second unit is 75% of the benzene that enters the second unit coming, that is, the bottom stream from the first unit. (a) Determine the flow rates for all of the streams leaving both units. (b) Determine the compositions of the bottom streams from both units.

4.3.2 Filtration Processes

Slurry consisting of $CaCO_3$ precipitate in a solution of NaOH and H_2O is washed with an equal mass of a dilute solution of 5 wt% NaOH in H_2O. The washed slurry, which leaves the unit, contains 2 kg solution per 1 kg of solid ($CaCO_3$). The clear solution (clear solution contains no $CaCO_3$ precipitate) withdrawn from the unit can be assumed to have the same concentration as the solution withdrawn with the solids. The feed slurry contains equal mass fractions of all components. On the basis of 100 kg/h of feed slurry, draw and label the process flowchart and calculate the composition of the clear solution.

4.3.3 Concentration of Orange Juice

Fresh orange juice is concentrated before freezing by the process illustrated in the flowchart below. Ninety percent of the water entering the evaporator in stream 2 is evaporated. Assume that only water evaporates and that none of the volatile components of the orange juice evaporates (Figure P4.3). Calculate the total mass flow rates of streams 2, 3, 4, and 5.

4.3.4 Separation of NaCl and KCl Mixture

A solution containing 15.0 wt% NaCl, 5.0 wt% KCl, and the balance water is fed to the separation process shown below. The feed rate is 18.4 t/h. The evaporator product (P) stream composition is 16.8 wt% NaCl, 21.6 wt% KCl, and the balance is water. The recycle (R) stream contains 20 wt% NaCl, the balance is KCl, and water. The bottom products of the evaporator and the crystallizer are pure and consist of dry KCl and NaCl, respectively. Calculate the total flow rate and the weight percentage composition of all streams indicated on the flowchart (Figure P4.4).

4.3.5 Sulfur Removal System

The raw feed to a sulfur removal system contains 15 mol% CO_2, 5% H_2S, and 1.41% CO_2. The balance is CH_4. The original absorber design placed a

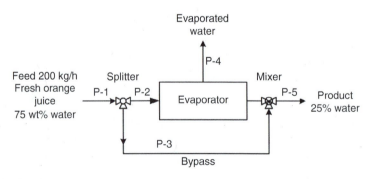

FIGURE P4.3
Evaporator process block diagram.

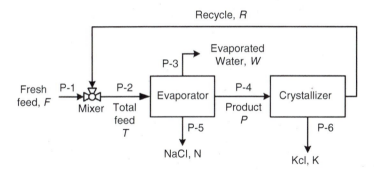

FIGURE P4.4
Process block diagram of evaporator–crystallizer process.

maximum flow limit of 820 kmol/h and yielded a product stream with only 1% H_2S and 0.3% CO_2. The feed to the unit is 1000 kmol/h. The excess feed flow is bypassed. To maintain the same product quality, the absorber was operated with an enhanced removal of the sulfur-containing species. Draw and label the process flowchart. Find the flow rates and compositions of all unknown streams.

4.3.6 Separation of DMF–Nitrogen Mixture

A separator is designed to remove exactly 75% of the DMF that is fed to it. The DMF is removed from the bottom of the separator. The exit composition of stream leaving the separator should contain maximum 10% DMF and the balance is nitrogen. To achieve this goal, a recycle stream is used, where part of the exit of the separator is back recycled. Draw and label a process flowchart. Calculate the flow rate and compositions of all unknown streams. Compute the fraction of the separator exit stream that must be recycled.

4.3.7 Separation of Benzene–Toluene Mixture

A distillation column separates a mixture of 100 kg/h of 50% benzene and the balance is toluene. The distillate at the top of the column was found to contain 95% benzene, and the bottom products contain 96% toluene. The vapor flow rate from the top of the column is 80 kg/h. The condenser condenses vapor completely. A portion of the condenser product is recycled to the column as reflux. The rest of the condensate is withdrawn as product. Calculate all unknown streams flow rates and compositions?

4.3.8 Separation of Potassium Nitrate

Given the process shown, find the recycle flow in kilograms per hour, the production rate of potassium nitrate, and the recycle ratio (Figure P4.8).

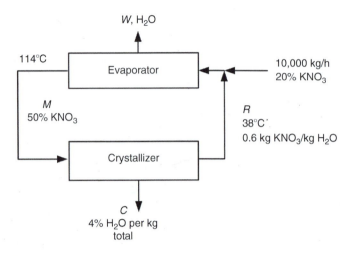

FIGURE P4.8
Process flow sheet of evaporator–crystallizer system.

Hint: You are asked to find three things: (i) the recycle flow (labeled R on the drawing), (ii) the production rate (labeled C on the drawing), and (iii) the recycle ratio, which will be calculated as R/10,000, if we do not change the basis. The sketch is already done, so we need to label the variables. Let us call the fresh feed F. If we look over the compositions, we will note that they are not consistently represented. Two are given as percentage KNO_3, one as percentage H_2O, and one as pounds of KNO_3 per pound of H_2O. We would usually prefer all the numbers to be in percent, so let us do the problem using percent nitrate. Shifting the water percentage to nitrate is easy; just subtract from 100. The other stream (R) requires a calculation. Do we really need to calculate the mass fraction of nitrate in R? Since we are asked two questions about the stream, it seems almost certain. To do this calculation, look at the composition given and choose a basis for computing the composition (we can change the basis for the rest of the problem).

4.3.9 Production of Instant Coffee

Coffee beans contain components that are soluble (S) in water and others that are insoluble (I). Instant coffee is produced by dissolving the soluble portion in boiling water in large percolators, then feeding the coffee to a spray dryer in which the water is evaporated, leaving the soluble coffee as a dry powder. The insoluble portion of the coffee beans (the spent grounds) passes through several drying operations, and the dried grounds are either burned or used as landfill. The solution removed from the grounds in the first stage of the drying operation may be fed to the spray dryer to join the effluent from the percolators. A flowchart of this process is shown in Figure P4.9. The symbols S and I denote the soluble and insoluble

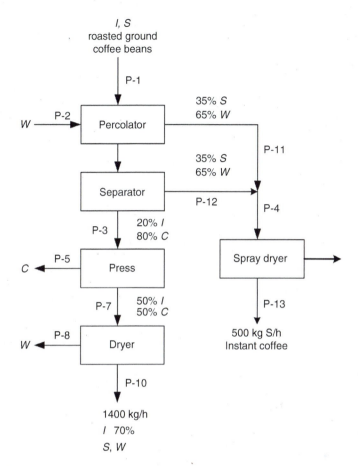

FIGURE P4.9
Block flow diagram of the synthesis of instant coffee.

components of the coffee beans, W is water, and C is a solution containing 35% S and 65% W by mass. Calculate the flow rates in kilograms per hour of each of streams 1 through 8.

Further Readings

1. Reklaitis, G.V. (1983) *Introduction to Material and Energy Balances,* John Wiley & Sons, New York.
2. Felder, R.M. and R.W. Rousseau (1978) *Elementary Principles of Chemical Processes,* John Wiley, New York.
3. Himmelblau, D.M. (1996) *Basic Principles and Calculations in Chemical Engineering,* 6th edn, Prentice-Hall, New Jersey.

4. Whirwell, J.C. and R.K. Toner (1969) *Conservation of Mass and Energy*, Blaisdell, Waltham, Massachusetts.
5. Shaheen, E.I. (1975) *Basic Practice of Chemical Engineering*, Houghton Mifflin, Boston.
6. Hougen, O.A., K.M. Watson, and R.A. Ragatz (1954) *Chemical Process Principles: Material and Energy Balances*, 2nd edn, John Wiley, New York.

5

Material Balances in Reactive Processes

At the End of This Chapter You Should Be Able to

1. Understand how to apply material balances involving chemical reactions.
2. Understand what is single-pass and overall conversion.
3. Identify limiting and excess reactant.
4. Recognize the extent of reaction.
5. Define flue gas, stack gas in combustion problems, theoretical air (oxygen), required air (oxygen), and excess air (oxygen).
6. Solve material balance problems containing chemical reactions using any of the three available solving techniques: molecular balance, extent of reaction method, and atomic balance.

5.1 Amount of Substance in Moles

A very useful way to express the amount of a substance is the mole:

- 1 mol $= 6.023 \times 10^{23}$ particles (i.e., atoms or molecules).
- Units frequently used are mol (g-mol), kmol (kg-mol), and lb-mol.
- Moles $=$ mass/molecular weight (MW).

5.1.1 Why Use the Mole?

- Reactions occur between molecules not kilograms of molecules, for example,

$$C + O_2 \rightarrow CO_2.$$

- We could work in "number of molecules," but quantities would be very large.

- We, therefore, use the "mole," which is a fixed number of molecules, equal to Avogadro's number (6.023×10^{23}).
- Because each molecule has a different mass, the mole converts into a different mass (g, kg, lb).

Example 5.1 Moles in chemical reactions

1. How many moles of atomic hydrogen and oxygen would be released in 1 mol of H_2O, if the latter was broken up into its constituent parts?
2. An average rain drop contains 0.05 g water, how many moles are there in this rain drop? What would be the mass of air in the same number moles, if we assume air has an MW of 29?
3. How many kilograms of H_2 can be obtained by electrolysis of 1 kg of water?
4. Balance the equation for glucose oxidation (i.e., determine α, β, and γ)

$$C_6H_{12}O_6 + \alpha O_2 \rightarrow \beta CO_2 + \gamma H_2O$$

Solution

1. The number of moles of atomic hydrogen and oxygen that would be released if 1 mol of water is broken up into its constituent parts can be calculated as follows:

   Consider 1 mol of water broken up into its constituent parts.

$$H_2O \rightarrow 2H + O$$

   There are two hydrogen atoms and one oxygen, atom in a water molecule. Thus, 1 mol of water will release 2 mol of atomic hydrogen and 1 mol of atomic oxygen if broken up.

2. The number of moles in an average rain drop contains 0.05 g water, which can be calculated knowing the MW of water (H_2O), that is, $(2 \times 1) + (1 \times 16) = 18$ g/mol.

   The number of moles is therefore

$$(n) = m/\text{MW} = 0.05 \text{ g}/(18 \text{ g/mol}) = 2.78 \times 10^{-3} \text{ mol}.$$

   The mass of air in the same number of moles of the rain drop can be calculated as follows. The MW of air is 29 g/mol. The mass of air in a similar number of moles of that of water is given by

$$\text{Grams of air in } 2.78 \times 10^{-3} \, \cancel{\text{mol}} \left| \frac{29 \text{ g}}{\cancel{\text{mol}}} \right. = 0.08 \text{ g}$$

3. Moles of hydrogen that result from electrolysis of 1 kg of water can be obtained as follows:
   Electrolysis is the use of electrical energy to turn water into H_2 and O_2.

   The relevant reaction is $H_2O \rightarrow H_2 + 1/2O_2$

   Thus, 1 mol of H_2O yields 1 mol of H_2.
   The moles in 1 kg of H_2O is $n = m/MW = 1$ kg/(18 kg/kmol).
   Thus, (1/18) kmol of H_2O yields (1/18) kmol of H_2.
   Mass of (1/18) kmol H_2 = number of moles of H_2 multiplied by the MW of hydrogen.

$$\text{Mass of (1/18) kmol } H_2 = \frac{1}{18} \text{ kmol} \left| \frac{2 \text{ kg}}{1 \text{ kmol}} \right| = \frac{1}{9} \text{ kg}$$

4. Balance the equation for glucose oxidation to determine α, β, and γ

$$C_6H_{12}O_6 + \alpha O_2 \rightarrow \beta CO_2 + \gamma H_2O$$

α, β, and γ are determined by balancing the number of each atom:

C: $6 = \beta$, H: $12 = 2\gamma \Rightarrow \gamma = 6$, O: $6 + 2\alpha = 2\beta + \gamma \Rightarrow \alpha = 6$

Thus, the balanced equation is

$$C_6H_{12}O_6 + 6O_2 \rightarrow 6CO_2 + 6H_2O$$

5.2 General Material Balance

There are two types of material balances: differential and integral.

5.2.1 Differential Balance

A differential balance is a balance at one particular instant in time and deals with rates (for mass balances: mass per time [kg/s]).Typically, use a general balance, where it may be written more precisely in a mathematical form as

$$\frac{dM}{dt} = \dot{M_I} - \dot{M_O} + \dot{G} - \dot{C}$$

TABLE 5.1

Equality Required for Input and Output of a Steady-State Process

Type of Balance	Without Chemical Reaction	With Chemical Reaction
Total mass	Yes	Yes
Total moles	Yes	No
Mass of a chemical compound	Yes	No
Moles of a chemical compound	Yes	No
Moles of an atomic species	Yes	Yes

where

dM/dt denotes the rate of change of the material M

$\dot{G}$ and $\dot{C}$ denote the rate of generation and consumption, respectively,

and the over dots denote flow rates.

There are some special cases:

For steady state: In $-$ out $+$ generation $-$ consumption $= 0$

$$\dot{M}_I - \dot{M}_O + \dot{G} - \dot{C} = 0$$

For no reaction: in $=$ out; $\dot{M}_I = \dot{M}_O$.

5.2.2 Integral Balance

An integral balance deals with the entire time of the process at once; so it uses amounts rather than rates: e.g., mass not mass/time. At steady state, none of the process variables change with time (if we ignore small random fluctuations). Under unsteady-state conditions, the process variables change with time. (One class of unsteady-state processes consists of oscillatory phenomena, where process variables change with time in a regular way. All other unsteady processes may be called "transient," meaning that the process variables continuously evolve over time.) The stoichiometric equation of the reaction imposes constraints on the relative amounts of reactants and products in the input and output streams. The simple relation "input equals the output" holds for steady-state processes under the following circumstances (Table 5.1).

5.3 Stoichiometry Basics

There are few definitions frequently used in solving material balance problems that involve chemical reactions. Such definitions are stoichiometric equations, stoichiometric coefficients, and stoichiometric ratios.

5.3.1 Stoichiometric Equation

It is an equation that relates the relative number of molecules or moles of reactants and products (but not mass) that participate in a chemical reaction. To be valid, the equation must be balanced. For example, the following stoichiometric equations are not balanced:

$$C_2H_5OH + O_2 \rightarrow CO_2 + H_2O \text{ (H and C atoms are not balanced)}$$

$$(NH_4)_2Cr_2O_7 \rightarrow Cr_2O_3 + N_2 + H_2O \text{ (O and H atoms are not balanced)}$$

The following equations are balanced because the number of atoms (C, H, O, and N) are the same on both sides of the equations:

$$C_2H_5OH + 3O_2 \rightarrow 2CO_2 + 3H_2O$$

$$(NH_4)_2Cr_2O_7 \rightarrow Cr_2O_3 + N_2 + 4H_2O$$

5.3.2 Stoichiometric Coefficients (ν_i)

These are the values preceding each molecular species, i, in a balanced stoichiometric equation. Values are defined to be positive for products and negative for reactants. For the reaction,

$$2SO_2 + O_2 \rightarrow 2SO_3$$

$$\nu_{SO_2} = -2,\ \nu_{O_2} = -1,\ \nu_{SO_3} = 2$$

5.3.3 Stoichiometric Ratio

It is ratio of stoichiometric coefficients in a balanced stoichiometric equation. For the oxidation of sulfur dioxide,

$$2SO_2 + O_2 \rightarrow 2SO_3$$

The following stoichiometric ratio is employed in solving material balance problems that involve chemical reaction:

$$\frac{2 \text{ mol } SO_3 \text{ generated}}{1 \text{ mol } O_2 \text{ consumed}}$$

Two reactants, A and B, are in stoichiometric proportion if the ratio (moles of A present)/(moles of B present) equals their stoichiometric ratio determined from the balanced stoichiometric equation.

5.4　Limiting and Excess Reactants

The reactant that is completely consumed when a reaction is run to completion is known as the limiting reactant. The other reactants are termed as excess reactants. To find the limiting reactant:

1. Balance the stoichiometric equation.
2. Take the ratio of the reactant feed rate to reactant stoichiometric coefficients. The limiting reactant is the reagent or the reactant that has the lowest ratio of the reactants molar feed rate to the reactant stoichiometric coefficient, that is,

$$\left(\frac{\dot{n}_i^0 \ (\text{molar flow rate of component } i \text{ in the feed})}{\nu_i \ (\text{stoichiometric cofficient of component } i)} \right)$$

The excess reactants are the ones that are left over. The fractional excess is the ratio of the amount by which the feed exceeds stoichiometric requirements (in moles) divided by the stoichiometric requirement (in moles). The fractional excess of the reactant is the ratio of the excess to the stoichiometric requirement:

$$\text{Fractional excess of A} = \frac{(n_A)_{\text{feed}} - (n_A)_{\text{stoich}}}{(n_A)_{\text{stoich}}} \ \text{or} \ \left(= \frac{(\dot{n}_A)_{\text{feed}} - (\dot{n}_A)_{\text{stoich}}}{(\dot{n}_A)_{\text{stoich}}} \right)$$

where $(n_A)_{\text{feed}}$ is the number of moles of an excess reactant, A, present in the feed to a reactor and $(n_A)_{\text{stoich}}$ is the stoichiometric requirement of A, or the amount needed to react completely with the limiting reactant. Percentage excess of A is 100 times the fractional excess.

Example 5.2　Limiting reactant

For the following the cases, determine which reactant is limiting and which is in excess as well as the percent excess for that component when

1. 2 mol of nitrogen (N_2) react with 4 mol of hydrogen (H_2) to form ammonia (NH_3) via the reaction ($MW_{N_2} = 28$, $MW_{H_2} = 2$, $MW_{NH_3} = 17$)

$$N_2 + 3H_2 \rightarrow 2NH_3$$

2. 100 kg ethanol (C_2H_5OH) reacts with 100 kg of acetic acid (CH_3COOH) to form ethyl acetate

$$C_2H_5OH + CH_3COOH \rightarrow CH_3COOHC_2H_5 + H_2O$$

3. 64 g methanol (CH_3OH) reacts with 0.5 mol of oxygen (O_2) to form formaldehyde.

$$CH_3OH + \frac{1}{2}O_2 \rightarrow HCHO + H_2O$$

Solution

1. The feed rate to stoichiometric ratio of both reactants is as follows:

$$N_2 + 3H_2 \rightarrow 2NH_3$$

$$\frac{2.0 \text{ lb-mol}}{1} \quad \frac{4 \text{ lb-mol}}{3}$$

$$2.0 \qquad \boxed{1.33}$$

This means that hydrogen (H_2) is the limiting reactant. Then the percent excess of nitrogen (N_2) can be calculated from the percent excess equation:

$$\% \text{ excess of } N_2 = \frac{(N_2)_{feed} - (N_2)_{stoich}}{(N_2)_{stoich}} \times 100\%$$

$$(N_2)_{stoich} = 4 \text{ lb-mol } H_2 \left| \frac{1 \text{ lb-mol } N_2}{3 \text{ lb-mol } H_2} \right| = \frac{4}{3} \text{ lb-mol } N_2$$

$$\% \text{ excess of } N_2 = \frac{2 - \frac{4}{3} \text{ lb-mol } N_2}{\frac{4}{3} \text{ lb-mol } N_2} \times 100\% = 50\%$$

2. $MW_{C_2H_5OH} = 46$, $MW_{CH_3COOH} = 60$, $MW_{CH_3COOHC_2H_5} = 88$, $MW_{H_2O} = 18$ kg/kmol

First you need to convert the masses given to moles. Thus,

$$C_2H_5OH: 100 \text{ kg} \left(\frac{1 \text{ kmol}}{46 \text{ kg}} \right) = 2.17 \text{ kmol } C_2H_5OH$$

$$CH_3COOH: 100 \text{ kg} \left(\frac{1 \text{ kmol}}{60 \text{ kg}} \right) = 1.67 \text{ kmol } CH_3COOH$$

The feed rate to stoichiometric ratio of both reactants is as follows:

$$C_2H_5OH + CH_3COOH \rightarrow CH_3COOHC_2H_5 + H_2O$$

$$\frac{2.17 \text{ kmol}}{1} \quad \frac{1.67}{1}$$

$$2.17 \qquad \boxed{1.67}$$

The chemical species with the lowest ratio is the limiting reactant. This means that acetic acid (CH_3COOH) is the limiting reactant. The percent excess of ethanol (C_2H_5OH) is

$$\% \text{ excess of A} = \frac{(n_A)_{feed} - (n_A)_{stoich}}{(n_A)_{stoich}} \times 100\%$$

$$\% \text{ excess} = \frac{2.17 - 1.67}{1.67} \times 100\% = 30.0\%$$

3. $MW_{CH_3OH} = 32$ g/mol, $MW_{O_2} = 32$ g/mol, $MW_{HCHO} = 30$ g/mol, $MW_{H_2O} = 18$ g/mol

First you need to convert the masses given to moles. Thus,

$$CH_3OH: 64 \text{ g}\left(\frac{1 \text{ mol}}{32 \text{ g}}\right) = 2 \text{ mol } CH_3OH$$

The feed to stoichiometric ratio is as follows:

$$CH_3OH + \frac{1}{2}O_2 \rightarrow HCHO + H_2O$$

$$\frac{2}{1} \quad \frac{0.5}{1/2}$$

$$2 \qquad \boxed{1}$$

This means oxygen (O_2) is the limiting reactant. The percent by which methanol is in excess is

$$\% \text{ excess of } CH_3OH = \frac{(n_{CH_3OH})_{feed} - (n_{CH_3OH})_{stoich}}{(n_{CH_3OH})_{stoich}} \times 100\%$$

$$(n_{CH_3OH})_{stoich} = (0.5 \text{ mol } O_2)_{feed}\left(\frac{1 \text{ mol } CH_3OH \text{ consumed}}{1/2 \text{ mol } O_2 \text{ consumed}}\right)$$

$$= 1 \text{ mol } CH_3OH$$

$$\% \text{ excess of } CH_3OH = \frac{(2 \text{ mol})_{feed} - (1 \text{ mol})_{stoich}}{(1 \text{ mol})_{stoich}} \times 100\% = 100\%$$

5.5 Fractional Conversion

Chemical reactions do not occur instantaneously, but rather proceed quite slowly. Therefore, it is not practical to design a reactor for complete conversion of the limiting reactant. Instead, the reactant is separated from the reactor outlet stream and recycled back to the reactor inlet. The fractional conversion of a reactant is the ratio of the amount reacted to the amount fed:

$$x_A = \frac{(n_A)_{reacted}}{(n_A)_{fed}} = \frac{(n_A)_{fed} - (n_A)_{out}}{(n_A)_{fed}}$$

The percentage conversion is

$$\text{Percent conversion} = \frac{(n_A)_{reacted}}{(n_A)_{fed}} \times 100\% = \frac{(n_A)_{fed} - (n_A)_{out}}{(n_A)_{fed}} \times 100\%$$

5.6 Methods of Solving Material Balances Involving Chemical Reactions

Three methods are discussed for solving material balance problems involving chemical reactions: extent of reaction, element balance, and component balance. Each approach provides the same results, but one method may be more convenient than the other for a given calculation, so you should be comfortable with all methods.

5.6.1 Extent of Reaction Method

The extent of reaction (ξ or $\dot{\xi}$) is the amount in moles (or molar flow rate) converted in a given reaction. The extent of reaction is a quantity that characterizes the reaction and simplifies our calculations.

For a continuous process at steady state:

$$\dot{n}_i = \dot{n}_{i0} + \nu_i \dot{\xi}$$

where $\dot{n}_{i0}$ and $\dot{n}_i$ are the molar flow rates of species i in the feed and outlet streams, respectively.

For a batch process:

$$n_i = n_{i0} + \nu_i \xi$$

where n_{i0} and n_i are the initial and final molar amounts of species i, respectively. The extent of reaction ξ (or $\dot{\xi}$) has the same units as n (or $\dot{n}$).

Summary:

$$\dot{n}_i = \dot{n}_{i0} + \nu_i \dot{\xi} \begin{cases} \dot{n}_i & \text{molar flow rate of component } i \text{ out of the system} \\ \dot{n}_{i0} & \text{molar flow rate of component } i \text{ entering the system} \\ \nu_i & \text{stoichiometric coefficient} + \text{for product} - \text{for reactant} \end{cases}$$

Example 5.3 Extent of reaction method

Ethylene oxide is produced by the reaction of ethylene with oxygen. The feed to the reactor contains 5 mol ethylene, 3 mol oxygen, and 2 mol ethylene oxide. Draw and label the process flow sheet. Write the material balance equations as a function of the extent of reaction.

Solution

The process flow sheet is shown in Figure E5.3.

$$2C_2H_4 + O_2 \leftrightarrow 2C_2H_4O$$

$$\dot{n}_i = \dot{n}_i^0 + \nu_i \dot{\xi}$$

1. General compound balance

$$C_2H_4: n_{C_2H_4} = n_{C_2H_4}^0 - 2\xi = 5 - 2\xi$$

$$O_2: n_{O_2} = n_{O_2}^0 - \xi = 3 - \xi$$

$$C_2H_4O: n_{C_2H_4O} = n_{C_2H_4O}^0 + 2\xi = 2 + 2\xi$$

$$\text{Total: } n = n^0 - \xi = 10 - \xi$$

5.6.2 Element or Atomic Balance Method

Element balances have no generation or consumption terms, and the mass balance is simplified to input equals output for continuous, steady-state processes. The element balance is based on the number of moles of that element regardless of the number of moles of the compound. The number of

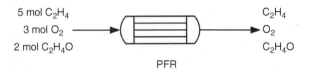

FIGURE E5.3
Extent of reaction method.

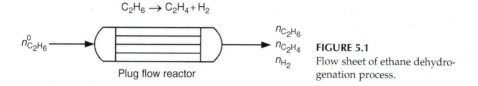

$$C_2H_6 \rightarrow C_2H_4 + H_2$$

$n^0_{C_2H_6}$ → Plug flow reactor → $n_{C_2H_6}$ $n_{C_2H_4}$ n_{H_2}

FIGURE 5.1
Flow sheet of ethane dehydrogenation process.

moles of each compound must be multiplied by the stoichiometric number for the element.

In the ethane dehydrogenation process (Figure 5.1), there are 2 mol of carbon atom for every mole of ethane. Thus, the element balances become

$$\text{C balance: } 2n^0_{C_2H_6} = 2n_{C_2H_4} + 2n_{C_2H_6}$$

$$\text{H balance: } 6n^0_{C_2H_6} = 4n_{C_2H_4} + 6n_{C_2H_6} + 2n_{H_2}$$

5.6.3 Molecular or Component Balance Approach

When performing molecular or component balances, consumption and generation terms need to be considered. Therefore, the general mass balance for steady-state flow processes becomes

$$m_{acc} = 0 = \sum m_{in} - \sum m_{out} + m_{gen} - m_{cons}$$

In the sulfur dioxide oxidation process (Figure 5.2) suppose 5 mol of O_2 is consumed, then the number of moles of SO_3 generated is

$$(5 \text{ mol of } O_2 \text{ consumed}) \frac{1 \text{ mol } SO_3 \text{ generated}}{1/2 \text{ mol } O_2 \text{ consumed}} = 10 \text{ mol } SO_3 \text{ produced}$$

Example 5.4 Extent of reaction, atomic balance, and molecular balance methods

Ammonia is burned to form nitric oxide. The fractional conversion is 0.5. The inlet molar flow rates of NH_3 and O_2 are 5 mol/h each. Calculate the exit component molar flow rates using

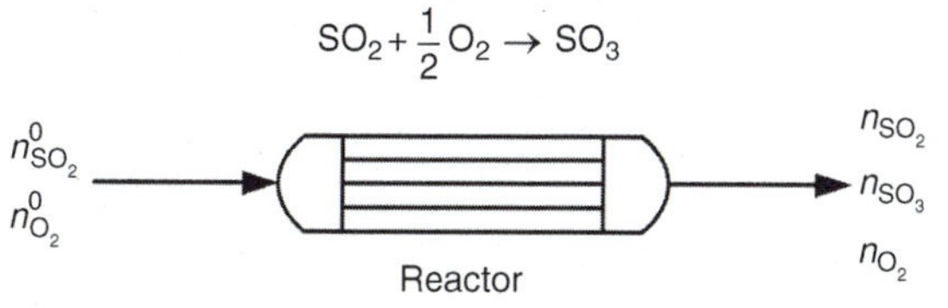

$$SO_2 + \frac{1}{2}O_2 \rightarrow SO_3$$

$n^0_{SO_2}$ $n^0_{O_2}$ → Reactor → n_{SO_2} n_{SO_3} n_{O_2}

FIGURE 5.2
Flow sheet of molecular compound balances.

a. Extent of reaction method
b. Atomic balance approach
c. Compound balance approach

Solution

The process flow sheet is shown in Figure E5.4.

$$4NH_3 + 5O_2 \rightarrow 4NO + 6H_2O$$

1. Extent of reaction method (ξ)

$$4NH_3 + 5O_2 \rightarrow 4NO + 6H_2O$$

Thus, $\nu_{NH_3} = -4$, $\nu_{O_2} = -5$, $\nu_{NO_2} = 4$, and $\nu_{H_2O} = 6$

The material balance can be written using the extent of reaction method as follows:

$$n_i = n_i^0 + \nu\xi$$

Balance on NH$_3$: $n_{NH_3} = n_{NH_3}^0 - 4\xi$

Balance on O$_2$: $n_{O_2} = n_{O_2}^0 - 5\xi$

Balance on NO: $n_{NO} = n_{NO}^0 + 4\xi$

Balance on H$_2$O: $n_{H_2O} = n_{H_2O}^0 + 6\xi$

Total number of moles at the outlet of the reactor:

$$n = n^0 + (-4 - 5 + 4 + 6)\xi = n^0 + \xi$$
$$n = n^0 + \xi$$

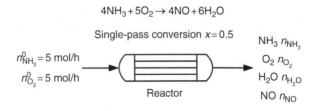

$$4NH_3 + 5O_2 \rightarrow 4NO + 6H_2O$$

Single-pass conversion $x = 0.5$

$n_{NH_3}^0 = 5$ mol/h

$n_{O_2}^0 = 5$ mol/h

Reactor

NH$_3$ n_{NH_3}

O$_2$ n_{O_2}

H$_2$O n_{H_2O}

NO n_{NO}

FIGURE E5.4
Process flow sheet of ammonia oxidation.

Inlet molar feed rates:

$$n^0_{NH_3} = 5 \text{ mol/h}, \; n^0_{O_2} = 5 \text{ mol/h}, \; n^0_{NO} = 0, \; n^0_{H_2O} = 0$$

$$\text{Conversion} = x = \frac{n^0_{O_2} - n_{O_2}}{n^0_{O_2}}$$

Solving the set of equations using *E–Z* solve gives the following results:

$$\xi = 0.5$$

$$n_{NH_3} = 3 \text{ mol/h}$$

$$n_{O_2} = 2.5 \text{ mol/h}$$

$$n_{H_2O} = 3 \text{ mol/h}$$

$$n_{NO} = 2 \text{ mol/h}$$

2. Element or atomic balance approach method

$$\text{Balance on N: } 5 = n_{NH_3} + n_{NO}$$

$$\text{Balance on O: } 2(5) = 2(n_{O_2}) + n_{H_2O} + n_{NO}$$

$$\text{Balance on H: } 3(5) = 3n_{NH_3} + n_{NO}$$

$$\text{Relations: conversion} = 0.5 = (5 - n_{O_2})/5$$

Solving the set of equations using *E–Z* solve leads to the following results:

$$n_{NH_3} = 3 \text{ mol/h}$$

$$n_{O_2} = 2.5 \text{ mol/h}$$

$$n_{H_2O} = 3 \text{ mol/h}$$

$$n_{NO} = 2 \text{ mol/h}$$

3. Compound or molecular balance approach

$$4NH_3 + 5O_2 \rightleftarrows 4NO + 6H_2O$$

Molecular balances can be done as the following:

1. Moles of O_2 (the limiting reactant) consumed or reacted

$$\text{Conversion} = X = 0.5 = \frac{\text{Moles reacted}}{\text{Moles in the feed}} = \frac{\text{Mole reacted}}{5}$$

Moles of O_2 reacted or consumed $= 0.5 \times 5 = 2.5$ mol
Moles of O_2 exiting the reactor $= 5 - 2.5 = 2.5$ mol

2. Moles of NH_3 consumed $= 2.5$ mol O_2 consumed $\times \frac{4 \text{ mol consumed}}{5 \text{ mol of } O_2 \text{ consumed}}$
Moles of NH_3 consumed $= 2$ mol
Moles of NH_3 leaving the reactor $=$ in $-$ consumed $= 5 - 2 = 3$ mol

3. Moles $\quad H_2O \quad$ generated $= 2.5 \quad$ mol $\quad O_2 \quad$ consumed $\times$
$\frac{6 \text{ mol generated}}{5 \text{ mol of } O_2 \text{ consumed}} = 3$ mol

4. Moles $\quad$ of $\quad$ NO $\quad$ generated $= 2.5 \quad$ mol $\quad O_2 \quad$ consumed $\times$
$\frac{4 \text{ NO generated}}{5 \text{ mol of consumed}} = 2$ mol

Results using molecular balance or compound balance leads to the following:

$$NH_3 \quad n_{NH_3} = 3 \text{ mol/h}$$

$$O_2 \quad n_{O_2} = 2.5 \text{ mol/h}$$

$$H_2O \quad n_{H_2O} = 3 \text{ mol/h}$$

$$NO \quad n_{NO} = 2 \text{ mol/h}$$

Conclusion

You have learned three methods for solving material balance problems involving chemical reactions. The first and the third method need the use of stoichiometric equations (i.e., the extent method and molecular balance method). The atomic balance method does not need the use of the stoichiometric reaction equation. All of the three methods lead to the same results.

Example 5.5   Extent of reaction, fractional conversion, and yield

Ethylene oxide (C_2H_4O) is produced by the reaction of ethylene (C_2H_4) with oxygen. The feed to the reactor contains 5 mol of ethylene, 3 mol of oxygen, and 2 mol of ethylene oxide. Set up the compound balances in general, and then use these to solve the problems suggested below.

Solution

The flow sheet of the process is shown in Figure E5.5.
    Material balances:
    The following reaction is taking place:

$$2C_2H_4 + O_2 \rightarrow 2C_2H_4O$$

Use the extent of reaction method to find moles of exit components.

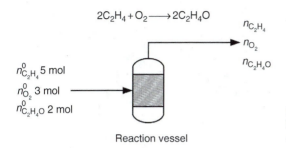

$$2C_2H_4 + O_2 \longrightarrow 2C_2H_4O$$

$n_{C_2H_4}$
n_{O_2}
$n_{C_2H_4O}$

$n^0_{C_2H_4}$ 5 mol
$n^0_{O_2}$ 3 mol
$n^0_{C_2H_4O}$ 2 mol

Reaction vessel

FIGURE E5.5
Flow sheet of ethylene oxide production.

$$C_2H_4: \dot{n}_{C_2H_4} = \dot{n}^0_{C_2H_4} - 2\dot{\xi}$$

$$O_2: \dot{n}_{O_2} = \dot{n}_{O_2} - \dot{\xi}$$

$$C_2H_4O: \text{n}_{C_2H_4O} = \text{n}^0_{C_2H_4O} + 2\dot{\xi}$$

Note that everything in the above equations is expressed in terms of a single unknown $\dot{\xi}$. If we know the moles in and out for any one component, we can solve for the extent of reaction and complete all of the material balances. Now that we have formulated the material balances for this problem in general, let us look at some specific problems using these balances.

(a) Suppose the amount of C_2H_4 coming out of the reactor is 5 mol, how much reaction has occurred?

From the balance for C_2H_4 above, we see that 5 mol $= 5$ mol $- 2$ ξ, so $\xi = 0$ mol. Thus, no reaction has occurred.

(b) Suppose $n_{C_2H_4} = 6$ mol, how much reaction has occurred?

From the balance for C_2H_4, we obtain 6 mol $= 5$ mol $- 2$ ξ, or $\xi = -0.5$ mol. Thus, the reaction proceeds to the left by 0.5 mol.

(c) Suppose $n_{C_2H_4} = 0$ mol, how much reaction has occurred?

From the balance for C_2H_4, we obtain 0 mol $= 5$ mol $- 2$ ξ, or $\xi = 2.5$ mol. Thus, the reaction proceeds to the right by 2.5 mol.

(d) Which component is the limiting reagent, and what is the theoretical yield?

To find the limiting reagent, we compare the ratio of moles of reactants in the feed to their stoichiometric coefficients. In this case, $n^0_{C_2H_4} = 5$ mol and $n^0_{O_2} = 3$ mol so

$$\frac{n^0_{C_2H_4}}{\nu_{C_2H_4}} = \frac{5}{2} < \frac{n^0_{O_2}}{\nu_{O_2}} = \frac{3}{1}$$

since the ratio of the molar feed rate to stoichiometric coefficient of C_2H_4 is less than that of O_2. There is too little C_2H_4, therefore C_2H_4 is the limiting reactant. The theoretical yield is the maximum

amount of product that could be produced if the reaction proceeded until the entire limiting reagent had been used up. Because
C_2H_4 is the limiting reagent, we use its material balance, and plug
in 0 mol for the outlet quantity. We already did this in (c) above
and found that this meant that $\xi = 2.5$ mol. We then use this value
of x in the material balance for the product to find the theoretical
maximum yield:

$$\text{Amount of } C_2H_2O \text{ produced} = n_{C_2H_4O} - n^0_{C_2H_4O} = 2\xi = 5 \text{ mol}$$

(e) Suppose $n_{C_2H_4} = 2$ mol, what is the fractional conversion?
 Again we use the limiting reagent to determine the fractional
 conversion. The fractional conversion is defined by

$$x = \frac{(n_{Lro} - n_{Lr})}{n_{Lro}} = \frac{(\text{moles used})}{(\text{moles in the feed})} = \frac{(5-2)}{5} = \frac{3}{5} \text{ or } 0.6$$

where the subscript Lr refers to the limiting reactant. The fractional conversion will often be given as a subsidiary relation in
material balances involving reactions. If the fractional conversion
is known, then one can use its value to determine the amount of
the limiting reagent leaving the reactor using

$$n_{Lr} = (1 - x) \times n_{Lro}$$

For example, if $x = 0.6$, then $n_{Lr} = (1 - 0.6)(5 \text{ mol}) = 2$ mol, as was
the case above.

Example 5.6   Acrylonitrile production

Acrylonitrile (C_3H_3N) is produced by the reaction of propylene, ammonia,
and oxygen:

$$C_3H_6 + NH_3 + \frac{3}{2}O_2 \rightarrow C_3H_3N + 3H_2O$$

$$A + B + \frac{3}{2}C \rightarrow D + 3E$$

The feed contains 10.0 mol% propylene (C_3H_6), 12.0 mol% ammonia (NH_3),
and 78.0 mol% air. A fractional conversion of 30.0% of the limiting reactant
is achieved. Taking 100 mol of feed as a basis, determine which reactant is
limiting, the percentage by which each of the other reactants is in excess,
and the molar amounts of all product gas constituents for 30% conversion of
the limiting reactant.

Solution

The flow sheet of the acrylonitrile production process is shown in Figure E5.6a.

Basis: 100 mol of feed

System: reaction vessel

Limiting reactant has lowest $\dfrac{n_i}{v_i}$ ratio

$$
\begin{array}{ccc}
C_3H_6 & : NH_3 & : O_2 \\
\dfrac{10 \text{ mol}}{1} & : \dfrac{12 \text{ mol}}{1} & : \dfrac{0.78(100)0.21}{3/2} \\
\boxed{10} & 12 & 10.9
\end{array}
$$

The lowest ratio corresponds to the limiting reactant, in this case propylene

$$
\% \text{ excess of } NH_3 = \frac{(12 - 10)}{10} 100\% = 20\%
$$

$$
\% \text{ excess of oxygen} = \frac{(10.9 - 10)}{10} 100\% = 9\%
$$

$$
C_3H_6 + NH_3 + \frac{3}{2}O_2 \rightarrow C_3H_3N + 3H_2O \ \xi
$$

$$
A + B + \frac{3}{2}C \rightarrow D + 3E
$$

1. Mole balance using the extent of reaction method

$$
n_A = n_{A0} - \xi = 10 - \xi
$$

$$
n_B = n_{B0} - \xi = 12 - \xi
$$

$$
n_C = n_{C0} - \frac{3}{2}\xi = (0.21 \times 78) - \frac{3}{2}\xi
$$

$$
n_D = n_{D0} + \xi = 0.0 + \xi
$$

$$
n_E = n_{E0} + 3\xi = 0.0 + 3\xi
$$

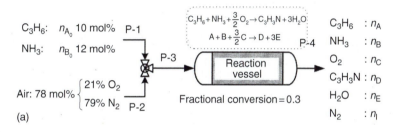

(a)

FIGURE E5.6a
Flow sheet of acrylonitrile production process (Extent of reaction method).

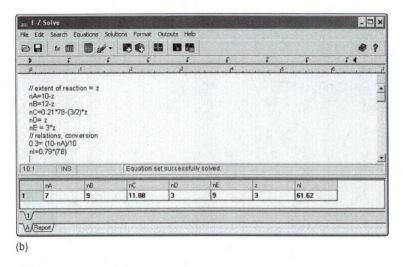

(b)

FIGURE E5.6b
Solution of the set of equations using *E–Z* solve.

Conversion of the limiting reactant

$$0.3 = \frac{10 - n_A}{10}$$

Solve the above set of material balance equations using *E–Z* solve as shown in Figure E5.6b.

2. **Mole balance using atomic balance approach**

In the atomic balance method, there is no need for the stoichiometric equation. The problem process flow sheet is shown in Figure E5.6c.

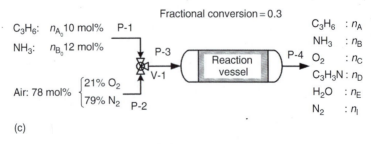

(c)

FIGURE E5.6c
Flow sheet of acrylonitrile production (atomic balance).

Solution

Basis: 100 mol for feed
Atomic balance equations (C, H, O, N)
Degree of freedom analysis (DFA) (atomic balance):

DFA	System: Reactor
Number of unknowns	6 (n_A, n_B, n_C, n_D, n_E, n_I)
Number of independent equations	4 (C, H, O, N)
Number of relations	2 (conversion, n_I is inert)
DF $= 6 - 4 - 2$	0

Setting the atomic balance equations: input $=$ output

$$C: 3 \times 10 = 3n_A + 3n_D$$

$$H: 6(10) + 3(12) = 6n_A + 3n_D + 2n_E$$

$$O: 2 \times 0.21(78) = 2n_C + n_E$$

$$N: 2 \times 0.79(78) + 12 = n_B + n_D + 2n_I$$

$$x = 0.3 = \frac{10 - n_A}{10}$$

The degree of freedom is zero, since we have four equations, two relations, and six unknowns. Solving the set of equations gives the following results:

Answer: $n_A = 7$, $n_B = 9$, $n_C = 11.88$, $n_D = 3$, $n_E = 9$, $n_I = 61.62$

In conclusion, atomic and extent of reaction methods give the same results.

3. **Material balance using molecular balance approach**

   We again have to use the stoichiometric equation

$$C_3H_6 + NH_3 + \frac{3}{2}O_2 \rightarrow C_3H_3N + 3H_2O$$

$$A + B + \frac{3}{2}C \rightarrow D + 3E$$

   Accumulation $=$ input $-$ output $+$ generation $-$ consumption
       The process flow sheet is shown in Figure E5.6d.
   Basis: 100 mol of feed

$$Conversion = 0.3 = (10 - n_A)/10 \Rightarrow n_A = 7 \text{ mol}$$

   NH_3 molecular balance:

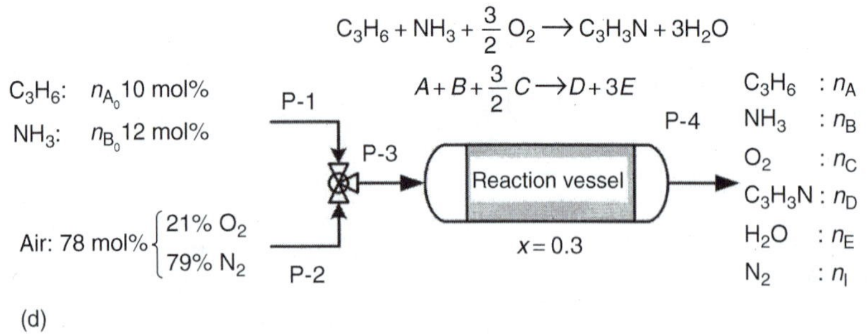

$$C_3H_6 + NH_3 + \frac{3}{2} O_2 \rightarrow C_3H_3N + 3H_2O$$

(d)

FIGURE E5.6d
Process flow sheet for acrylonitrile process (Molecular balance).

$$0.0 = \text{in} - \text{out} + \text{generation} - \text{consumption}$$

$$0.0 = 12 \text{ mol} - n_B + 0.0 - \left(\frac{1 \text{ mol NH}_3 \text{ consumed}}{1 \text{ mol C}_3H_6 \text{ consumed}} \right)$$

$$\times 3 \text{ mol C}_3H_6 \text{ consumed}$$

$$n_B = 9 \text{ mol}$$

O_2 molecular balance: $0.0 = \text{in} - \text{out} + \text{generation} - \text{consumption}$

$$0.0 = 0.21(78) \text{ mol in} - n_C + 0.0 - \left(\frac{3/2 \text{ mol O}_2 \text{ consumed}}{1 \text{ mol C}_3H_6 \text{ consumed}} \right)$$

$$\times 3 \text{ mol C}_3H_6 \text{ consumed}$$

$$n_C = 11.88 \text{ mol}$$

C_3H_3N molecular balance: $0.0 = \text{in} - \text{out} + \text{generation} - \text{consumption}$

$$0.0 = 0.0 - n_D + \left(\frac{1 \text{ mol C}_3H_3N \text{ generated}}{1 \text{ mol C}_3H_6 \text{ consumed}} \right) \times 3 \text{ mol C}_3H_6 \text{ consumed}$$

$$n_D = 3 \text{ mol}$$

H_2O molecular balance: $0.0 = \text{in} - \text{out} + \text{generation} - \text{consumption}$

$$0.0 = 0.0 - n_E + \left(\frac{3 \text{ mol H}_2O \text{ generated}}{1 \text{ mol C}_3H_6 \text{ consumed}} \right) \times 3 \text{ mol C}_3H_6 \text{ consumed} \Rightarrow n_E$$

$$= 9 \text{ mol}$$

5.7 Multiple Reactions and Extent of Reaction

Generally, the syntheses of chemical products do not involve a single reaction but rather multiple reactions. The goal is to maximize the production of the desirable product and minimize the production of unwanted by-products. For example, ethylene is produced by the dehydrogenation of ethane:

$$C_2H_6 \rightarrow C_2H_4 + H_2$$
$$C_2H_6 + H_2 \rightarrow 2CH_4$$
$$C_2H_4 + C_2H_6 \rightarrow C_3H_6 + CH_4$$

This leads to the following definitions:

$$\text{Yield} = \frac{\text{moles of desired product formed}}{\begin{array}{c}\text{moles formed if there were no side reactions}\\ \text{and limiting reactant reacted completely}\end{array}}$$

$$\text{Selectivity} = \frac{\text{moles of desired product formed}}{\text{moles of undesired product formed}}$$

The concept of extent of reaction can also be applied to multiple reactions, with each reaction having its own extent. If a set of reactions takes place in a batch or continuous, steady-state reactor, we can write

$$n_i = n_{i0} + \sum_j \nu_{ij}\xi_j$$

where
ν_{ij} is the stoichiometric coefficient of substance i in reaction j.
ξ_j is the extent of reaction for reaction j.

For a single reaction, the above equation reduces to the equation reported in a previous section (see Section 5.6.1).

Example 5.7 Oxidation reaction

Consider the pair of reactions in which ethylene is oxidized either to ethylene oxide (desired) or to carbon dioxide (undesired). Express the moles (or molar flow rates) of each of the five species in terms of the extent of reaction in the following reactions:

$$C_2H_4 + \frac{1}{2}O_2 \rightarrow C_2H_4O$$
$$C_2H_4 + 3O_2 \rightarrow 2CO_2 + 2H_2O$$

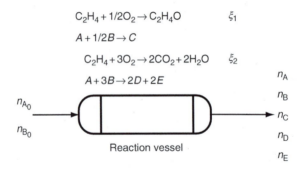

FIGURE E5.7
Flow sheet of multiple reaction process.

Solution

The process flow sheet is shown in Figure E5.7.

Assign extent of reaction for each reaction:

$$C_2H_4 + \frac{1}{2}O_2 \rightarrow C_2H_4O \; \xi_1$$

$$A + \frac{1}{2}B \rightarrow C$$

$$C_2H_4 + 3O_2 \rightarrow 2CO_2 + 2H_2O \; \xi_2$$

$$A + 3B \rightarrow 2D + 2E$$

Mole balance using the extent of reaction approach:

$$C_2H_4: n_A = n_{A0} - \xi_1 - \xi_2$$

$$O_2: n_B = n_{B0} - \frac{1}{2}\xi_1 - 3\xi_2$$

$$C_2H_4O: n_C = \xi_2$$

$$CO_2: n_D = 2\xi_2$$

$$H_2O: n_E = 2\xi_2$$

Example 5.8   Production of ethylene

The following two multiple reactions take place in a continuous reactor at steady state.

$$C_2H_6 \rightarrow C_2H_4 + H_2$$

$$C_2H_6 + H_2 \rightarrow 2CH_4$$

The feed contains 85.0 mol% ethane (C_2H_6), and the balance is inert (I). The fractional conversion of ethane is 0.501, and the fractional yield of ethylene (C_2H_4) is 0.471. Calculate the molar composition of the product gas and the selectivity of ethylene for methane production.

Solution

Assign extent of reaction for each reaction, ξ_1 for the first reaction and extent ξ_2 for the second reaction and assign symbols to simplify the solution.

$$C_2H_6 \rightarrow C_2H_4 + H_2 \quad \xi_1$$
$$A \rightarrow B + C$$
$$C_2H_6 + H_2 \rightarrow 2CH_4 \quad \xi_2$$
$$A + C \rightarrow 2D$$

The process flow sheet is shown in Figure E5.8a.

Basis: 100 mol of feed

$$C_2H_6: n_A = 85 - \xi_1 - \xi_2$$
$$C_2H_4: n_B = 0.0 + \xi_2$$
$$H_2: n_C = 0.0 + \xi_1 - \xi_2$$
$$CH_4: n_D = 0.0 + 2\xi_2$$

$$x_A = 0.501 = \frac{85 - n_A}{85} \Rightarrow n_A = 42.415 \text{ mol}$$

$$\text{Yield}_B = 0.471 = \frac{n_B}{85} \Rightarrow n_B = 40.035 \text{ mol}$$

The problem can now be easily solved. Knowing the number of moles of ethylene produced will give the value of the extent of reaction 2 (ξ_2). Knowing the number of moles of ethane (A) and extent of reaction 2 and substituting these values in the ethane balance equation give the value of extent of reaction 1 (ξ_1). Once the extents of reactions are obtained substitute

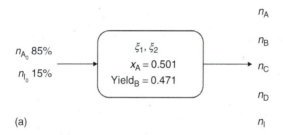

(a)

FIGURE E5.8a
Flow sheet of ethylene production.

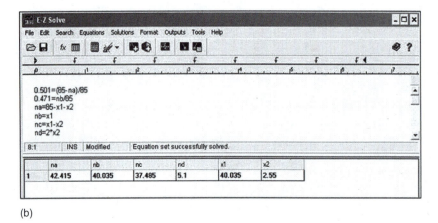

(b)

FIGURE E5.8b
Results with *E–Z* solve.

these values in the rest of the equations to get the number of moles of exit products. You can also use *E–Z* solve as shown in Figure E5.8b where *x* represents the extent of reaction.

5.8 Degree of Freedom Analysis for Reactive Processes

5.8.1 Molecular Species Balances and Extent of Reaction

$$n_{DF} = \left\{ \begin{array}{c} \text{number of} \\ \text{unknowns} \end{array} \right\} + \left\{ \begin{array}{c} \text{number of} \\ \text{independent} \\ \text{chemical} \\ \text{reactions} \end{array} \right\} - \left\{ \begin{array}{c} \text{number of} \\ \text{independent} \\ \text{molecular} \\ \text{species} \\ \text{balances} \end{array} \right\} - \left\{ \begin{array}{c} \text{number} \\ \text{of other} \\ \text{equations} \\ \text{relating} \\ \text{variables} \end{array} \right\}$$

The DFA for molecular species balances and extent of reaction balance is almost the same.

5.8.2 Atomic Species Balances

$$n_{DF} = \left\{ \begin{array}{c} \text{number of} \\ \text{unknowns} \end{array} \right\} - \left\{ \begin{array}{c} \text{number of} \\ \text{independent} \\ \text{atomic} \\ \text{species} \\ \text{balances} \end{array} \right\} - \left\{ \begin{array}{c} \text{number of} \\ \text{molecular} \\ \text{balances on} \\ \text{independent} \\ \text{nonreactive} \\ \text{species} \end{array} \right\} - \left\{ \begin{array}{c} \text{number} \\ \text{of other} \\ \text{relations} \\ \text{relating} \\ \text{variables} \end{array} \right\}$$

5.9 Independent Chemical Reactions

Chemical reactions are independent, if the stoichiometric equation of any one of them cannot be obtained by adding and subtracting multiples of the stoichiometric equations of others.

Consider the following equations:

$$A \rightarrow 2B \text{ (1)}$$
$$B \rightarrow C \quad \text{(2)}$$
$$A \rightarrow 2C \text{ (3)}$$

These three reactions are not all independent, since $(3) = (1) + 2 \times (2)$.

5.10 Independent Species Balances

Generally, there are n balances for n independent species. But if two species are in the same ratio to each other wherever they appear in the process, balances on these species will not be independent equations. For example, air consists of 21 mol% O_2 and 79 mol% N_2. In ammonia synthesis, reactants $(N_2:H_2)$ in certain cases enter the reactor on stoichiometric basis (1:3).

5.11 Chemical Equilibrium

Reactions do not proceed instantly; in fact, predicting the speed at which a reaction occurs is very important. Furthermore, reactions do not necessarily happen independently; in fact, often the reverse of the reaction we are interested in also takes place. Chemical equilibrium is reached when the rates of the forward and reverse reactions are equal to each other (i.e., compositions no longer change with time). While we will not calculate rates, we need to know what affects them (because this will affect the equilibrium). Things that we must consider that affect reaction rates and hence equilibrium are temperature and concentration.

Example 5.9 Equilibrium reaction
Consider the reaction of methane with oxygen.

$$2CH_4 + O_2 \rightleftarrows 2CH_3OH$$

We are given that at equilibrium the compositions of the components satisfy

$$K(T) = \frac{y_{CH_3OH}^2}{y_{CH_4}^2 y_{O_2}}$$

If you are given the feed compositions ($n_{CH_4}^0$, $n_{O_2}^0$, and $n_{CH_3OH}^0$) and the equilibrium constant $K(T)$, how do you determine the equilibrium compositions?

Solution

Utilize equilibrium expressions to determine equilibrium compositions and extents of reaction.

$$n_{CH_3OH} = n_{CH_3OH}^0 + 2\xi$$

$$n_{CH_4} = n_{CH_4}^0 - 2\xi$$

$$n_{O_2} = n_{O_2}^0 - \xi$$

$$n = n^0 - \xi$$

$$K(T) = \frac{y_{CH_3OH}^2}{y_{CH_4}^2 y_{O_2}} = \frac{\left(\frac{n_{CH_3OH}^0 + 2\xi}{n^0 - \xi}\right)^2}{\left(\frac{n_{CH_4}^0 - 2\xi y_{O_2}}{n^0 - \xi}\right)^2 \left(\frac{n_{O_2}^0 - \xi}{n^0 - \xi}\right)}$$

Example 5.10 Methane oxidation process

Methane (CH_4) and oxygen react in the presence of a catalyst to form formaldehyde (HCHO). In a parallel reaction, methane is oxidized to carbon dioxide and water:

$$CH_4 + O_2 \rightarrow HCHO + H_2O$$
$$CH_4 + 2O_2 \rightarrow CO_2 + 2H_2O$$

The feed of the reactor contains equimolar amounts of methane and oxygen. Assume a basis of 100 mol feed/s.

(a) Draw and label a flowchart. Use a degree of freedom analysis based on extents of reaction to determine how many process variables must be specified for the remaining variable values to be calculated.

(b) Use ξ to derive expressions for the product stream component flow rates in terms of the two extents of reaction, ξ_1 and ξ_2.

(c) The fractional conversion of methane is 0.900 and the fraction yield of formaldehyde is 0.855. Calculate the molar composition of the reactor output stream and the selectivity of formaldehyde production relative to carbon dioxide production.

Solution

Basis: 100 mol/s of feed to reactor

(a)

$$CH_4 + O_2 \rightarrow HCHO + H_2O \quad \xi_1$$
$$A + B \rightarrow C + D$$
$$CH_4 + 2O_2 \rightarrow CO_2 + 2H_2O \quad \xi_2$$
$$A + 2B \rightarrow E + 2D$$

The process flow sheet of reactor with multiple reactions is shown in Figure E5.10.

Degree of freedom analysis

Number of unknowns: $5 + 2$ (n_A, n_B, n_C, n_D, n_E, ξ_1, ξ_2)

Number of independent equations: 5 (equal number of components)

Number of available relations: 0

DF $= 7 - 5 - 0 = 2$

(b)

Material balance equations:

$$CH_4: n_A = 50 - \xi_1 - \xi_2$$
$$O_2: n_B = 50 - \xi_1 - 2\xi_2$$
$$HCHO: n_C = \xi_1$$

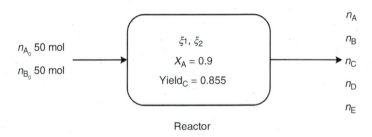

FIGURE E5.10
Flow sheet of reaction involving multiple reactions.

$$H_2O: n_D = 0.0 + \xi_1 + 2\xi_2$$

$$CO_2: n_E = \xi_2$$

(c)

The fractional conversion of methane is 0.900.

$$0.9 = \frac{50 - n_A}{50}$$

Yield of formaldehyde is 0.855

$$Yield_C = 0.855 = \frac{n_C}{50}$$

5.12 Combustion Reactions

Combustion is the rapid reaction of a fuel with oxygen to produce energy. Combustion is a very important industrial chemical reaction. Fuels include coal (C, H, S and others), fuel oil (high MW hydrocarbons and some S), gaseous fuel (natural gas—mostly methane), or liquefied petroleum gas (propane and/or butane). Maximum energy is produced when fuel is completely oxidized. The product gas is called stack gas or flue gas. Complete combustion results in all C oxidized to CO_2, all H oxidized to H_2O, and all S oxidized to SO_2. In incomplete combustion, C is oxidized to CO and CO_2

$$C_4H_{10} + \frac{13}{2}O_2 \rightarrow 4CO_2 + 5H_2O \text{ complete combustion of butane}$$

$$C_4H_{10} + \frac{9}{2}O_2 \rightarrow 4CO + 5H_2O \text{ incomplete combustion of butane}$$

5.12.1 Theoretical and Excess Air

For obvious economic reasons, air (79% N_2, 21% O_2) is the source of oxygen in combustion reactions. Combustion reactions are always conducted with excess air, thus ensuring good conversion of the expensive fuel.

Theoretical oxygen is the moles or molar flow rate of O_2 required for complete combustion of all the fuel. Theoretical air is the quantity of air that contains the theoretical oxygen.

$$\text{Theoretical air} = \frac{1}{0.21} \times \text{theoretical } O_2$$

Excess air is the amount of air fed to the reactor that exceeds the theoretical air.

$$\text{Percent excess air} = \frac{(\text{moles air})_{\text{fed}} - (\text{moles air})_{\text{theoretical}}}{(\text{moles air})_{\text{theoretical}}} \times 100\%$$

Example 5.11 Coal combustion process

Consider the coal combustion process as shown in Figure E5.11. Calculate the flow rate of all streams and their compositions. Assuming all the coal is consumed, note that the given feed composition is in mass fraction, calculate the following:

a. Percent excess air.
b. Ratio of water vapor and dry gas.
c. Selectivity of CO_2 relative to CO.
d. Yield of C to CO_2.
e. If only 900 ppm of SO_2 is allowed in the dry gas, find the ratio of kilogram SO_2 removed per kilogram of coal fed to the furnace.

Solution

The combustion process flow sheet is shown in Figure E5.11.

(a) Set 100 mol/min dry gas product as the basis.

$$N_2 \left(100 \frac{\text{mol}}{\text{min}} \right) (0.7775) = 77.75 \text{ mol } N_2/\text{min}$$

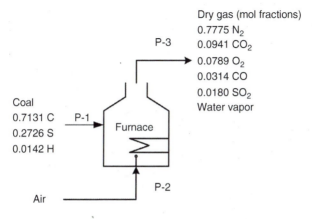

Dry gas (mol fractions)
0.7775 N_2
0.0941 CO_2
0.0789 O_2
0.0314 CO
0.0180 SO_2
Water vapor

Coal
0.7131 C
0.2726 S
0.0142 H

FIGURE E5.11
Flow sheet of combustion process.

C atomic balance:

$$\left(100\frac{mol}{min}\right)(0.0941 + 0.0314) = 12.55 \text{ mol C/min}$$

Assuming all the coal is consumed. You need to calculate coal composition in mole fraction. To convert mass fraction to mole fraction do the following:

Assume basis of 100 g coal. The mass of each component:

$$C: 0.7131(100 \text{ g}) = 71.31 \text{ g}$$

$$S: 0.2726(100 \text{ g}) = 27.26 \text{ g}$$

$$H: 0.0142(100 \text{ g}) = 1.42 \text{ g}$$

Number of moles of each atom = mass of atom/MW

$$C: 71.31 \text{ g C}/(12 \text{ g/mol}) = 5.94 \text{ mol C}$$

$$S: 27.26 \text{ g S}/(32 \text{ g/mol}) = 0.85 \text{ mol S}$$

$$H: 1.42 \text{ g H}/(1 \text{ g/mol}) = 1.42 \text{ mol H}$$

Total number of moles = 8.21 mol total
Mole fractions = mole of each atom/total number of moles
Mole fraction of C: 5.94/8.21 = 72.35% mol C
Feed mole fraction: C: 0.7235, S: 0.1035, H: 0.1730

Reactions that took place inside the furnace:

$$C + O_2 \rightarrow CO_2$$

$$S + O_2 \rightarrow SO_2$$

$$2H + \frac{1}{2}O_2 \rightarrow H_2O$$

Amount reacted for the complete combustion from 100 mol dry gas as the basis

$$0.7235(\text{coal used}) = 12.55 \text{ mol} \rightarrow 17.35 \text{ mol feed/min}$$

$$(0.1730)(17.35) = 3.0 \text{ mol H in feed/min}$$

Reaction	Number of Atoms in the Feed	Moles of O_2 Required for Complete Combustion
$C + O_2 \rightarrow O_2$	12.55 mol C	12.55
$S + O_2 \rightarrow SO_2$	1.80 mol S	1.8
$2H + \frac{1}{2}O_2 \rightarrow H_2O$	3.0 mol H	0.75
		15.10 mol O_2 needed theoretically

Amount of N_2 represents amount or (air/O_2) fed

$$\left(\frac{77.75 \text{ mol } N_2}{\text{min}}\right)\left(\frac{0.21 \text{ mol } O_2}{0.79 \text{ mol } N_2}\right) = 20.67 \text{ mol } O_2 \text{ in feed/min}$$

Excess air (equal to excess O_2)

$$\frac{20.67 - 15.10}{15.10} \times 100\% = 36.9\% \text{ excess air}$$

(b) Ratio of water vapor to dry gas:

Assume all H_2O in the product gas is from the reactions.

$$\text{From } 3.0 \text{ mol } H \rightarrow 1.5 \text{ mol } H_2O \text{ formed}$$

$$\frac{1.5 \text{ mol } H_2O}{100 \text{ mol dry gas}} = 0.015 \text{ mol } H_2O/\text{mol dry gas}$$

(c) Selectivity of CO_2 relative to CO:

$$\frac{9.41 \text{ mol } CO_2}{3.14 \text{ mol } CO} = 2.997 \text{ mol } CO_2/\text{mol } CO$$

(d) Yield of C to CO_2:

$$\frac{9.41 \text{ mol } CO_2}{12.55 \text{ mol } C} = 0.75 \text{ mol } CO_2/\text{mol } C$$

(e) 900 ppm SO_2 (dry gas) allowed:

Have $(0.0180 \ O_2)(1 \times 10^6) = 18{,}000$ ppm SO_2 (dry gas basis)

18000 ppm SO_2 − 900 ppm = 17,100 ppm of SO_2 is to be removed

$$\text{MW } SO_2 = 32 + 32 = 64$$

$$\frac{17{,}100}{18{,}000} \times 100\% = 95\% \text{ or } 0.95 \text{ of } SO_2 \text{ needs to be removed}$$

$$(100 \text{ mol dry gas})\left(\frac{0.0171 \text{ mol } SO_2 \text{ to remove}}{\text{mol dry gas}}\right)$$

$$\left(\frac{64 \text{ g } SO_2}{\text{mol } SO_2}\right)\left(\frac{\text{kg}}{1000 \text{ g}}\right) = 0.10944 \text{ kg } SO_2 \text{ to remove}$$

The mass of coal fed in kilograms: $(12.55)(12) + (1.80)(32) + (3.0)(1.0)$

$$= \left(\frac{211.2}{1000}\right) \text{ kg} = 0.2112 \text{ kg}$$

$0.10944 \text{ kg } SO_2/0.2112 \text{ kg coal} = 0.518 \text{ kg } SO_2/\text{kg coal}$

Example 5.12   Fermentation bioprocess

Acetobacter aceti bacteria convert ethanol to acetic acid under aerobic conditions. A continuous fermentation process for vinegar production is proposed using nonviable *A. aceti* cells, which are immobilized on the surface of gelatin beads. The production target is 1.8 kg/h acetic acid; however, the maximum acetic acid concentration tolerated by the cells is 10%. Air is pumped into the fermentor at a rate of 200 g-mol/h.

(a) What is the minimum amount of ethanol required?
(b) What is the minimum amount of water that must be used to dilute ethanol to avoid acid inhibition?
(c) What is the composition of the fermentor off-gas?

Solution
The process flowchart is shown in Figure E5.12

$$C_2H_5OH + O_2 \rightarrow CH_3COOH + H_2O$$

Number of moles of acetic acid produced:

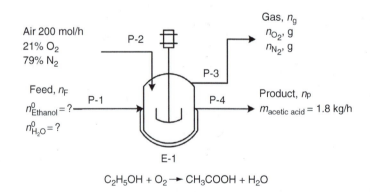

FIGURE E5.12
Flow sheet of acetic acid fermentation process.

$$1.8 \; \frac{kg}{h} \left| \frac{1}{60 \; kg/kmol} \right. = 0.03 \; \frac{kmol}{h} \text{acetic acid } (CH_3COOH) \text{ produced}$$

Material balance equations (extent of reaction)

$$n_{CH_3COOH} = 0.0 + \xi \Rightarrow 0.03 = 0 + \xi \Rightarrow \xi = 0.03 \; mol$$

$$n_{H_2O} = n_{H_2O}^0 + \xi$$

$$n_{C_2H_5OH} = n_{C_2H_5OH}^0 - \xi$$

$$n_{O_2} = n_{O_2}^0 - \xi$$

(a) The minimum amount of ethanol required is 0.03 kmol/h, calculated based on the mole balance equation of oxygen:

$$n_{C_2H_5OH} = n_{C_2H_5OH}^0 - \xi \Rightarrow 0 = n_{C_2H_5OH}^0 - 0.03 \Rightarrow n_{C_2H_5OH}^0 = 0.03 \; kmol/h$$

(b) The minimum amount of water required to avoid acid inhibition:

$$0.10 = kg \text{ acetic acid}/(kg \text{ water} + kg \text{ acetic acid})$$

$$0.10 = 1.8 \; kg/h(kg \text{ of water} + 1.8 \; kg/h \text{ of acetic acid})$$

$$\text{Mass in kg of water out} = 16.2 \; kg/h$$

$$\text{Moles of water out } (n_{H_2O}) = 16.2 \; kg/h/(18 \; kg/kmol) = 0.9 \; kmol$$

Using the mole balance equation for water

$$n_{H_2O} = n_{H_2O}^0 + \xi \Rightarrow 0.9 \; kmol = n_{H_2O}^0 + 0.03 \; kmol/h$$

$$\text{Moles of inlet water required to avoid inhibition} = n_{H_2O}^0$$

$$= 0.87 \; kmol/h$$

(c) Oxygen balance: Fermentor off-gases

$$n_{O_2} = n_{O_2}^0 - \xi = 0.21(0.2 \; kmol/h) - 0.03 = 0.012 \; kmol/h$$

Moles of nitrogen out = moles of nitrogen inlet (inert) = 0.79 (0.2)
$$= 0.158 \; kmol/h$$

Composition of oxygen (mole fraction) = 0.012/(0.012 + 0.158) = 0.07

Nitrogen mole fraction = 0.93

5.13 Problems

5.13.1 Incomplete Combustion of Butane

Butane (C_4H_{10}) is fed at a rate of 100 mol/s to a boiler with 50% excess air. 50% of the butane is consumed, and the product gas contains 10 mol CO_2 per mol CO. Calculate the molar composition of the stack gas.

(Answer: $n_{C_4H_{10}} = 50$, $n_{CO} = 18.2$, $n_{CO_2} = 181.8$, $n_{H_2O} = 250$, $n_{O_2} = 659$)

5.13.2 Complete Combustion of Butane

Butane is burned completely in a process fed with 25.0% excess air. Determine the composition in mole fractions of the (complete) combustion product stream to three decimal places.

(Answer: $y_{N_2} = 0.742$, $y_{O_2} = 0.04$, $y_{CO_2} = 0.097$, $y_{H_2O} = 0.121$)

5.13.3 Methane Combustion

Methane in the amount of 55 mol per min is combusted with dry air at 90°C and 770 mmHg. A partial analysis of the stack (exit) gas reveals that the mole fraction of nitrogen is 0.8477 on a dry basis and the molar flow rate of the dry stack gas is 683 mol/min. The fractional conversion of methane is 97.5%, and the molar ratio of CO_2 to CO is ~9.9. On the basis of the given information determine the percent excess air fed to the combustion chamber.

(Answer: Percent excess air $= 40\%$)

5.13.4 Burning Ethyl Ketone with Excess Air

A gaseous mixture containing 80 mol percent methane and the balance methyl ethyl ketone (C_4H_8O) is burned with 50% excess air. 95% of the methane and 75% of the methyl ethyl ketone burn. However, some of the methane that burns forms CO. The exhaust gas contains 0.5 mol percent CO on a dry basis. Calculate the molar composition of the exhaust gas.

(Answer: $y_i = \frac{n_i}{n_{Total}}$, $y_{CH_4} = 0.002$, $y_{C_4H_8O} = 0.0024$, $y_{N_2} = 0.741$, $y_{CO} = 0.0045$, $y_{CO_2} = 0.0618$, $y_{O_2} = 0.085$, $y_{H_2O} = 0.103$)

5.13.5 Roasting of Iron Pyrite

In one of the earliest processes used to make sulfuric acid, an ore containing iron pyrite (FeS_2) is roasted (burned) with air. The following reactions take place in the furnace (s denotes a solid species and g a gaseous species):

$$FeS_2 \ (s) + 15/2 \ O_2 \ (g) \rightarrow Fe_2O_3(s) + 4 \ SO_3 \ (g)$$
$$2FeS_2 \ (s) + 11/2 \ O_2 \ (g) \rightarrow Fe_2O_3 \ (s) + 4 \ SO_2 \ (g)$$

(In later stages of the process, not to be considered here, the SO_2 is further oxidized to SO_3, which is then absorbed in water to produce the H_2SO_4 product.) A solid ore containing 82 wt% FeS_2 and 18 wt% inerts is fed to the furnace. Dry air is fed in 40% excess of the amount theoretically required to oxidize all of the sulfur in the ore to SO_3. A pyrite conversion of 85% is obtained, with 40% of the FeS_2 converted to form sulfur dioxide, and the rest forming sulfur trioxide. Two streams leave the roaster: a gas stream containing SO_3, SO_2, O_2, and N_2, and a solid stream containing unconverted pyrite, ferric oxide (Fe_2O_3), and inerts. Using 100 kg/min of ore as a basis, calculate the rate of Fe_2O_3 production (kg/min) in the outlet solid stream, and the total molar flow and composition of the outlet gas stream.

(Answer: Amount of Fe_2O_3 produced $= 290.5$ mol $\times$ 159.69 g/mol$\times$ $\frac{kg}{1000 \, g} = 46.4$ kg)

$$\text{Total molar gas flow} = 16302.6 \text{ mol}$$

$$\text{Molar fraction of SO}_2 = \frac{464.8}{16302.6} = 0.079$$

$$\text{Molar fraction of SO}_3 = \frac{657.2}{16302.6} = 0.043$$

$$\text{Molar fraction of O}_2 = \frac{1642.1}{16302.6} = 0.100$$

$$\text{Molar fraction of N}_2 = \frac{13498.6}{16302.6} = 0.828$$

5.13.6 Water–Gas Shift Reaction

One hundred kilomolecules per hour of a stream containing 20 mol% CO, 20 mol% H_2O, and 60 mol% H_2 is combined with additional steam and fed to a water–gas shift reactor to produce more hydrogen. The equilibrium constant of the water–gas shift reaction is 200, and the mole fraction of hydrogen at the outlet is 0.75. Write and balance the chemical equation. How much additional steam is fed to the water–gas shift reactor? (Answer: $\xi = 18.8 \ \frac{kmol}{h}$, $n_2 = 5.09 \ \frac{kmol}{h}$)

5.13.7 Production of Sulfuric Acid

Sulfuric acid (H_2SO_4) is formed by first burning sulfur in air to produce sulfur dioxide (SO_2):

$$S + O_2 \rightarrow SO_2$$

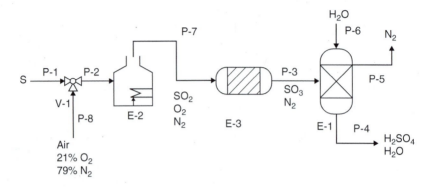

FIGURE P5.13.7
Process flow sheet for sulfuric acid production.

The sulfur dioxide then is catalytically reacted with more air to form sulfur trioxide:

$$SO_2 + \frac{1}{2}O_2 \rightarrow SO_3$$

Finally, the sulfur trioxide is combined with water to form sulfuric acid:

$$SO_3 + H_2O \rightarrow H_2SO_4$$

The flow diagram for this process is shown in Figure P5.13.7. The components in each stream are as shown in the flow diagram. The concentrated sulfuric acid produced is 98% sulfuric acid and 2 wt% water. This process is to produce 200,000 kg/day of the concentrated acid. How much air, water, and sulfur must enter the process in order to meet this production demand? Use the following MWs when solving this problem.

$$N_2 = 28, \ O_2 = 32, \ S = 32 \ g/mol$$

$$SO_2 = 64, \ SO_3 = 80, \ H_2SO_4 = 98 \ g/mol$$

(Answer: Thus, the flow rate of the air is $316,000 + 96,000 = 412,000$ kg/day. The flows are $S = 64,000$ kg/day, $H_2O = 4,000$ kg/day, and air $= 412,000$ kg/day)

Further Readings

1. Reklaitis, G.V. (1983) *Introduction to Material and Energy Balances*, John Wiley & Sons, New York.
2. Felder, R.M. and R.W. Rousseau (1978) *Elementary Principles of Chemical Processes*, John Wiley, New York.

3. Himmelblau, D.M. (1974) *Basic Principles and Calculations in Chemical Engineering*, 3rd edn, Prentice-Hall, New Jersey.
4. Whirwell, J.C. and R.K. Toner (1969) *Conservation of Mass and Energy*, Blaisdell, Waltham, Massachusetts.
5. Cordier, J.-L., B.M. Butsch, B. Birou, and U. von Stockar (1987) The relationship between elemental composition and heat of combustion of microbial biomass. *Appl. Microbiol. Biotechnol. 25*, 305–312.
6. Atkinson, B. and F. Mavituna (1991) *Biochemical Engineering and Biotechnology Handbook*, 2nd edn, Macmillan, Basingstoke.

6

Multiple Systems Involving Reaction, Recycle, and Purge

At the End of This Chapter You Should Be Able to

1. Draw and label a process flow sheet for problems involving recycle, bypass, and purge.
2. Explain the purpose of a recycle stream, a bypass stream, and a purge stream.
3. Use the single-pass and overall conversion in solving recycle problems involving chemical reactions.
4. Write a set of independent material balances for a complex process involving chemical reactions and more than one unit.
5. Solve problems involving multiple systems.

6.1 Reaction with Product Separation and Recycle

A couple of new issues arise, when considering recycle in processes that include reactions, which are conversions, if the reactant is recycled, and accumulations. A recycle stream is introduced to recover and reuse unreacted reactants. Two definitions of reactant conversion are used in the analysis of chemical reactors with product separation and recycle of unconsumed reactants. The overall conversion ignores the fact that recycle is occurring and is calculated purely on the basis of the difference between the process inputs and the product outside the recycle stream (Figure 6.1).

$$\text{Overall conversion} = \frac{\text{reactant input to process} - \text{reactant output from process}}{\text{reactant input to process}}$$

The single-pass conversion is based on the difference between the reactor input and output, inside the recycle loop.

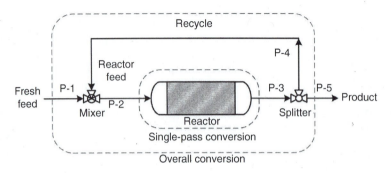

FIGURE 6.1
Single-pass and overall conversion in the presence of recycle.

$$\text{Single-pass conversion} = \frac{\text{reactant input to reactor } - \text{ reactant output from reactor}}{\text{reactant input to reactor}}$$

6.2 Reaction with Recycle and Purge

A problem that can occur in processes that involve recycle is that a material that enters the process in the feed stream or is generated in the reactor may remain entirely in the recycle stream rather than being carried out in the product stream. To prevent this buildup, a portion of the recycle stream is withdrawn as a purge stream. In the process flowchart, a purge point is a simple splitter. The recycle stream before and after the purge point has the same composition, and only one independent material balance can be written. If a reaction produces a component that remains entirely within the recycle, then we are in trouble! Why? How might this happen? A purge stream is one that bleeds material out of a recycle stream so that accumulation is avoided (see Figure 6.2). The recycle stream allows operation of the

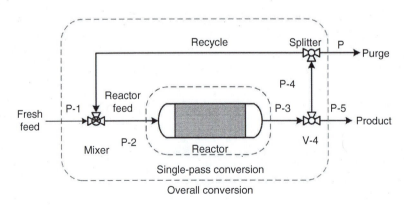

FIGURE 6.2
Single-pass and overall conversion in the presence of recycle and purge.

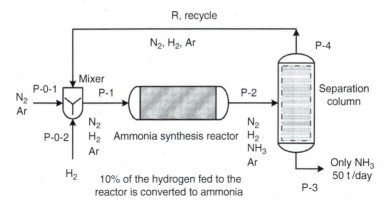

FIGURE 6.3
Process flow sheet for ammonia synthesis.

reactor at low single-pass conversion, and has high overall conversion for the system.

6.2.1 Flow Sheet for Reaction with Recycle

Figure 6.3 shows a process flow sheet where unreacted compounds are recycled for further processing.

6.2.2 Flow Sheet for Reaction with Recycle and Purge

Figure 6.4 shows a process flow sheet with recycle and purge. The purge stream is one that splits material out of a recycle stream so as to avoid accumulation within the reactor. An industrial example is the fluidized bed reactor used for polyethylene production (UNIPOL process, see Figure 2.24b). Nitrogen (N_2) in the feed in the polyethylene reactor is inert and is

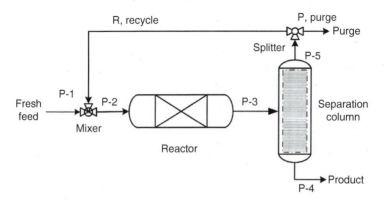

FIGURE 6.4
Process flow sheet with recycle and purge streams.

used as a diluent carrier gas for the ethylene and comonomers. If there is no purge the unreacted nitrogen will be accumulated within the reactor and will affect reactor operating conditions, and hence reactor performance.

Example 6.1   Methanol production

Methanol is produced by the reaction of carbon dioxide and hydrogen:

$$CO_2 + 3H_2 \rightarrow CH_3OH + H_2O$$

The fresh feed to the process contains hydrogen, carbon dioxide, and 0.400 mol% inerts (I). The reactor effluent passes to a condenser that removes essentially all the methanol and water formed but none of the reactants or inerts. The latter substances are recycled to the reactor. To avoid buildup of the inerts in the system, a purge stream is withdrawn from the recycle. The feed to the reactor (not the fresh feed to the process) contains 28.0 mol% CO_2, 70.0 mol% H_2, and 2.00 mol% inerts. The single-pass conversion of hydrogen is 60.0%. Calculate the molar flow rates and molar compositions of the fresh feed, the total feed to the reactor, the recycle stream, and the purge stream for a production rate of 155 mol CH_3OH/h.

Solution

The methanol production process-labeled flow sheet is shown in Figure E6.1a. First, performing the degree of freedom (DF) analysis will narrow down the process unit or system to start with.

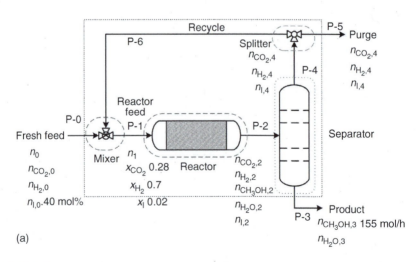

(a)

FIGURE E6.1a
Methanol production process flow sheet.

| | Systems | | | |
DF Analysis	Mixing	Reactor	Condenser	Overall Process
Number of unknowns	7	6+1	9	7+1
Number of independent equations	3	5	5	5
Number of auxiliary relations	0	1	0	0
DF	4	1	4	3

The possible balance can be done around the reactor while taking 100 mol of feed to the reactor (not the fresh feed) as the basis. Taking this basis will reduce the reactor DF to 0 and hence the problem is solvable.

Basis: 100 mol of feed to the reactor
System: reactor

$$CO_2 + 3H_2 \rightarrow CH_3OH + H_2O$$

Material balance using the extent of reaction approach:

$$n_{CO_2,2} = 0.28(100) - \xi$$

$$n_{H_2,2} = 0.7(100) - 3\xi$$

$$n_{CH_3OH,2} = 0.0 + \xi$$

$$n_{H_2O,2} = 0.0 + \xi$$

$$n_{I,2} = n_{I,1} = 0.02(100)$$

The conversion can be calculated as

$$0.6 = \frac{0.7(100) - n_{H_2,2}}{0.7(100)}$$

Using *E–Z* solve for solving the set of algebraic material balance equations (Figure E6.1b), the following answers are obtained:

$$n_{CO_2,2} = 14 \text{ mol}, \quad n_{H_2,2} = 28 \text{ mol}, \quad n_{CH_3OH,2} = 14 \text{ mol}$$

$$n_{H_2O,2} = 14 \text{ mol}, \quad n_{I,2} = 2 \text{ mol}, \quad \xi = 14 \text{ mol}$$

Balance around the condenser. Note that the production rate of methanol will not be 155 mol/h because we have taken the unknown stream as the basis. The production rate for the new assumed basis will be different from 155 kmol/h. Scaling up will be considered later:

$$n_{CO_2,2} = n_{CO_2,4} = 14 \text{ mol}$$

$$n_{H_2,2} = n_{H_2,4} = 28 \text{ mol}$$

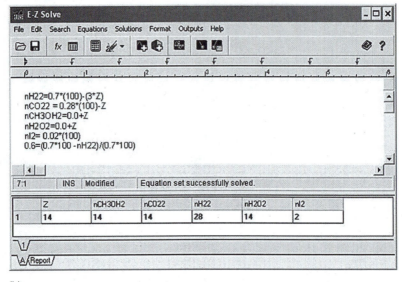

(b)

FIGURE E6.1b
Results using E–Z solve.

$$n_{I,2} = n_{I,4} = 2 \text{ mol}$$

$$n_{CH_3OH,2} = n_{CH_3OH,3} = 14 \text{ mol}$$

$$n_{H_2O,2} = n_{H_2O,3} = 14 \text{ mol}$$

The recycle stream and the purge stream have the same composition but not the same amount; hence the composition of the recycle and purge streams are

$$x_{CO_2,4} = x_{CO_2,5} = x_{CO_2,6} = 14/(14 + 28 + 2) = 0.32$$

$$x_{H_2,4} = x_{H_2,5} = x_{H_2,6} = 28/(14 + 28 + 2) = 0.635$$

$$x_{I,4} = x_{I,5} = x_{I,6} = 2/(14 + 28 + 2) = 0.045$$

Balance around the mixing point.
Total balance: $n_0 + n_6 = 100$
Inert balance: $0.004 \times n_0 + 0.045 \times n_6 = 0.02 \times 100$
The answer can be rounded off to $n_0 = 61$ mol and $n_6 = 39$ mol.
Other components can be determined as follows:

$$n_{CO_2,0} + 0.32 \times 39 = 0.28 \times 100$$

$$n_{H_2,0} + 0.635 \times 39 = 0.7 \times 100$$

$$n_{I,0} + 0.045 + 39 = 0.02 \times 100$$

$$n_{CO_2,0} = 15.5 \text{ mol}, \quad n_{H_2,0} = 45.25 \text{ mol}, \quad n_{I,0} = 0.25 \text{ mol}$$

Scaling

Amount of fresh feed required to produce 155 mol of methanol is equal to

$$155 \text{ mol CH}_3\text{OH produced} \times \frac{61 \text{ mol of fresh feed}}{14 \text{ mol of CH}_3\text{OH produced}}$$

$$= 675.357 \text{ mol of fresh feed}$$

A fresh feed of 675.357 mol/h is required to produce 155 mol/h.

6.3 Reaction and Multiple-Unit Steady-State Processes

The following points must be considered when solving problems where multiple-unit steady-state processes involve chemical reactions:

1. Some systems contain reactions and some will not. For subsystems with reactions (generally the reactor and the overall system).
   - Use individual component flows around the reactor and not the compositions.
   - Include stoichiometry and generation/consumption on molar balances.
   - In subsystems without reaction (such as mixers, splitters, and separators):
     - Input = output (moles are conserved, no generation/consumption terms).
     - More flexibility as to stream specification (component flows or composition).

2. Reactions combined with product separation and recycle:
   - Recycle stream is introduced to recover and reuse unreacted reactants.
   - Recycle allows operation of reactor at low single-pass conversion, and has a high overall conversion for the system.

3. Subsystems involving reactions with recycle and purge:
   - Purge is introduced to prevent buildup of inerts (or incompletely separated products) in the system.
   - Purge point is a simple splitter. That is, only one independent material balance exists.

Example 6.2 Ammonia synthesis reaction

Nitrogen gas and hydrogen gas are fed to a reactor in stoichiometric quantities that react to form ammonia. The conversion of nitrogen to ammonia is 25%. How much nitrogen is required to make 100 t/day of ammonia?

Solution

The process-labeled flow sheet of ammonia synthesis process is depicted in Figure E6.2.

$$N_2 + 3H_2 \rightarrow 2NH_3$$

Note that the inlet mole of nitrogen and hydrogen is in stoichiometric quantities:

$$\frac{N_2 \text{ in}}{H_2 \text{ in}} = \frac{1}{3}$$

1. The basis of 100 t/day of ammonia is given. Since we have a chemical reaction, let us convert this to moles:

$$100\frac{t}{day}\ \frac{10^3 \text{ kg}}{t}\ \frac{\text{kgmol NH}_3}{17 \text{ kg NH}_3} = 5882 \text{ kgmol NH}_3$$

2. From the problem statement we know that 25% of the nitrogen entering the reactor has been converted to ammonia. Knowing the definition of conversion to be

$$\text{Conversion} = X_A = \frac{\text{mol N}_{2\ \text{converted}}}{\text{mol N}_{2\ \text{in}}} = \frac{\text{mol N}_{2\ \text{in}} - \text{mol N}_{2\ \text{out}}}{\text{mol N}_{2\ \text{in}}}$$

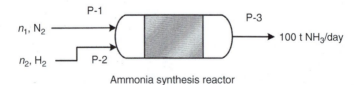

Ammonia synthesis reactor

Conversion $= 0.25$

FIGURE E6.2
Ammonia synthesis reactor.

$$0.25 = \frac{5882 \frac{\text{kgmol NH}_3}{\text{day}} \frac{\text{kgmol N}_2}{2 \text{ kgmol NH}_3}}{x_{N_2} \dot{n}_1}$$

Since stream 1 is pure nitrogen, $x_{N_2} = 1.0$.

$$\dot{n}_1 = 11{,}764 \frac{\text{kgmol N}_2}{\text{day}}$$

$$\dot{m}_1 = 11{,}764 \frac{\text{kgmol N}_2}{\text{day}} \frac{28 \text{ kg N}_2}{\text{kgmol N}_2} \frac{\text{t}}{10^3 \text{kg}} = 329 \frac{\text{t N}_2}{\text{day}}$$

$$\dot{n}_2 = \frac{3 \text{ kmol H}_2 \text{ fed}}{1 \text{ kmol N}_2 \text{ fed}} \times \left(11{,}764 \frac{\text{kmol N}_2 \text{ fed}}{\text{day}} \right) \Rightarrow \dot{n}_2 = 35{,}292 \frac{\text{kmol H}_2 \text{ fed}}{\text{day}}$$

$$\dot{m}_2 = 35{,}292 \frac{\text{kmol H}_2 \text{ fed}}{\text{day}} \left| \frac{2 \text{ kg H}_2}{1 \text{ kmol}} \right| \frac{1 \text{ t}}{1000 \text{ kg}} = 70.584 \frac{\text{t H}_2}{\text{day}}$$

Example 6.3 Ammonia reactors with recycle stream

Since the single-pass operation in Figure E6.2 is wasteful, we decide to separate the product stream and recycle the unwanted gases back to the reactor. Now how much nitrogen is required to make 100 t/day of ammonia? What is the flow rate of the recycle stream?

Solution

The process flowchart is shown in Figure E6.3.

1. Let us use 100 t/day of ammonia as the basis. In addition, since we have a chemical reaction, convert mass to moles as we did in

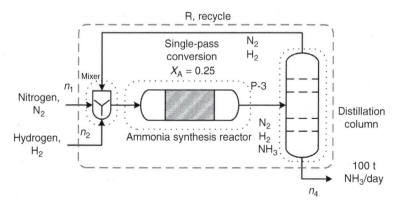

FIGURE E6.3
Ammonia synthesis with recycle.

TABLE E6.3

DF Analysis for the Ammonia Synthesis with Recycle Flow Sheet

DF Analysis	System within Control Volume		
	Reactor	Distillation Column	Overall Process
Number of unknowns	$7 + 1(\xi)$	5	$2 + 1(\xi)$
Number of independent equations	3	3	3
Number of auxiliary relations	1	—	0
DF	4	2	0

Figure E6.2, to get 5882 kmol NH_3/day. We can now draw a number of different control volumes and write a number of equations to solve this problem. Here is one way to go about it. Although we want to look at the recycle stream in this case, it may be easier to start with an overall balance (see the control volume outlined above). The other way is to perform a DF analysis for all available systems and start with the system with 0 DF. The DF analysis reveals that starting with the overall process is the best choice (Table E6.3).

2. Since the control volume above (overall process) contains a chemical reaction, the number of moles may not be conserved and atoms may switch from one molecule to another.

Mass in = mass out
Atoms in = atoms out
Overall atomic balance for nitrogen:

$$x_{N_2}\dot{n}_1 + x_{N_2}\dot{n}_2 = x_{N_2}\dot{n}_4, \ x_{N_2}\dot{n}_2 = 0$$

$$x_{N_2}\dot{n}_1 = x_{N_2}\dot{n}_4$$

$$1.0\dot{n}_1 = (1.0)5882 \ \frac{\text{kgmol } NH_3}{\text{day}} \ \frac{\text{kgmol } N_2}{2 \ \text{kgmol } NH_3} = 2941 \ \frac{\text{kgmol } N_2}{\text{day}}$$

$$2941 \ \frac{\text{kgmol } N_2}{\text{day}} \ \frac{28 \ \text{kg } N_2}{\text{kgmol } N_2} \ \frac{t}{10^3 \text{kg}} = 82.3 \ \frac{t \ N_2}{\text{day}}$$

$$2941 \ \frac{\text{kgmol } N_2}{\text{day}} \ \frac{28 \ \text{kg } N_2}{\text{kgmol } N_2} \ \frac{t}{10^3 \text{kg}} = 82.3 \ \frac{t \ N_2}{\text{day}}$$

3. From the problem statement we know that 25% of the nitrogen entering the reactor has been converted to ammonia. Knowing the definition of conversion to be

$$\text{Conversion} = X_A = \frac{\text{mol } N_2 \text{ reacted}}{\text{mol } N_2 \text{fed}} = \frac{\text{mol } N_2 \text{ fed} - \text{mol } N_2 \text{ out}}{\text{mol } N_2 \text{ fed}}$$

$$0.25 = \frac{x_{N_2}\dot{n}_1 + x_{N_2}\dot{n}_R - x_{N_2}\dot{n}_3}{x_{N_2}\dot{n}_1 + x_{N_2}\dot{n}_R}$$

Rearranging, we get

$$0.25 = \frac{2941 \frac{\text{kgmol } N_2}{\text{day}} + x_{N_2}\dot{n}_R - x_{N_2}\dot{n}_3}{2941 \frac{\text{kgmol } N_2}{\text{day}} + x_{N_2}\dot{n}_R} \qquad (6.1)$$

We now have two unknowns and one equation. To solve the problem we will need one more equation. Another independent equation will be a N_2 balance around the separator. Since no chemical reactions occur in the separator, moles in the separator are conserved and we can write a balance equation for molecular nitrogen.

$$x_{N_2}\dot{n}_3 = x_{N_2}\dot{n}_4 + x_{N_2}\dot{n}_R, \quad x_{N_2}\dot{n}_4 = 0$$

$$x_{N_2}\dot{n}_3 = x_{N_2}\dot{n}_R \qquad (6.2)$$

Now, we have two equations and two unknowns. Substitute Equation 6.2 into Equation 6.1 and solve

$$0.25 = \frac{2941 \frac{\text{kgmol } N_2}{\text{day}} + x_{N_2}\dot{n}_R - x_{N_2}\dot{n}_R}{2941 \frac{\text{kgmol } N_2}{\text{day}} + x_{N_2}\dot{n}_R}$$

$$0.25 = \frac{2941 \frac{\text{kgmol } N_2}{\text{day}}}{2941 \frac{\text{kgmol } N_2}{\text{day}} + x_{N_2}\dot{n}_R}$$

$$x_{N_2}\dot{n}_R = 8823 \frac{\text{kgmol } N_2}{\text{day}} \quad \text{in recycle stream}$$

$$8823 \frac{\text{kgmol } N_2}{\text{day}} \frac{28 \text{ kg } N_2}{\text{kgmol } N_2} \frac{t}{10^3 \text{ kg}} = 247 \frac{t \, N_2}{\text{day}}$$

We also know that hydrogen was fed stoichiometrically, so the amount of hydrogen is three times greater than the amount of nitrogen in every stream:

$$3x_{N_2}\dot{n}_R = x_{H_2}\dot{n}_R$$

$$3 \times 8823 \frac{\text{kgmol } N_2}{\text{day}} = 26{,}469 \frac{\text{kgmol } H_2}{\text{day}} \quad \text{in recycle stream}$$

$$26{,}469 \frac{\text{kgmol } H_2}{\text{day}} \frac{2\text{kg } N_2}{\text{kgmol } N_2} \frac{t}{10^3 \text{ kg}} = 52.9 \frac{t \, H_2}{\text{day}}$$

$$52.9 \frac{t \, H_2}{\text{day}} + 247 \frac{t \, N_2}{\text{day}} = 300 \frac{t}{\text{day}}$$

Example 6.4 Ammonia reaction with inert in the feed stream

Suppose that the nitrogen stream in Figure E6.3 has 0.2% argon (Ar) as an impurity. If we do only recycling, then Ar concentration will build up and eventually shut down the reaction. To avoid this, we purge a portion of the recycle stream so that the level of Ar in the recycle stream is maintained at 7.0%. Now, how much nitrogen is required to make 100 t/day of ammonia? What is the flow rate of the recycle stream? What is the flow rate of the purge stream?

Solution

This problem is a little more complicated than the problems above and is solved in detail below. Here is an outline of the solution procedure:

(1) Draw the process flow sheet, and then choose a basis.
(2) Perform balances around: Overall system, separator, reactor, and splitter.

The process flow sheet is shown in Figure E6.4a.
The choice of the system to start with is facilitated by performing DF analysis as shown in Table E6.4.1.

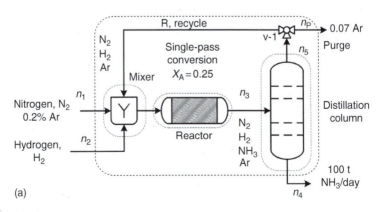

(a)

FIGURE E6.4a
Ammonia synthesis process with Ar in the nitrogen stream.

TABLE E6.4.1

DF Analysis for Ammonia Synthesis with Purge Stream

DF Analysis	System within Control Volume			
	Reactor	Separator	Splitter	Overall Process
Number of unknowns	$9+1(\xi)$	7	3	$3+1(\xi)$
Number of independent equations	4	4	1	4
Number of auxiliary relations	1(conversion)	—	—	—
DF	6	3	2	0

Here is one way to tackle this problem. Again, choose a basis of 100 t/day of ammonia, which converts to 5882 kgmol/day of ammonia. Next, draw the diagram.

Next, we know some information about some of the streams from common sense and simple calculations. For example, we know that hydrogen is fed stoichiometrically so for every mole of nitrogen in the system, 3 mol of hydrogen is present:

$$x_{N_2} + x_{H_2} + x_{Ar} + x_{NH_3} = 1.0$$

$$x_{NH_3} = 0 \text{ in stream R}$$

$$x_{Ar} = 0.07 \text{ in stream R}$$

$$3x_{N_2} = x_{H_2}$$

Solving: $x_{N_2} + 3x_{N_2} + 0.07 = 1.0$, $x_{N_2} = 0.2325$, $3x_{N_2} = x_{H_2} = 0.697$

Since stream 5 splits into streams R and P, we know that the compositions of streams R and P are the same as stream 5. Now we can start drawing balances to solve for the streams. Since we do not know anything about stream 3 yet, it might be best to balance it later. Let us start with balances around the whole system (Figure E6.4b) since we know the compositions of streams 1, 2, 4, and P as the DF is 0.

In the overall balance, we know the compositions of all four streams (1, 2, 4, P) but we only know the molar flow rates of stream 4. This leaves us with three unknowns. Therefore, you need three independent equations to solve for the molar flow rates of 1, 2, and P. Mass and atoms are always conserved, but because the control volume contains a chemical reaction, the atoms may move between molecules and moles may not be conserved. So we most certainly cannot write an overall mole balance. We can however write individual or overall atom balances. We have three types of atoms: N, H, and Ar, so we can write three independent equations (i.e., one for each atomic component).

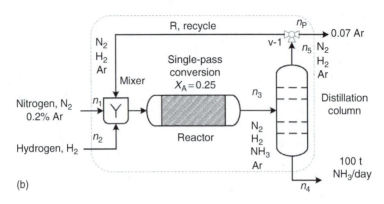

FIGURE E6.4b
Ammonia synthesis process with recycle and purge (system: overall process).

Overall atomic balances for the system

$$\text{Ar balance } x_{\text{Ar}}\dot{n}_1 + x_{\text{Ar}}\dot{n}_2 = x_{\text{Ar}}\dot{n}_4 + x_{\text{Ar}}\dot{n}_\text{P}$$

$$\text{N balance } x_\text{N}\dot{n}_1 + x_\text{N}\dot{n}_2 = x_\text{N}\dot{n}_4 + x_\text{N}\dot{n}_\text{P}$$

$$\text{H balance } x_\text{H}\dot{n}_1 + x_\text{H}\dot{n}_2 = x_\text{H}\dot{n}_4 + x_\text{H}\dot{n}_\text{P}$$

Simplification of the above equations
Argon (Ar) atomic balance

$$x_{\text{Ar}}\dot{n}_1 + x_{\text{Ar}}\dot{n}_2 = x_{\text{Ar}}\dot{n}_4 + x_{\text{Ar}}\dot{n}_\text{P}$$

$$x_{\text{Ar}}\dot{n}_2 = 0 \text{ and } x_{\text{Ar}}\dot{n}_4 = 0$$

$$x_{\text{Ar}}\dot{n}_1 = x_{\text{Ar}}\dot{n}_\text{P}$$

$$0.002\dot{n}_1 = 0.07\dot{n}_\text{P} \qquad (6.3)$$

Nitrogen (N) atomic balance

$$x_{\text{N}_2}\dot{n}_1 + x_{\text{N}_2}\dot{n}_2 = x_{\text{NH}_3}\dot{n}_4(\text{N}_2/2\text{NH}_3) + x_{\text{N}_2}\dot{n}_\text{P}$$

$$x_{\text{N}_2}\dot{n}_2 = 0$$

$$x_{\text{N}_2}\dot{n}_1 = x_{\text{NH}_3}\dot{n}_4(\text{N}_2/2\text{NH}_3) + x_{\text{N}_2}\dot{n}_\text{P}$$

$$0.998\dot{n}_1 = 5882 \text{ kgmol NH}_3(\text{N}_2/2\text{NH}_3) + 0.233\dot{n}_\text{P} \qquad (6.4)$$

$$x_{\text{N}_2}\dot{n}_1 + x_{\text{N}_2}\dot{n}_2 = x_{\text{NH}_3}\dot{n}_4(\text{N}_2/2\text{NH}_3) + x_{\text{N}_2}\dot{n}_\text{P}$$

$$x_{\text{N}_2}\dot{n}_2 = 0$$

$$x_{\text{N}_2}\dot{n}_1 = x_{\text{NH}_3}\dot{n}_4(\text{N}_2/2\text{NH}_3) + x_{\text{N}_2}\dot{n}_\text{P}$$

$$0.998\dot{n}_1 = 5882 \text{ kgmol NH}_3(\text{N}_2/2\text{NH}_3) + 0.233\dot{n}_\text{P} \qquad (6.5)$$

Hydrogen (H) atomic balance

$$x_{H_2}\dot{n}_1 + x_{H_2}\dot{n}_2 = x_{NH_3}\dot{n}_4(3H_2/2NH_3) + x_{H_2}\dot{n}_P$$

$$x_{H_2}\dot{n}_2 = x_{NH_3}\dot{n}_4(3H_2/2NH_3) + x_{H_2}\dot{n}_P$$

$$1.000n_2 = 5882 \text{ kgmol NH}_3(3H_2/2NH_3) + 0.697n_P \qquad (6.6)$$

Substituting Equation 6.3 into Equation 6.4 and solving

$$0.998\dot{n}_1 = 5882 \text{ kgmol NH}_3(N_2/2NH_3) + 0.233\left(\frac{0.002}{0.07}\right)\dot{n}_1$$

$$\dot{n}_1\left[0.998 - 0.233\left(\frac{0.002}{0.07}\right)\right] = 5882 \text{ kgmol NH}_3\frac{N_2}{2NH_3} \Rightarrow \dot{n}_1 = 2967\frac{\text{kgmol}}{\text{day}}$$

Substituting $\dot{n}_1$ in Equation 6.3 and substituting these results in Equation 6.6 gives

$$1.000n_2 = 5882 \text{ kgmol NH}_3\frac{3H_2}{2NH_3} + 0.697\left(84.8\frac{\text{kgmol}}{\text{day}}\right)$$

$$1.000n_2 = 5882 \text{ kgmol NH}_3\frac{3H_2}{2NH_3} + 0.697\left(84.8\frac{\text{kgmol}}{\text{day}}\right)$$

$$0.002\left(2967\frac{\text{kgmol}}{\text{day}}\right) = 0.07\dot{n}_P \Rightarrow \dot{n}_P = 84.8\frac{\text{kgmol}}{\text{day}}$$

Now, we need to solve for streams 3, 5, and R. Let us start with streams 3 and R. A drawn control volume around the separator is shown in Figure E6.4c. The border of the control volume cross streams 3 and R, so we can

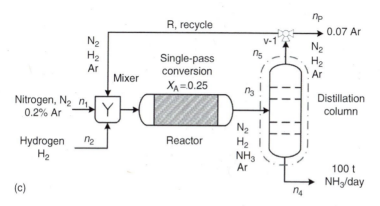

(c)

FIGURE E6.4c
Ammonia synthesis process (system: distillation column).

write several balances. Since the control volume does not contain a chemical reaction, moles are conserved and we can write balances for molecular species rather than just atoms. Since we can write molecular balances, we have four components: N_2, H_2, Ar, NH_3—and we can write an independent equation with each one.

N_2 balance

$$x_{N_2} \dot{n}_3 = x_{N_2} \dot{n}_4 + x_{N_2} \dot{n}_P + x_{N_2} \dot{n}_R$$

$$x_{N_2} \dot{n}_4 = 0$$

$$x_{N_2} \dot{n}_3 = 0.233 \left(84.8 \; \frac{\text{kgmol}}{\text{day}} + \dot{n}_R \right) \qquad (6.7)$$

This equation has three unknowns, so we will need four additional independent pieces of information or equations to solve.

H_2 balance

$$x_{H_2} \dot{n}_3 = x_{H_2} \dot{n}_4 + x_{H_2} \dot{n}_P + x_{H_2} \dot{n}_R, \quad x_{H_2} \dot{n}_4 = 0$$

$$x_{H_2} \dot{n}_3 = 0.697 \left(84.8 \; \frac{\text{kmol}}{\text{day}} + \dot{n}_R \right) \qquad (6.8)$$

Ar balance

$$x_{Ar} \dot{n}_3 = x_{Ar} \dot{n}_4 + x_{Ar} \dot{n}_P + x_{Ar} \dot{n}_R, \quad x_{Ar} \dot{n}_4 = 0$$

$$x_{Ar} \dot{n}_3 = 0.07 \left(84.8 \; \frac{\text{kmol}}{\text{day}} + \dot{n}_R \right) \qquad (6.9)$$

NH_3 balance

$$x_{NH_3} \dot{n}_3 = x_{NH_3} \dot{n}_4 + x_{NH_3} \dot{n}_P + x_{NH_3} \dot{n}_R$$

$$x_{NH_3} \dot{n}_4 = 5882 \; \frac{\text{kgmol NH}_3}{\text{day}}; \quad x_{NH_3} \dot{n}_P = 0; \quad x_{NH_3} \dot{n}_R = 0$$

$$x_{NH_3} \dot{n}_3 = \dot{n}_4 = 5882 \; \frac{\text{kgmol NH}_3}{\text{day}} \qquad (6.10)$$

Many new unknowns have been added including all of the mole fractions in stream 3: x_{N_2}, x_{H_2}, x_{Ar}, x_{NH_3}. Solving the problem needs many new equations. Let us look back at Equation 6.7:

N_2 balance

$$x_{N_2} \dot{n}_3 = x_{N_2} \dot{n}_4 + x_{N_2} \dot{n}_P + x_{N_2} \dot{n}_R, \quad x_{N_2} \dot{n}_4 = 0$$

$$x_{N_2} \dot{n}_3 = 0.233 \left(84.8 \frac{\text{kgmol}}{\text{day}} + \dot{n}_R \right) \tag{6.11}$$

What we really need is an equation with this same set of unknowns. We can use the equation for the conversion of nitrogen:

$$X_A = 0.25 = \frac{(\text{mol } N_2)_{\text{in}} - (\text{mol } N_2)_{\text{out}}}{(\text{mol } N_2)_{\text{in}}}$$

$$X_A = 0.25 = \frac{x_{N_2} \dot{n}_1 + x_{N_2} \dot{n}_2 + x_{N_2} \dot{n}_R - x_{N_2} \dot{n}_3}{x_{N_2} \dot{n}_1 + x_{N_2} \dot{n}_2 + x_{N_2} \dot{n}_R}$$

$$X_A = 0.25 = \frac{2961 \frac{\text{kgmol } N_2}{\text{day}} + 0.233 \dot{n}_R - x_{N_2} \dot{n}_3}{2961 \frac{\text{kgmol } N_2}{\text{day}} + 0.233 \dot{n}_R} \tag{6.11a}$$

Now substitute Equation 6.11 into Equation 6.11a

$$X_A = 0.25 = \frac{2961 \frac{\text{kgmol } N_2}{\text{day}} + 0.233 \dot{n}_R - 0.233 \left(84.8 \frac{\text{kgmol}}{\text{day}} + \dot{n}_R \right)}{2961 \frac{\text{kgmol } N_2}{\text{day}} + 0.233 \dot{n}_R}$$

$$0.25 \left(2961 \frac{\text{kgmol } N_2}{\text{day}} + 0.233 \dot{n}_R \right) = 2961 \frac{\text{kgmol } N_2}{\text{day}} - 0.233 \left(84.8 \frac{\text{kgmol}}{\text{day}} \right)$$

$$\dot{n}_R = 37{,}785 \frac{\text{kgmol}}{\text{day}}$$

$$x_{N_2} \dot{n}_3 = 0.233 \left(84.8 \frac{\text{kgmol}}{\text{day}} + \dot{n}_R \right) = 8824 \frac{\text{kgmol } N_2}{\text{day}}$$

$$x_{Ar} \dot{n}_3 = 0.07 \left(84.8 \frac{\text{kgmol}}{\text{day}} + \dot{n}_R \right) = 2651 \frac{\text{kgmol Ar}}{\text{day}}$$

Now let us fill in our table with what we have calculated so far. Now that we know the amount of N_2 in stream 3, we also know the amount of hydrogen in stream 3. It is three times the number of moles of N_2 present or 26,472 kgmol/day. We now know the molar flow rate of each component in stream 3 so we can calculate the mole fraction and total flow rate of stream 3. We just need to know the overall flow of stream 5. Calculating the flow rate of stream 5 can be done by taking the splitter as the system as shown in the control volume around the splitter as follows (Figure E6.4d):

You can probably calculate the balance in your head, but let us do it anyway. There is no chemical reaction in this balance, so moles are conserved. We can write an overall balance:

$$\dot{n}_5 = \dot{n}_R + \dot{n}_P$$

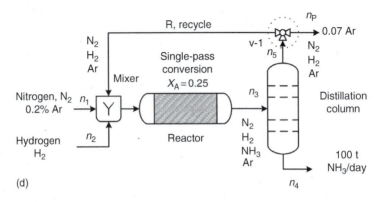

(d)

FIGURE E6.4d
Ammonia synthesis process with recycle and purge (system: splitter).

$$\dot{n}_5 = 37{,}785\frac{\text{kgmol}}{\text{day}} + 84.8\frac{\text{kgmol}}{\text{day}}$$

$$\dot{n}_5 = 37{,}870\frac{\text{kgmol}}{\text{day}}$$

The final results that include all stream flow rates and compositions are shown in Table E6.4.2.

Example 6.5   Ammonia synthesis process

A stream of 99.8% nitrogen and 0.2% argon is combined with stoichiometric hydrogen and fed to a reactor to produce 50 t/day of ammonia. 10% of the hydrogen fed to the reactor is converted to ammonia. The products are then separated in a condenser at −38°C and 3 atm. The liquid ammonia is drawn from the bottom of the condenser as product. The uncondensed hydrogen,

TABLE E6.4.2

Final Results of Ammonia Synthesis Process with Recycle and Purge

Stream Number	N₂		H₂		NH₃		Ar		Total
	x_i	n_i	x_i	n_i	x_i	n_i	x_i	n_i	n_i
1	0.998	2961	0	0	0	0	0.002	5.9	2967
2	0	0	1.000	8882	0	0	0	0	8882
3	0.201	8824	0.604	26,472	0.134	5882	0.060	2651	43,829
4	0	0	0	0	1.000	5882	0	0	5882
5	0.233	8824	0.697	26,395	0	0	0.07	2651	37,870
R	0.233	8804	0.697	26,336	0	0	0.07	2645	37,785
P	0.233	19.8	0.697	59.1	0	0	0.07	5.9	84.8

nitrogen, and ammonia are recycled and combined with the nitrogen and hydrogen stream entering the reactor. A portion of the recycle stream is purged to keep the mole fraction of Ar in the recycle stream at 0.07. (Antoine's coefficients for ammonia are: $A = 7.55466$, $B = 1002.711$, and $C = 247.885$.)

a. What is the molar flow of the fresh feed?
b. What is the composition of the recycle stream?
c. What is the composition of the purge stream?
d. What is the flow rate of the purge stream?
e. What is the flow rate of the recycle stream?
f. What is the flow rate of stream 2?
g. What is the composition of stream 2?

Solution
Basis: 1 mol of nitrogen stream (i.e., $N_2 + Ar$)
  The process flow sheet is shown in Figure E6.5a.
  Parts a–d

$$50 \text{ t } NH_3 \frac{2000 \text{ lb}_m}{t} \frac{454 \text{ g}}{lb_m} \frac{mol}{17 \text{ g}} = 2671 \frac{kmol \text{ } NH_3}{day}$$

Composition in fresh feed (0.998 N_2 and 0.002 Ar)

$$0.998 \text{ } N_2 \frac{3H_2}{N_2} = 2.994 \text{ } H_2 \Rightarrow n_{Total} = 3.994$$

Normalizing

$$x_{N_2,1} = \frac{0.998}{3.994} = 0.250, \quad x_{H_2,1} = \frac{2.994}{3.994} = 0.7495, \quad x_{Ar,1} = \frac{0.002}{3.994} = 0.0005$$

Use Raoult's law to find the composition of recycle and purge streams

$$\log P_{NH_3}^{sat} [mmHg] = A - \frac{B}{T[°C] + C}$$

$$T = -38°C, \quad A = 7.55466, \quad B = 1002.711, \quad C = 247.885$$

$$P_{NH_3}^{sat} = 599 \text{ mmHg}$$

$$y_1 P = x_1 P_1^{sat}$$

$$y_{NH_3,P} = \frac{(1.0)(599 \text{ mmHg})}{3 \text{ atm} \dfrac{760 \text{ mmHg}}{\text{atm}}} \Rightarrow y_{NH_3,P} = y_{NH_3,R} = 0.263$$

$$0.07 + x_{N_2,P} + 3x_{N_2,P} + 0.263 = 1.0$$

$$x_{N_2,P} = x_{N_2,R} = 0.167, \; x_{H_2,P} = x_{H_2,R} = 0.500, \; x_{Ar,P} = x_{Ar,R} = 0.070$$

System: Overall balance (Figure E6.5a)
Argon–nitrogen atomic balance

$$x_{Ar,1} n_1 = x_{Ar,P} n_P$$

$$0.0005 n_1 = 0.07 n_P$$

Nitrogen atomic balance

$$0.250 n_1 \frac{2N}{N_2} = 2671 \text{ kmol NH}_3 \frac{N}{NH_3} + 0.263 n_P \frac{N}{NH_3} + 0.167 n_P \frac{2N}{N_2}$$

$$0.5 n_1 = 2671 + 0.430 n_P$$

We are solving two equations and two unknowns

$$0.5 n_1 = 2671 + 0.430 \left(\frac{0.0005}{0.07} n_1 \right)$$

$$n_1 = 5375 \frac{\text{kmol}}{\text{day}}, \; n_P = 38 \frac{\text{kmol}}{\text{day}}$$

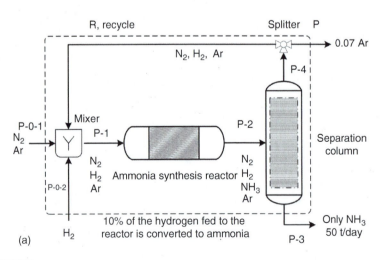

(a)

FIGURE E6.5a
Flow sheet of ammonia synthesis process (system: overall process).

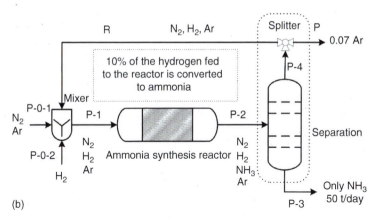

(b)

FIGURE E6.5b
Ammonia synthesis process (system: separator + splitter).

Parts e and f
Solve for the flow of recycle and reactor effluent streams
System: Separator + splitter (Figure E6.5b)
Nitrogen (N_2) molecular balance

$$x_{N_2,2}n_2 = x_{N_2,P}n_P + x_{N_2,R}n_R$$

$$x_{N_2,2}n_2 = 0.167(38 + n_R)$$

System: Reactor + mixer (Figure E6.5c)
Nitrogen (N_2) molecular balance

$$x_{N_2,1}n_1 + x_{N_2,R}n_R = x_{N_2,2}n_2 + 0.1(x_{N_2,1}n_1 + x_{N_2,R}n_R)$$

$$x_{N_2,2}n_2 = 0.9(x_{N_2,1}n_1 + x_{N_2,R}n_R)$$

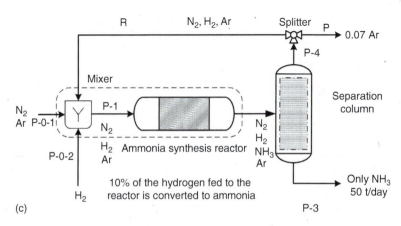

(c)

FIGURE E6.5c
Ammonia synthesis process (system: reactor + mixer).

Solving two equations and two unknowns

$$0.167(38 + n_R) = 0.9(x_{N_2,1}n_1 + x_{N_2,R}n_R)$$

$$6.346 + 0.167n_R = 0.9\{0.250(5374) + 0.167n_R\} \Rightarrow n_R = 72{,}027 \text{ kmol/day}$$

Example 6.6   Methanol production

Methanol is produced by the reaction of CO with H_2. A portion of the methanol leaving the reactor is condensed, and the unconsumed CO and H_2 and the uncondensed methanol are recycled back to the reactor. The reactor effluent flows at a rate of 275 mol/min and contains 21.8 wt% H_2, 33.9 wt% CO, and 44.3 wt% methanol. The mole fraction of methanol in the recycle stream is 0.004. Calculate the molar flow rates of CO and H_2 in the fresh feed, and the production rate of methanol.

Solution

Note that the mass fractions were given. We must convert these to mole fractions to solve the problem (see Figure E6.6).

DF Analysis	Mixing	Reactor	Condenser	Overall
Number of unknowns	7	3 + 1	3	3 + 1
Number of independent equations	3	3	3	3
Auxiliary relations	0	0	0	0
DF	4	1	0	1

Basis: 1 min of operation
Sytstem: Condenser
CO: 24.8 mol $= n_3$
H_2: 222.2 mol $= n_4$

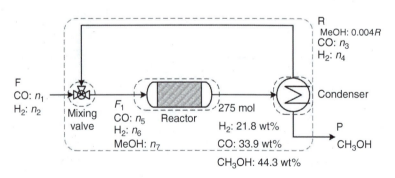

FIGURE E6.6
Methanol production process.

CH_3OH: $28.0 \text{ mol} = P + 0.004R$
Total: $275 \text{ mol} = P + R$
The last two equations can be solved simultaneously to obtain
$P = 27.3 \text{ mol}$ and $R = 247.7 \text{ mol}$
Now that P has been found we can do the block balances.

System: Overall process
   We will use element balances around the block. Element balances are generally advised when doing balances around the block. Component balances around the block can be done, but do not forget to include the reactor and the consumption/generation terms.
C balance $n_1 = P = 27.3 \text{ mol}$

H balance $2n_2 = 4\,P = 4 \times P \Rightarrow n_2 = 2 \times P = 54.6 \text{ mol}$

Now scale to the requested conditions:

$$F = 81.9 \text{ mol/min}, \quad n_1 = 27.3 \text{ mol/min}, \quad n_2 = 54.6 \text{ mol/min}, \quad P = 27.3 \text{ mol/min}$$

Example 6.7   Hydrogen production process
A reactor is designed to take a feed of 200 mol/min of propane and 50% excess steam and convert them to CO and H_2 with a 65% conversion rate. Determine the mole fractions of the reactor products (Figure E6.7).

Solution
Basis: 1 min operation = 200 mol C_3H_8 fed to reactor
DF analysis
Number of unknowns: $5 + 1$
Number of independent equations: 4
Number of relations: 2
$DF = 6 - 4 - 2 = 0$
Mole balance

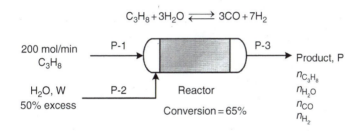

FIGURE E6.7
Hydrogen production process.

$$n_{C_3H_8} = (n_{C_3H_8})_0 - \xi$$

$$n_{H_2O} = (n_{H_2O})_0 - 3\xi$$

$$n_{CO} = (n_{CO})_0 + 3\xi$$

$$n_{H_2} = (n_{H_2})_0 + 7\xi$$

$$n_T = n_{T_0} + 6\xi$$

6.3.1 Auxiliary Relationship

1. Now we find the limiting reagent, which is obviously C_3H_8 since the problem states that there is an excess of water. We can then use one of the subsidiary relations to determine the amount (number of moles) of steam fed to the reactor. We find the number of moles of steam required for complete conversion by the limiting reagent; the actual amount fed will be 50% more than this amount. This is done in the following equations:

$$50\% \text{ excess steam} \Rightarrow 0.5 = (n_2^0 - n_{2,\text{required}})/n_{2,\text{required}}$$

$$n_{2,\text{req}} = (3)(200 \text{ mol})/(1) = 600 \text{ mol}$$

$$n_{2_0} = (0.5)(600) + 600 = (1.5)(600) = 900 \text{ mol}$$

2. Now we can use the second subsidiary relation to find the number of moles of C_3H_8 that exit the reactor unreacted. From the definitions already used we write that the fraction unconverted (which is one minus the fractional conversion) is

$$1 - x = 0.35 = n_1/n_{1_0}, \text{ thus } n_1 = 0.35n_{1_0} = (0.35)(200) = 70 \text{ mol}$$

Substituting in above mole balance equations yields

$$C_3H_8: 70 = 200 - \xi$$

$$H_2O: n_{H_2O} = 900 - 3\xi, \text{ CO}: n_{CO} = 3\xi, \text{ H}_2: n_{H_2} = 7\xi, \text{ total } n_T = 100 + 6\xi$$

Solving for C_3H_8 yields $\xi = 130$ mol:

$$n_{C_3H_8} = 70 \text{ mol}, n_{H_2O} = 510 \text{ mol}, n_{CO} = 390 \text{ mol}, n_{H_2} = 910 \text{ mol}, n_T = 1880$$

Example 6.8   Multiple-reaction process

The following reversible reactions occur in a reactor:

$$A + 2B \leftrightarrow C + D$$

$$2D + B \leftrightarrow E$$

$$E + B \leftrightarrow A$$

Suppose 10 mol of A, 50 mol of B, and 5 mol of D are fed to the reactor and the product stream is sampled and found to contain 30 mol% C, 20 mol% A, and 5 mol% E. How many moles leave the reactor and what is the percent conversion of A and B?

Solution

$$A + 2B \leftrightarrow C + D \quad \xi_1$$
$$2D + B \leftrightarrow E \quad \xi_2$$
$$E + B \leftrightarrow A \quad \xi_3$$

Basis: 65 mol of feed
Mole balance

$$A: n_A = 10 - \xi_1 + \xi_3$$
$$B: n_B = 50 - 2\xi_1 - \xi_2 - \xi_3$$
$$C: n_C = \xi_1$$
$$D: n_D = 5 + \xi_1 - 2\xi_2$$
$$E: n_E = 0 + \xi_2 - \xi_3$$
$$n_T = 65 - \xi_1 - 2\xi_2 - \xi_3$$

Note that we have information about the inlet and outlet for three of the compounds, and that we have three degrees of freedom to solve for. We simply set up the three equations for the mole fractions in terms of the balances and solve the three equations for the three unknowns.

$$x_c = \xi_1/(65 - \xi_1 - 2\xi_2 - \xi_3) = 0.3 \text{ so } 19.5 - 0.3\xi_1 - 0.6\xi_2 - 0.3\xi_3 = \xi_1$$
$$1.3\xi_1 + 0.6\xi_2 + 0.3\xi_3 = 19.5$$

$$x_A = (10 - \xi_1 + \xi_3)/(65 - \xi_1 - 2\xi_2 - \xi_3) \quad \text{so} \quad 13 - 0.2\xi_1 - 0.4\xi_2 - 0.2\xi_3$$
$$= 10 - \xi_1 + \xi_3$$
$$0.8\xi_1 - 0.4\xi_2 - 1.2\xi_3 = -3$$

$$x_E = (\xi_2 - \xi_3)/(65 - \xi_1 - 2\xi_2 - \xi_3) = 0.05 \quad \text{so} \quad 3.25 - 0.05\xi_1 - 0.1\xi_2 - 0.05\xi_3$$
$$= \xi_2 - \xi_3$$
$$0.05\xi_1 + 1.1\xi_2 - 0.95\xi_3 = 3.25$$

Solving these three equations for the three extents of reaction gives

$$\xi_1 = 9.828 \text{ mol}, \quad \xi_2 = 8.017 \text{ mol}, \quad \text{and} \quad \xi_3 = 6.379 \text{ mol}$$

We can now substitute these values into the remaining material balances to find the fractional conversion:

$$n_T = 65 - \xi_1 - 2\xi_2 - \xi_3 = 32.76 \text{ mol}$$

$$n_A = 0.2(20 \text{ mol}) = 4.0 \text{ mol of A}$$

$$x_A = (n_{A_0} - n_A)/n_{A_0} = (10 - 4)/10 = 60\%$$

Likewise for B,

$$n_B = 50 - 2\xi_1 - \xi_2 - \xi_3 = 12.33 \text{ mol of B}$$

$$x_B = (n_{B_0} - n_B)/n_{B_0} = (50 - 12.33)/50 = 75.3\%$$

Example 6.9   Methanol production

In the process of methanol production 100 mol of 32 mol% CO, 64 mol% H_2, and 4 mol% of N_2 is fed to the reactor. The product from the reactor is condensed where liquid methanol is separated. Unreacted gases are recycled in a ratio of 5:1 moles recycle to fresh feed. Parts of the recycled gases are purged. Calculate all unknown stream flow rates and compositions (Figure E6.9).

Basis: 100 mol/h fresh feed
Five moles of recycle: 1 mol of fresh feed
Reactor feed: $5(100) + 100 = 600 \text{ mol/h} = n_r$
N_2 balance at feed mixing point: $(100)(0.04) + 500(x_{CN}) = 600(0.31)$
$x_{CN} = 0.148 \text{ mol } N_2/\text{mol recycle}$

Overall balances:

$$N_2: 100(0.04) = n_P(0.148) \rightarrow n_P = 27.03 \text{ mol purged/h}$$

$$C: 100(0.32) = x_{C_c} n_P + n_m = x_{C_c}(27.03) + n_m = 32$$

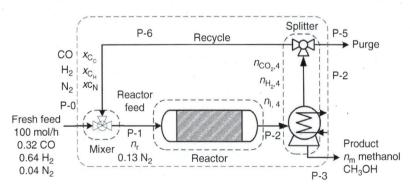

FIGURE E6.9
Methanol production process.

$$H: 2(64) = 128 = 4(n_m) + 2(27.03)(1 - 0.148 - x_{C_c})$$
$$128 = 4n_m + 54.06(0.852 - x_{C_c})$$

Solve for $n_m = 24.3$ mol CH_3OH, $x_{C_c} = 0.284$

6.4 Problems

6.4.1 Chemical Reactor Analysis

In the production of propylene oxide, CO_2 (at high pressure) is used as solvent since the reactant gases (H_2, O_2, and C_3H_6) and the propylene oxide products are soluble in CO_2, and CO_2 does not react with these gases (See Figure P6.4.1). A Pt/Pd/Ti catalyst promotes the propylene oxide synthesis. A by-product, propane, is also produced in this system. The desired reaction is

C_3H_6 (propylene) $+ H_2 + 1/2O_2 \rightarrow C_3H_8O$ (propylene oxide)

The by-product reaction is

$C_3H_6 + H_2 \rightarrow C_3H_8$ (propane)

In the test system, 35 mol/h air (79 mol% N_2, 21 mol% O_2) are mixed with 4 mol/h C_3H_6, 1.26 mol/h H_2, and 20,000 mol/h CO_2. The mixture is fed to a continuous reactor and the reactor output is monitored. The percent single-pass conversion of C_3H_6 is 90.5%. The fractional selectivity of C_3H_6 to C_3H_8O is 0.771.

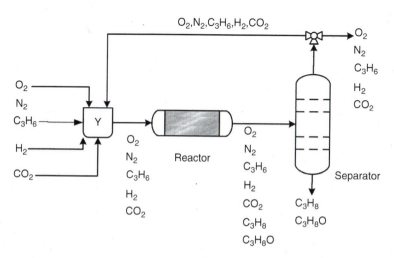

FIGURE P6.4.1
Process flow sheet for the production of propylene oxide.

1. What is the limiting reactant in this system? Justify your answer.

2. Calculate the mole percentage of each component in the effluent stream.

3. You notice that the process wastes unreacted reactants, and you want to recycle components as well as the solvent. Draw a block flow diagram of a process that recycles CO_2, O_2, N_2, H_2, and C_3H_6. Individual air, CO_2, H_2, and C_3H_6 streams are mixed before their entry into the reactor. C_3H_8O and C_3H_8 are separated from the reactor effluent and are removed. Assume this separator is perfect. You want to recycle the remaining components such that the reactor can operate at steady state. Draw the block flow diagram and indicate how many additional specifications (if any) will be required to solve for flow rates and compositions of all streams in the system. Do not solve for flows or compositions.

(Answer: $\xi_1 = 0.29$, $\xi_2 = 0.09$, $\dot{n}_{Pe,out} = 3.62$, $\dot{n}_{H,out} = 0.88$, $\dot{n}_{O,out} = 7.21$, $\dot{n}_{N,out} = 27.65$, $\dot{n}_{C,out} = 20,000$, $\dot{n}_{Po,out} = 0.29$, and $\dot{n}_{Pa,out} = 0.09$)

6.4.2 Laundry Detergent Synthesis Process

Butanal (C_4H_8O), used to make laundry detergents, is synthesized by the reaction of propylene (C_3H_6) with CO and H_2.

$$C_3H_6 + CO + H_2 \rightarrow C_4H_8O$$

In an existing process, 180 kmol/h of C_3H_6 (P) is mixed with 420 kmol/h of a mixture containing 50% CO and 50% H_2 and with a recycle stream containing propylene and then fed to a reactor, where a single-pass conversion of propylene of 30% is achieved. The desired product butanal (B) is removed in one stream, unreacted CO and H_2 are removed in a second stream, and unreacted C_3H_6 is recovered and recycled. Calculate the production rate of butanal (kmol/h) and the flow rate of the recycle stream (kmol/h) (Figure P6.4.2).

(Answer: $\dot{n}_P$, fed to reactor = 600 kmol/h. From a balance on the mixer, we can find that the recycle rate must be (600–180) = 420 kmol/h)

6.4.3 Butanal Production

A company contacts you and claims that they can supply you with a cheaper source of propylene, which could replace your current supply. Unfortunately, the cheaper source is contaminated with propane at a 5:95 ratio (propane:propylene). The cheaper source is economically attractive if the butanal production rate is maintained at the current rate and if an overall conversion of 0.90 can be achieved. Under your reactor conditions, propane is an inert, and it is too expensive to separate propane from

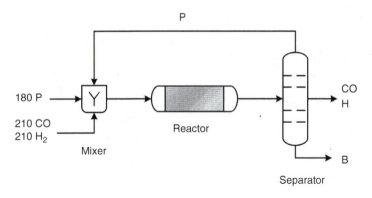

FIGURE P6.4.2
Laundry detergent process.

propylene, so you decide to install a purge stream. Assume the $CO + H_2$ stream remains the same (420 kmol/h, 50 mol% CO) as does the single-pass conversion of propylene in the reactor (0.3). Given this, show how the block flow diagram must be modified to accommodate the cheaper source of propylene, and calculate the following

(i) Flow rate of the contaminated propylene stream to the process (kmol/h).
(ii) Mole percentage inert and the total flow rate (kmol/h) of the purge stream.
(iii) Mole percentage inert and the total flow rate (kmol/h) of the feed to the reactor.

A splitter and purge are now essential. We no longer know the incoming flow rate but we fix the butanal production rate at 180 kmol/h, from Figure P6.4.3.

(Answer: (i) $\dot{n}_{P,fed} = 200$ kgmol/h, (ii) 34.5 mol%, and (iii)17.8 mol% inert)

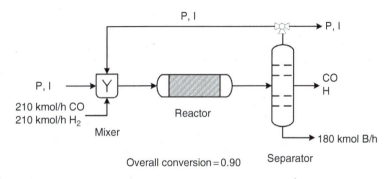

FIGURE P6.4.3
Flows heet of butane production process.

6.4.4   Hydrodealkylation Process

Hydrodealkylation is a process where short organic side chain are removed from larger molecules using hydrogen. During hydrodealkylation of toluene ($C_6H_5CH_3$), both benzene (C_6H_6) and methane (CH_4) are produced.

$$C_6H_5CH_3 + H_2 \rightarrow C_6H_6 + CH_4$$

In an industrial process 100 lb-mol/h of pure toluene (liquid) is fed along with 200 lb-mol/h of pure hydrogen (gas). These streams are combined with the recycle stream and fed to a reactor. In the reactor, 75% of the toluene fed to the reactor is consumed. The effluent from the reactor enters a vapor–liquid separator. All of the hydrogen and methane (gas) leave this separator in the vapor phase; all of the toluene and benzene (liquid) leave the separator in the liquid stream. The liquid stream then enters a second separator in which all of the benzene leaves in one stream and all of the toluene leaves in the other stream. The benzene stream leaves the process as a product stream. The toluene stream is completely recycled by mixing it with the toluene fed to the process. The vapor/liquid separator is split into two equal streams. One of these streams is combined with the hydrogen fed to the system that results in the gaseous stream entering the reactor. The other stream is sent to another process for further processing.

(a) Draw a process flow diagram for this process.

(b) How much benzene is produced (lb-mol/h)? (Answer: 200 lb-mol/h.)

(c) What is the composition of the vapor stream sent to the other process for further processing? (Answer: 50 mol% CH_4 and 50 mol% H_2.)

(d) How much toluene is actually fed to the reactor? (Answer: 133.3 lb-mol/h.)

(e) What is the composition of the vapor stream fed to the reactor? (Note: the answer is not 100% hydrogen)? (Answer: Flow of H_2 in $= 200 + 100 = 300$ lb-mol/h, flow of CH_4 in $= 100$ lb-mol/h.)

6.4.5   Uranium and Zirconium as Nuclear Fuels

One route toward separating uranium (U) and zirconium (Zr) contained in spent nuclear fuel involves the reaction of these elements with HCl. In the simplified flow sheet shown in Figure P6.4.5, 10 mol/h of 90% Zr and 10% U mixture is reacted with a HCl stream containing some water to produce metal chlorides via the reactions

$$U + 3HCl \rightarrow UCl_3 + 3/2H_2$$
$$Zr + 4HCl \rightarrow ZrCl_4 + 2H_2$$

The U and Zr are completely converted to chlorides, provided the total HCl fed to the reactor is twice the amount required by the reaction stoichiometry. The UCl_3 produced is a solid and hence is readily removed from the remaining gaseous reaction products. The vapors from this first reactor are passed to a second reactor where the $ZrCl_4$ vapor is reacted with steam to produce solid ZrO_2 according to the reaction: $ZrCl_4 + 2H_2O \rightarrow ZrO_2 + 4HCl$.

The reaction goes to completion and the ZrO_2 solid is separated from the gaseous reactor products. The remainder of the process is concerned with the concentration of the residual HCl for recycling back to the first reactor. The gases leaving the second reactor are sent to an absorber which uses 90% H_2O and 10% HCl solutions as a solvent to absorb HCl. This results in an absorber liquid containing 50% HCl and 50% H_2O and an overhead gas containing 90% H_2 and 10% HCl. The absorber liquid is sent to a stripper, which boils off most of the HCl to produce 90% HCl and 10% H_2O recycle vapor. The remaining stripper bottoms are cooled and recycled to the absorber for use as solvent. Calculate all flows and compositions, assuming all specified compositions are in mole percentage (see Figure P6.4.5).

(Answer: $n_8 = 80.926$ mol/h and $n_9 = 161.852$ mol/h)

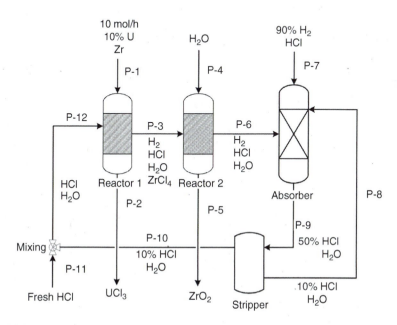

FIGURE P6.4.5
Separation process of uranium and zirconium from nuclear fuel.

Further Readings

1. Felder, R.M. and R.W. Rousseau (1999) *Elementary Principles of Chemical Processes*, 3rd edn, John Wiley, New York.
2. Reklaitis, G.V. (1983) *Introduction to Material and Energy Balances*, John Wiley & Sons, New York.
3. Himmelblau, D.M. (1996) *Basic Principles and Calculations in Chemical Engineering*, 6th edn, Prentice-Hall, New Jersey.

7

Energy Balance without Reaction

At the End of This Chapter You Should Be Able to

1. Explain kinetic and potential energy, closed and open systems, extensive and intensive properties, internal energy, heat and work, and adiabatic and isothermal cases.

2. Convert energy from one set to another.

3. Apply and simplify the general energy balance equation to solve particular problems.

4. Calculate enthalpy changes from heat capacity equations.

5. Become familiar with the steam table.

6. Become familiar to use the psychrometric chart.

7.1 Enthalpy and Energy Balances

Energy is required for breaking and forming chemical bonds inside chemical reactors. Every chemical process involves the transfer of energy. For instance, combustion is required for power generation, energy is required for volatilization, and energy is removed at condensers in distillation columns. Evaporators and condensers are required for phase change.

Energy balances are used to

1. Determine the amount of energy that flows into or out of each process unit.

2. Calculate the net energy requirement for the process.

3. Assess ways of reducing energy requirements in order to improve process profitability.

7.1.1 How Does Energy Move across Systems?

Energy crosses open and closed systems in different forms. Systems can be operated under isothermal or adiabatic conditions. An open system is one where material moves across its boundaries. In an open system, one way to change the energy in the system is by the movement of material. A closed system is one where no mass crosses the system boundaries. In a closed system, the two ways to change the energy is to supply heat as thermal energy or do work, that is, mechanical or electrical. An isothermal system is one where the temperature does not change with time and in space. This does not mean that no heat crosses the boundaries. An adiabatic system is one where no heat crosses the boundaries of the system, that is, the system is insulated. Energy has units of force times distance:

SI units: N·m ($=$ Joule)
CGS system: dyne·cm ($=$ erg)
American Engineering System: ft·lbf

7.2 Forms of Energy

Energy takes three forms: kinetic energy (E_k), potential energy (Ep), and internal energy (U).

7.2.1 Kinetic Energy (E_k)

Kinetic energy is the energy carried by a moving system because of its velocity:

$$\dot{E}_k = \frac{1}{2}\dot{m}v^2 \Rightarrow W = J/s = \left(\frac{kg}{s}\right)\left(\frac{m}{s}\right)^2$$

$\dot{m} =$ mass flow rate (kg/s), $v =$ velocity (m/s)

Example 7.1 Kinetic energy calculations
Water flows into a process unit through a 2 cm inside diameter (2 cm ID) pipe at a rate of 2.0 m^3/h. Calculate $\dot{E}_k$ for this stream in joules per second.

Solution

$$\dot{m} = \rho \dot{V} = \frac{1000 \text{ kg}}{\cancel{m^3}}\left|\frac{2 \cancel{m^3}}{\cancel{h}}\right|\frac{\cancel{h}}{3600 \text{ s}} = 0.56\frac{kg}{s}$$

$$v = \left(\frac{2.00\dfrac{m^3}{h}\left|\dfrac{h}{3600 \text{ s}}\right.}{3.14 \times (0.02 \text{ m})^2/4}\right) = 1.77\frac{m}{s}$$

$$\dot{E}_k = \frac{1}{2}\dot{m}v^2 = \frac{1}{2}\left(0.56\ \frac{\text{kg}}{\text{s}}\right)\left(1.77\ \frac{\text{m}}{\text{s}}\right)^2\left(\frac{1\ \text{N}}{\frac{\text{kg m}}{\text{s}^2}}\right)\left(\frac{1\ \text{J}}{1\ \text{N m}}\right) = 0.88\ \frac{\text{J}}{\text{s}}$$

7.2.2 Potential Energy (E_p)

Potential energy is the energy due to the position of the system in a potential field (e.g., earth's gravitational field, $g = 9.8\ \text{m/s}^2$).

$$E_p = mgz \Rightarrow J = [\text{kg}][\text{m/s}^2][\text{m}]$$

$$\dot{E}_p = \dot{m}gz \Rightarrow W = J/s = [\text{kg/s}][\text{m/s}^2][\text{m}]$$

m = mass (kg), $\dot{m}$ = mass flow rate (kg/s), z = height of object (m)

Example 7.2   Potential energy calculations

Crude oil is pumped at a rate of 15.0 kg/s from a point 220 m below the earth's surface to a point 20 m above ground level. Calculate the attendant rate of increase of potential energy.

Solution

$$\dot{E}_p = \dot{m}gz = 15.0\ \frac{\text{kg}}{\text{s}} \times 9.81\ \frac{\text{m}}{\text{s}^2} \times [20 - (-220)]\text{m}$$

$$= 35{,}316\ \frac{\text{kg m}^2}{\text{s s}^2} = \frac{\text{N m}}{\text{s}} = \frac{\text{J}}{\text{s}}$$

7.2.3   Internal Energy (U)

Internal energy is the sum of molecular, atomic, and subatomic energies of matter. It consists of all forms of energy other than kinetic and potential energy, including the energy due to the rotational and vibrational motion of molecules within the system, the interactions between molecules within the system, and the motion and interactions of electrons and nuclei within molecules. Internal energy (U) is related to enthalpy (H) via the relation: $\hat{H} = \hat{U} + P\hat{V}$.

U and H are the functions of temperature, chemical composition, and physical state (solid, liquid, or gas). They both weakly depend on pressure. U and H are relative quantities; their values must be defined with respect to their reference state.

Example 7.3   Internal energy

The specific internal energy of helium at 300 K and 1 atm is 3800 J/mol, and the specific molar volume at the same temperature and pressure is 24.63 L/mol. Calculate the specific enthalpy of helium at this temperature and pressure, and the rate at which enthalpy is transported by a stream of helium at 300 K and 1 atm with a molar flow rate of 250 kmol/h.

Solution

$$\hat{H} = \hat{U} + P\hat{V}$$

$$\hat{H} = 3899\,\frac{J}{mol} + (1\ \text{atm})\left(24.63\,\frac{L}{mol}\right)\left(\frac{1\ m^3}{1000\ L}\right)\left(\frac{1.01325 \times 10^5\ N/m^3}{1\ \text{atm}}\right)$$

$$= 6295.63\ J/mol$$

$$H = \dot{n} \times \hat{H}$$

$$H = 250\,\frac{k\ mol}{h} \times 6295.63\,\frac{J}{mol}\,\frac{1000\ mol}{k\ mol}\,\frac{kJ}{1000\ J} = 1.57 \times 10^6\ kJ/h$$

7.3   Intensive versus Extensive Variables

1. Extensive variables: variables that depend on the size of the system, for example, mass, number of moles, volume (mass or molar flow rate and volumetric flow rate), kinetic energy, potential energy, and internal energy
2. Intensive variables: variables that do not depend on the size of the system, for example, temperature, pressure, density, specific volume, and composition (mass or mole fraction)

7.4   Transfer of Energy

In a closed system, no mass is transferred across the system boundaries (i.e., batch system); energy may be transferred between the system and the surroundings in two ways:

1. Heat (Q) is the energy that flows due to a temperature difference between the system and its surroundings, and always flows from regions at high temperature to regions at low temperature. By

convention, heat is defined to be positive if it flows to a system (i.e., gained).

2. Work (W) is the energy that flows in response to any driving force (e.g., applied force, torque) other than temperature, and is defined as positive if it flows from the system (i.e., work done by the system). In chemical processes, work may, for instance, come from pumps, compressors, moving pistons, and moving turbines. Heat or work only refers to energy that is being transferred to or from the system.

Example 7.4 Power calculations

A certain gasoline engine has an efficiency of 30%, that is, it converts into useful work 30% of the heat generated by burning a fuel. If the engine consumes 0.80 L/h of gasoline with a heating value of 3.5×10^4 kJ/L, how much power does it provide? Express the answer both in kilowatts and horsepower (hp).

Solution

$$\text{Power provided} = 0.8 \frac{L}{h} \times \frac{h}{3600\ s} \times 3.5 \times 10^4 \frac{kJ}{L} \times 0.3 = 2.33 \frac{kJ}{s} = 2.33\ \text{kW}$$

7.5 First Law of Thermodynamics

The first law of thermodynamics states that energy can neither be created nor destroyed, but can be transformed. Its conservation equation is similar in form to the conservation equation of mass.

Accumulation = in − out + generation − consumption

If generation = 0 and consumption = 0, the energy balance for this case becomes

Accumulation = in − out (Figure 7.1)

7.5.1 Energy Balance on Closed Systems

In a closed system, although no mass crosses the boundaries, input energy and output energy are not necessarily equal to zero since energy can be transferred across the boundary. Therefore, the energy balance becomes

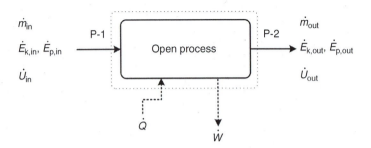

FIGURE 7.1
Overall energy balance.

$$\left\{ \begin{array}{c} \text{energy} \\ \text{flow} \\ \text{into} \\ \text{system} \end{array} \right\} - \left\{ \begin{array}{c} \text{energy} \\ \text{flow} \\ \text{out of} \\ \text{system} \end{array} \right\} + \left\{ \begin{array}{c} \text{generated} \\ \text{energy} \\ \text{flow in} \\ \text{system} \end{array} \right\} - \left\{ \begin{array}{c} \text{consumed} \\ \text{energy} \\ \text{flow in} \\ \text{system} \end{array} \right\}$$

$$= \left\{ \begin{array}{c} \text{accumulated} \\ \text{energy} \\ \text{in} \\ \text{system} \end{array} \right\} \tag{7.1}$$

7.5.2 Possible Simplifications on Energy Balance in a Closed System

$$\Delta U + \Delta E_k + \Delta E_p = Q - W$$

Possible simplifications:

- If $T_{\text{system}} = T_{\text{surroundings}}$, then $Q = 0$ since no heat is being transferred due to the lack of the driving force, that is, temperature difference is nil.
- If the system is perfectly insulated, then $Q = 0$ (system is adiabatic) since no heat is being transferred between the system and the surroundings.
- If the system is not accelerating/decelerating, then $\Delta E_k = 0$.
- If the system is not rising or falling, then $\Delta E_p = 0$.
- If energy is not transferred across the system boundary by a moving part (e.g., piston, impeller, rotor), then $W = 0$.
- If the system is at constant temperature (system is isothermal), no phase changes or chemical reactions are taking place, and only minimal pressure changes, then $\Delta U = 0$.

Example 7.5   Compression Process
Compressing a gas in an isolated cylinder.

Solution
Figure E7.5 is a schematic that shows the compression process. Assume that the piston is isolated; consequently, heat loss or gain from the surrounding is negligible ($Q = 0$). The system is stationary, and hence kinetic energy is zero and potential energy due to the movement of the piston is negligible. The equation is reduced to the following form:

$$\Delta U + \Delta E_k + \Delta E_p = Q - W$$

$$\therefore \Delta U = -W$$

Example 7.6   Energy balance in closed system
Write and simplify the closed system energy balance for each of the following processes, and state whether nonzero heat and work terms are positive or negative:

a. The contents of a closed flask are heated from 25°C to 80°C.

b. A tray filled with water at 20°C is put into a freezer. The water turns into ice at −5°C. (*Note*: When a substance expands it does work on its surroundings and when it contracts the surroundings do work on it.)

c. A chemical reaction takes place in a closed adiabatic (perfectly insulated) rigid container.

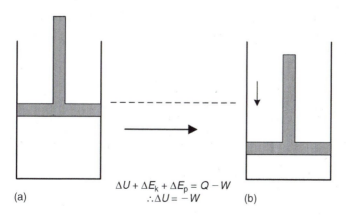

(a)

$$\Delta U + \Delta E_k + \Delta E_p = Q - W$$
$$\therefore \Delta U = -W$$

(b)

FIGURE E7.5
Compressing a gas in a piston cylinder.

d. Repeat part (c), only suppose the reactor is isothermal rather than adiabatic and that when the reaction was carried out adiabatically the temperature in the reactor increased.

Solution

Begin by defining the system.

a. No change in elevation $\rightarrow$ potential energy is zero.
   System is stationary $\rightarrow$ kinetic energy is zero.
   No moving parts or generated current.

$$\Delta U + \Delta \cancel{E}_k + \Delta \cancel{E}_p = Q - \cancel{W}, \Rightarrow \Delta U = Q$$

b. No height change $\rightarrow$ potential energy is zero.
   System is stationary $\rightarrow$ kinetic energy is zero.

$$\Delta U + \Delta \cancel{E}_k + \Delta \cancel{E}_p = Q - W \Rightarrow \Delta U = Q - W$$

c. No height change $\rightarrow$ potential energy is zero.
   System is stationary $\rightarrow$ kinetic energy is zero.
   No moving parts or generated current.
   Adiabatic $\rightarrow$ Q is zero.

$$\Delta U + \Delta \cancel{E}_k + \Delta \cancel{E}_p = \cancel{Q} - \cancel{W} \Rightarrow \Delta U = 0$$

d. No height change $\rightarrow$ potential energy is zero.
   System is stationary $\rightarrow$ kinetic energy is zero.
   No moving parts or generated current.

$$\Delta U + \Delta \cancel{E}_k + \Delta \cancel{E}_p = Q - \cancel{W} \Rightarrow \Delta U = Q$$

Example 7.7 Cylinder with a movable piston

A gas is contained in a cylinder fitted with a movable piston. The initial gas temperature is 20°C (Figure E7.7).

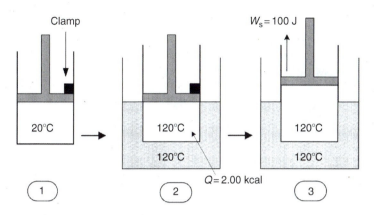

FIGURE E7.7
Cylinders with a movable piston.

The cylinder is placed in a constant temperature oil bath with the piston held in a fixed position by a clamp. Heat in the amount of 2.00 kcal is transferred to the gas, which equilibrates at 120°C (and a higher pressure). The piston is then released, and the gas does 100 J of work in moving the piston to its new equilibrium position. The final gas temperature is 120°C. Write the energy balance equation for each of the two stages of this process, and in each case, solve for the unknown energy term in the equation. Consider the gas in the cylinder to be the system, neglect the change in potential energy of the gas as the piston moves vertically, and assume the gas behaves ideally. Express all energies in joules.

Solution
Case 1 to 2
  No moving boundaries → work is zero
  No vertical displacement → potential energy is zero
  System is stationary → kinetic energy is zero

$$\Delta U + \Delta \cancel{E_k} + \Delta \cancel{E_p} = Q - \cancel{W}$$

$$\therefore \Delta U = Q$$

$$\Delta U = 2.00 \text{ kcal} \frac{1000 \text{ cal}}{\text{kcal}} \frac{1 \text{ J}}{0.239 \text{ cal}} = 8370 \text{ J}$$

Case 2 to 3
  No vertical displacement → potential energy is zero
  System is stationary → kinetic energy is zero
  Temperature is constant → internal energy is zero since it depends on temperature

$$\Delta \cancel{U} + \Delta \cancel{E_k} + \Delta \cancel{E_p} = Q - W \Rightarrow 0 = Q - W \Rightarrow Q = W = 100 \text{ J}$$

7.5.3 Energy Balance in Open Systems at Steady State

In open systems, material crosses the system boundary as the process occurs (e.g., continuous process at steady state). In an open system, work must be done to push input fluid streams at a pressure P_{in} into the system, and work is done by the output fluid streams at pressure P_{out} and flow rate on the surroundings as it leaves the system (Figure 7.2).

Net rate of work done by the system is

$$\dot{W}_f = \dot{W}_{out} - \dot{W}_{in} = P_{out}\dot{V}_{out} - P_{in}\dot{V}_{in}$$

For several input and output streams,

$$\dot{W}_f = \underbrace{\sum P_j \dot{V}_j}_{\substack{\text{output} \\ \text{streams}}} - \underbrace{\sum P_j \dot{V}_j}_{\substack{\text{input} \\ \text{streams}}}$$

The total rate of work $(\dot{W})$ done by a system on its surroundings is divided into two parts:

$$\dot{W} = \dot{W}_s + \dot{W}_f$$

where shaft work $(\dot{W}_s)$ is the rate of work done by the process fluid on a moving part within the system (e.g., piston, turbine, and rotor), and flow work $(\dot{W}_f)$ is the rate of work done by the fluid at the system outlet minus the rate of work done on the fluid at the system inlet. The general balance equation for an open continuous system (Figure 7.3) at steady state in the absence of generation/consumption term is

Energy input = energy output

$$\text{Energy input} = \dot{U}_{in} + \dot{E}_{k,in} + \dot{E}_{p,in} + P_{in}\dot{V}_{in}$$

$$\text{Energy output} = \dot{U}_{out} + \dot{E}_{k,out} + \dot{E}_{p,out} + P_{out}\dot{V}_{out}$$

$$\text{Energy transferred} = \dot{Q} - \dot{W}_s$$

FIGURE 7.2
Work done by a flowing fluid.

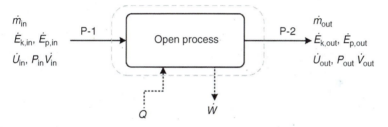

FIGURE 7.3
Overall energy balance of open systems.

Energy transferred = energy out − energy in

$$\dot{Q} - \dot{W}_s = \left(\dot{U}_{out} + \dot{E}_{k,out} + \dot{E}_{p,out} + P_{out}\dot{V}_{out}\right) - \left(\dot{U}_{in} + \dot{E}_{k,in} + \dot{E}_{p,in} + P_{in}\dot{V}_{in}\right)$$

$$\dot{Q} - W_s = \Delta\dot{U}_{in} + \Delta\dot{E}_k + \Delta\dot{E}_p + \Delta\left(P\dot{V}\right)$$

$$\dot{H} = \dot{U} + P\dot{V}$$

$$\Delta\dot{H} = \Delta\dot{U} + \Delta\left(P\dot{V}\right)$$

Rearranging the above equations leads to the first law of thermodynamics for an open system at steady state

$$\dot{Q} - \dot{W}_s = \Delta\dot{H} + \Delta\dot{E}_k + \Delta\dot{E}_p \text{ (where } \Delta = \text{output} - \text{input)}$$

7.5.4 Possible Simplifications on Energy Balance in an Open System

If $T_{system} = T_{surroundings}$, then $Q = 0$ since there is no heat being transferred due to the lack of the driving force, that is, temperature difference is nil.

- If the system is perfectly insulated, then $Q = 0$ (system is adiabatic) since no heat is being transferred between the system and the surroundings.
- If the system is not accelerating/decelerating, then $\Delta E_k = 0$.
- If the system is not rising or falling, then $\Delta E_p = 0$.
- If energy is not transferred across the system boundary by a moving part (e.g., piston, impeller and rotor), then $W = 0$.
- If the system is at constant temperature (system is isothermal), no phase changes or chemical reactions are taking place, with only minimal pressure changes, then $\Delta U \doteq 0$.

7.6 Enthalpy Calculations

The difference between energy and enthalpy is the flow work term. If we take $\hat{H} = F(T, P)$, and we can write $\hat{H}$ in terms of any two variables, we think about getting to our final state. First, by keeping T constant (rather than $\hat{V}$), and then letting P change, the flow work term can be changed from $\Delta(P\hat{V})$ to $\hat{V}\Delta P$ (the changes in $\hat{V}$ would have occurred in the constant P part). Keeping P constant and letting T change, we can get the expression for the constant P part as $\Delta\hat{H} = \int C_P dT \approx C_P \Delta T$ (at constant P).

7.7 Reference States and State Properties

It is not necessary to know the reference state to calculate $\Delta\hat{H}$ for the transition from one state to another. $\Delta\hat{H}$ from state 1 to state 2 equals $\hat{H}_2 - \hat{H}_1$ regardless of the reference state upon which $\hat{H}_1$ and $\hat{H}_2$ were based. If different tables are used, one must make sure they have the same reference state. $\hat{H}$ and $\hat{U}$ are state properties; their values depend only on the state of the species, temperature, and pressure and not on how the species reached its state. When a species passes from one state to another, both $\Delta\hat{U}$ and $\Delta\hat{H}$ for the process are independent of the path taken from the first state to the second one.

Example 7.8 Saturated methyl chloride

The following entries are taken from a data table for saturated methyl chloride (Table E7.8):

1. What reference state was used to generate the given enthalpies?
2. Calculate $\Delta\hat{H}$ and $\Delta\hat{U}$ for the transition of saturated vapor CH_3Cl from 10°C to −18°C.

TABLE E7.8

Properties of Saturated Methyl Chloride

State	T (°C)	P (atm)	$\hat{V}$ (m³/kg)	$\hat{H}$ (kJ/kg)
Liquid	−40	0.47	9.7×10^{-4}	0.00
Vapor	−18	1.29	0.310	456.00
Vapor	10	3.54	0.120	470.06

Solution

1. The reference is liquid at $-40°C$, because the enthalpy at this temperature is zero.

2. $\Delta\hat{H}$ and $\Delta\hat{U}$ for the transition of saturated CH_3Cl vapor from $10°C$ to $-18°C$ can be calculated as

$$\Delta\hat{H} = \hat{H}_{-18°C} - \hat{H}_{10°C} = 456.00 - 470.06 = -14.06 \text{ kJ/kg}$$

$$\Delta\hat{U} = \Delta\hat{H} - \Delta\left(P\hat{V}\right) = \Delta\hat{H} - \left[\left(P\hat{V}\right)_{-18} - \left(P\hat{V}\right)_{10}\right]$$

$$= -14.06 \frac{\text{kJ}}{\text{kg}} - \left(1.29 \times 0.312 - 3.54 \times 0.12\left[\text{atm} \times \frac{\text{m}^3}{\text{kg}}\right]\right)\left(\frac{101.325 \text{ kN/m}^2}{1 \text{ atm}}\right)$$

$$= -14.06 \frac{\text{kJ}}{\text{kg}} - \left(-0.022 \times 101.325 \frac{\text{kN} \cdot \text{m}}{\text{kg}}\right) = -11.80 \frac{\text{kJ}}{\text{kg}}$$

7.8 Use of Linear Interpolation in Steam Tables

Sometimes, you need an estimate of specific enthalpy, specific internal energy, or specific volume at a temperature and a pressure that is between tabulated values. In this case, one can use a linear interpolation. Use this equation to estimate y for an x between x_0 and x_1:

$$\frac{y - y_0}{y_1 - y_0} = \frac{x - x_0}{x_1 - x_0}$$

Example 7.9 Saturated steams

1. Determine the vapor pressure, specific internal energy, and specific enthalpy of saturated steam at $133.5°C$.

2. Show that water at $400°C$ and 10 bar is superheated steam and determine its specific volume, specific internal energy, and specific enthalpy relative to liquid water at the triple point, and its dew point.

Solution

1. Vapour pressure $= 299.6$ kpa, specific enthalpy $= 2725$ kJ/kg, specific internal energy $= 2543$ kJ/kg
2. At 10 bar the saturated temperature is 179.9°C and 400°C is higher than the saturated temperature at 10 bar, so the state of water is superheated steam.

   At 10 bar and 400°C

   Specific enthalpy $= 3263$ kJ/kg

   Specific internal energy $= 2957$ kJ/kg

   Specific volume $= 0.3066$ m³/kg

7.9 Enthalpy Change in Nonreactive Processes

Change in enthalpy can occur because of

1. Change in temperature;
2. Change in phase; or
3. Mixing of solutions and reactions.

7.9.1 Enthalpy Change as a Result of Temperature Change

Sensible heat is the heat transferred to raise or lower the temperature of a material in the absence of phase change. In the energy balance calculations, sensible heat change is determined by using a property of matter called the heat capacity at constant pressure, or just heat capacity (C_P). Units for C_P are (J/mol) K or (cal/g)°C. Appendix B lists C_P values for several organic and inorganic compounds. There are several methods for calculating enthalpy change using C_P values. When C_P is constant, the change in enthalpy of a substance due to change in temperature at constant pressure is

$$\Delta H = mC_P(T - T_{\text{ref}}) \qquad (7.2)$$

Heat capacities for most substances vary with temperature where the values of C_P vary for the range of the change in temperature. Heat capacities are tabulated as polynomial functions of temperature such as

$$C_P = a + bT + cT^2 + dT^3 \qquad (7.3)$$

Coefficients a, b, c, and d for a number of substances are given in Appendix B. In this case, the enthalpy change is

$$\Delta H = m \int_{T_{ref}}^{T} C_P \, dT \tag{7.4}$$

Examples 7.10   Constant heat capacity

What is the change in enthalpy of 100 g formic acid at 80°C; the heat capacity of formic acid is 0.524 cal/g°C?

Solution

$$\Delta H = mC_P(T - T_{ref})$$

$$\Delta H = (100 \text{ g}) \left(0.524 \frac{\text{cal}}{\text{g°C}}\right) (80°\text{C} - 25°\text{C}) = 5500 \text{ cal}$$

Example 7.11   Heat capacity as a function of temperature

Calculate the heat required to raise the temperature of 200 kg of nitrous oxide from 20°C to 150°C in a constant-volume vessel. The constant-volume heat capacity of N_2O in this temperature range is given by the following equation (T is in °C):

$$C_v \left(\frac{\text{kJ}}{\text{kg·°C}}\right) = 0.855 + 9.42 \times 10^{-4}T$$

Solution

$$Q - W = \Delta U = \dot{n}\Delta \hat{U} = \dot{n} \int_{20°\text{C}}^{150°\text{C}} C_v \, dT = \dot{n} \int_{20°\text{C}}^{150°\text{C}} (0.855 + 9.42 \times 10^{-4}T)dT$$

$W = 0.0$ (rigid vessel; no moving part)

$$Q = \dot{n} \int_{20°\text{C}}^{150°\text{C}} (0.855 + 9.42 \times 10^{-4}T)dT = 0.855T - 9.42 \times 10^{-4}\frac{T^2}{2}$$

$$= 200 \text{ kg} \left[0.855(150 - 20) - 9.42 \times 10^{-4}\frac{(150^2 - 20^2)}{2}\right] \left[\frac{\text{kJ}}{\text{kg·°C}}\right] = 20018.2 \text{ kJ}$$

7.9.2 Enthalpy Change because of Phase Changes

Phase changes, such as evaporation and melting, are accompanied by relatively large changes in internal energy and enthalpy, as bonds between molecules are broken and reformed. Heat transferred to or from a system, causing change of phase at constant temperature and pressure, is known as latent heat. The types of latent heats are latent heat of vaporization, which is the heat required to vaporize a liquid; latent heat of fusion, which is the heat required to melt a solid; and latent heat of sublimation, which is the heat required to directly vaporize a solid. Heat is released during condensation, and heat is required to vaporize a liquid or melt a solid. Table A.1 reports these two latent heats for substances at their normal melting and boiling points (i.e., at a pressure of 1 atm). Sensible heat refers to heat that must be transferred to raise or lower the temperature of a substance without change in phase. The quantity of sensible heat required to produce a temperature change in a system can be determined from the appropriate form of the first law of thermodynamics. The heat capacity at constant pressure, C_P, for most incompressible liquids and solids is equal; $C_P \approx C_v$ and for ideal gases, $C_P = C_v + R$.

Example 7.12   Heat exchanger

How much heat must be removed per hour from a heat exchanger designed to condense 60 kg/h of steam at 200°C and 1 bar to saturated liquid (Figure E7.12)?

Solution

Basis: 1 h

From the superheated steam table (Appendix C) for saturated liquids as a function of pressure, we find the following:

Inlet: (1 bar, 200°C), $\hat{H}_1 = 2875$ kJ/kg
Outlet: (1 bar, saturated liquid), $\hat{H}_1 = 417.5$ kJ/kg

$$m_{in} = m_{out} = 60 \text{ kg}$$

$$\Delta H + \Delta E_k + \Delta E_p = Q - W_S \text{ (general form of the energy balance)}$$

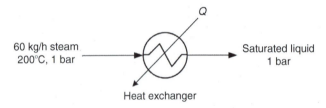

FIGURE E7.12
Heat exchanger.

Now we make the following simplifying assumptions for this condenser:

- No shaft work: $W_s = 0$.
- The inlet and outlet lines are not tremendously different in height: $\Delta E_p = 0$.
- There is little difference in the velocity of the fluid in the inlet and outlet lines (since we do not know anything about the diameters of the two lines, this is a good assumption. It would take a very large change in velocity to make a significant difference to the enthalpy.): $\Delta E_k = 0$.

The simplified form of the energy balance is therefore:

$$H_{out} - H_{in} = Q = m_{out}\hat{H}_{out} - m_{in}\hat{H}_{in} = m(\hat{H}_{out} - \hat{H}_{in})$$

$$Q = (60 \text{ kg})(417.5 - 2875) \text{ kJ/kg} = -147450 \text{ kJ/h}$$

Scale with basis: 1 h

$$Q = (-147{,}450 \text{ kJ/h})(1 \text{ h}) = -1.5 \times 10^5$$

$$Q = (-147{,}450 \text{ kJ/h})/(1 \text{ h}) = -1.5 \times 10^5 \text{ kJ}$$

Example 7.13 Heat added to a boiler

Liquid water is fed to a boiler at 24°C under a pressure of 10 bar and is converted at constant pressure to saturated steam. Calculate $\Delta\hat{H}(\text{kJ/kg})$ for this process and the heat input required for producing 15,000 m³/h of steam at the exiting conditions. Assume that the kinetic energy of the liquid entering the boiler is negligible and that the steam is discharged through a 15 cm ID (inner diameter) pipe (Figure E7.13).

Solution

$$Q - W_s = \dot{m}\Delta\hat{H} + \Delta K_E + \Delta K_P$$

$$\Delta\hat{H} = \hat{H}_2 - \hat{H}_1$$

$$= 2778 \frac{\text{kJ}}{\text{kg}} - 99.73 \frac{\text{kJ}}{\text{kg}} = 2678.27 \frac{\text{kJ}}{\text{kg}}$$

$$\dot{m}_2 = \dot{V} \times \rho = \dot{V} \times \frac{1}{\hat{V}} = 15{,}000 \frac{\text{m}^3}{\text{h}} \times 1 \left/ \left[0.1944 \frac{\text{m}^3}{\text{kg}} \right] \right. = 77160.5 \frac{\text{kg}}{\text{h}}$$

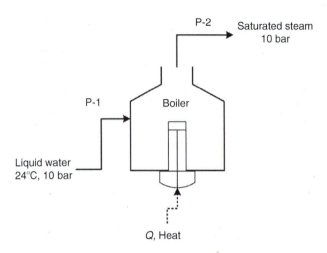

FIGURE E7.13
Flow sheet of a boiling process.

$$v_2 = \frac{\dot{V}_2}{\frac{\pi D^2}{4}} = \frac{15{,}000 \text{ m}^3/\text{h}}{\frac{\pi (0.15)^2}{4} \text{ m}^2} = 848{,}826 \frac{\text{m}}{\text{h}}$$

$$Q = \dot{m}\Delta\hat{H} + \Delta K_E$$

$$= \dot{m}\Delta\hat{H} + KE_2 - KE_1$$

$$= \dot{m}\Delta\hat{H} + \frac{1}{2}\dot{m}_2 v_2^2 - 0$$

$$= 77{,}160.5 \frac{\text{kg}}{\text{h}} \, 2678.27 \frac{\text{kJ}}{\text{kg}} + \frac{1}{2} 77{,}160.5 \frac{\text{kg}}{\text{h}} \times \left(848{,}826 \frac{\text{m}}{\text{h}}\right)^2$$

$$= 77{,}160.5 \frac{\cancel{\text{kg}}}{\text{h}} 2678.27 \frac{\text{kJ}}{\cancel{\text{kg}}} + \frac{1}{2} 77{,}160.5 \frac{\cancel{\text{kg}}}{\text{h}} \times (848{,}826)^2 \frac{\cancel{\text{m}^2}}{\cancel{\text{h}^2}} \frac{\cancel{\text{h}^2}}{(3600)^2 \, \cancel{s^2}} \frac{\cancel{N}}{\frac{\cancel{\text{kg}} \cdot \cancel{\text{m}}}{s^2}}$$

$$\times \frac{\cancel{J}}{\cancel{N} \times \cancel{\text{m}}} \frac{\text{kJ}}{1000 \, \cancel{J}} = 2 \times 10^8 \text{ kJ/h}$$

Example 7.14 Enthalpy of phase change

One thousand kilogram per hour of a liquid mixture of 70 wt% acetone and 30 wt% benzene is heated from 10°C to 50°C in a heat exchanger using steam. The steam enters the heat exchanger as a saturated vapor at 200°C of 90% quality, and exits as saturated liquid water at 200°C. How much steam is required in kilograms per hour? Be sure to use a linear interpolation to get the correct values from the steam tables.

Solution

Energy balance around the heat exchange

Enthalpy change for the heat exchanger is zero: $\Delta \dot{H} = 0$ (assume an adiabatic heat exchanger)

$$\dot{m}_{steam}(0.9)\Delta H^{vap}_{201.4°C} = \dot{n}_{acetone,benzene} \int_{10°C}^{50°C} C_{P,mix}\, dT$$

$$C_{P,mix} = \sum y_i C_{Pi} = 0.7 C_{P,acetone} + 0.3 C_{P,benzene}$$

$$C_{Pi} = a + bT + cT^2 + dT^3$$

Acetone (liquid): $a = 0.123$, $b = 18.6 \times 10^{-5}$, $c = 0$, $d = 0$
Benzene (liquid): $a = 0.1265$, $b = 23.4 \times 10^{-5}$, $c = 0$, $d = 0$

$$\int_{10°C}^{50°C} C_{P,mix}\, dT = \{0.7(0.123) + 0.3(0.1265)\}(50°C - 10°C)$$

$$+ \frac{\{0.7(18.6 \times 10^{-5}) + 0.3(23.4 \times 10^{-5})\}}{2} \left[(50°C)^2 - (10°C)^2\right]$$

$$\int_{10°C}^{50°C} C_{P,Mix}\, dT = 5.2025\ \frac{kJ}{mol}$$

$$\dot{m}_{steam}(0.9)\left(1933.2\ \frac{kJ}{kg}\right) = 1000\ \frac{kmol}{h}\left(\frac{10^3\ mol}{kmol}\right)\left(5.2025\ \frac{kJ}{mol}\right) \quad \dot{m}_{steam} = 2990\ kg/h$$

7.9.3 Enthalpy Change because of Mixing

The thermodynamic property of an ideal mixture is the sum of the contributions from the individual compounds.

Example 7.15 Mixing

Saturated steam at 1 atm and at a rate of 1150 kg/h is mixed with superheated steam available at 400°C and 1 atm to produce superheated steam at 300°C and 1 atm. Calculate the amount of superheated steam produced at 300°C, and the required volumetric flow rate of the 400°C steam.

Solution

The process flow sheet is shown in Figure E7.15.

$$\text{Mass balance: } \dot{m}_1 + \dot{m}_2 = \dot{m}_3 \Rightarrow 1150\ \frac{kg}{h} + \dot{m}_2 = \dot{m}_3 \tag{7.5}$$

$$\dot{m}_1 \hat{H}_1 + \dot{m}_2 \hat{H}_2 = \dot{m}_3 \hat{H}_3 \tag{7.6a}$$

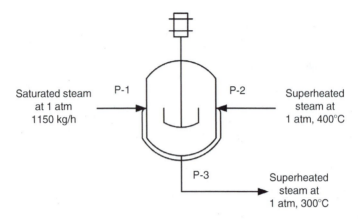

FIGURE E7.15
Flow sheet of a mixing process.

$$1150\,\frac{kg}{h}\left(2676\,\frac{kJ}{kg}\right) + \dot{m}_2\left(3278\,\frac{kJ}{kg}\right) = \dot{m}_3\left(3074\,\frac{kJ}{kg}\right) \qquad (7.6b)$$

Solving Equations 7.5 and 7.6 simultaneously gives $\dot{m}_3 = 3393.63$ kg/h.

7.10 Energy Balance on Bioprocesses

Bioprocesses are unlike many chemical processes. Bioprocesses are not particularly energy intensive. Fermentation and enzyme reactors are operated at temperatures and pressures close to ambient conditions, and energy input for downstream processing is minimized to avoid damaging heat-labile products. Nevertheless, energy effects are important because biological catalysts are very sensitive to heat and changes in temperature. In large-scale processes, heat released during biochemical reactions can cause cell death or denaturation of enzymes, if heat is not properly removed. The law of conservation of energy means that an energy accounting system can be set up to determine the amount of steam or cooling water required to maintain optimum process temperature.

Example 7.16 Heat exchanger
A liquid stream flows through a heat exchanger in which it is heated from 25°C to 80°C. The inlet and outlet pipes have the same diameter, and there is no change in elevation between these points.

Simplify the general energy balance equation.

Solution
System: heat exchanger

No change in kinetic energy since no change in velocity between inlet and exit.

No change in potential energy because there is no change in elevation.

No work because there are no moving parts or generated current.

$$\Delta \dot{H} + \Delta \dot{E}_k + \Delta \dot{E}_p = \dot{Q} - \dot{W}_s \Rightarrow \Delta \dot{H} = \dot{Q}$$

Examples 7.17 Cooling gluconic acid
In downstream processing of gluconic acid, a concentrated fermentation broth containing 20 wt% gluconic acid is cooled in a heat exchanger before crystallization. A liquid leaving an evaporator at the rate of 2000 kg/h at 90°C must be cooled to 6°C. Cooling is achieved by heat exchange with 2700 kg/h water initially at 2°C. If the final temperature of the cooling water is 50°C what is the rate of heat loss from the gluconic acid solution to the surroundings? Assume the heat capacity of gluconic acid is 0.35 (kcal/g)°C.

The process flow sheet is shown in Figure E7.17.

Solution

$$Q - W_s = \sum_{out} m_i \hat{H}_i - \sum_{in} m_i \hat{H}_i$$

$$Q - W_s = \left[m_w \hat{H}_w + m_c \hat{H}_c \right]_{out} - \left[m_w \hat{H}_w + m_c \hat{H}_c \right]_{in}$$

The enthalpy of water is found from the steam table as saturated liquid water.

$$Q - W_s = \left[\left(1600 \frac{kg}{h}\right) \hat{H}_w + \left(400 \frac{kg}{h}\right) \hat{H}_c \right]_{out} - \left[\left(1600 \frac{kg}{h}\right) \hat{H}_w + \left(400 \frac{kg}{h}\right) \hat{H}_c \right]_{in}$$

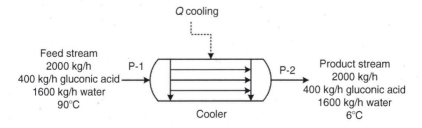

FIGURE E7.17
Flow sheet of a cooling system.

$$Q - W_s = \left[\left(1600 \ \frac{kg}{h} \right) \left(24.56 \ \frac{kJ}{kg} \right) + \left(400 \ \frac{kg}{h} \right) \hat{H}_c \right]_{out}$$
$$- \left[\left(1600 \ \frac{kg}{h} \right) \left(377.6 \ \frac{kJ}{kg} \right) + \left(400 \ \frac{kg}{h} \right) \hat{H}_c \right]_{in}$$

$$Q - W_s = \left(1600 \ \frac{kg}{h} \right) \left(24.56 - 377.6 \ \frac{kJ}{kg} \right)$$
$$+ \left(400 \ \frac{kg}{h} \right) \left(\hat{H}_c|_{out} - \hat{H}_c|_{in} \right)$$

$$Q - W_s = \left(1600 \ \frac{kg}{h} \right) \left(24.56 - 377.6 \ \frac{kJ}{kg} \right) \frac{h}{3600 \ s}$$
$$+ \left(400 \ \frac{kg}{h} \right) C_P(T_2 - T_1)$$

$$Q - W_s = \left(1600 \ \frac{kg}{h} \frac{h}{3600 \ s} \right) \left(24.56 - 377.6 \ \frac{kJ}{kg} \right)$$
$$+ \left(400 \ \frac{kg}{h} \frac{h}{3600 \ s} \right) 0.35 \ \frac{cal}{g^\circ C} \left| \frac{4.18 \ J}{1 \ cal} \right| \frac{1 \ kJ}{1000 \ J} \left| \frac{1000 \ g}{1 \ kg} (6^\circ C - 90^\circ C) \right.$$

$$Q - W_s = -150.68 \ \frac{kJ}{s} - 13.65 \ \frac{kJ}{s} = -164.33 \ \frac{kJ}{s}$$

$$W_s = 0 \ (\text{no shaft work})$$

$$Q = (Q_{loss} + Q_{coolant}) = -164.33 \ \frac{kJ}{s}$$

Heat gained by the cooling gained from the hot gluconic acid is

$$Q_{coolant} = m\Delta\hat{H} = 2700 \ \frac{kg}{h} \left(\hat{H}_w|_{@50^\circ C} - \hat{H}_w|_{@2^\circ C} \right)$$

$$Q_{coolant} = m\Delta\hat{H} = 2700 \ \frac{kg}{h} \left| \frac{h}{3600 \ s} \left(209.5 \ \frac{kJ}{kg} - 8.124 \ \frac{kJ}{kg} \right) = 151 \ \frac{kJ}{s} \right.$$

$$Q_{loss} + 151 \ \frac{kJ}{s} = -164.33 \ \frac{kJ}{s} \Rightarrow Q_{loss} = -315.33 \ \frac{kJ}{s}$$

Example 7.18 Turbine rotor
Water passes through the sluice gate of a dam and falls on a turbine rotor, which turns a shaft connected to a generator. The fluid velocity on both sides of the dam is negligible, and the water undergoes insignificant pressure and temperature changes between the inlet and outlet. Simplify the general energy balance equation.

Solution
System: water
  No change in temperature, enthalpy is zero.
  No change in velocity, kinetic energy is zero.
  The temperature of the system and surrounding is the same, heat transfer is zero.

$$\Delta \dot{H} + \Delta \dot{E}_k + \Delta \dot{E}_p = \dot{Q} - \dot{W}_s \Rightarrow \Delta \dot{E}_p = -\dot{W}_s$$

Example 7.19 Pumping of crude oil
Crude oil is pumped through a cross-country pipeline. The pipe inlet is 200 m higher than the outlet, the pipe diameter is constant, and the pump is located near the midpoint of the pipeline. Energy dissipated by friction in the line is transferred as heat through the wall. Simplify the general energy balance equation.

Solution
System: crude oil
  No change in kinetic energy as diameter is same.

$$\Delta \dot{H} + \Delta \dot{E}_k + \Delta \dot{E}_p = \dot{Q} - \dot{W}_s$$
$$\Delta \dot{H} + \Delta \dot{E}_p = \dot{Q} - \dot{W}_s$$

Example 7.20 Steam drives turbine
Five hundred kilograms per hour of steam drives a turbine. The steam enters the turbine at 44 atm and 450°C at a linear velocity of 60 m/s and leaves at a point 5 m below the turbine inlet at atmospheric pressure and a velocity of 360 m/s. The turbine delivers shaft work at a rate of 70 kW, and the heat loss from the turbine is estimated to be 104 kcal/h. Calculate the specific enthalpy change associated with the process.

Solution
System: turbine

$$\Delta \dot{H} + \Delta E_k + \Delta \dot{E}_p = \dot{Q} - \dot{W}_s$$

$$\Delta \dot{H} + \frac{1}{2}mv^2 + mg\Delta z = \dot{Q} - \dot{W}_s$$

$$Q = 70 \text{ kW} = 70 \ \frac{\text{kJ}}{\text{s}}$$

$$W_s = 10^4 \frac{\text{kcal}}{\text{h}} \frac{\text{h}}{3600 \text{ s}} \frac{1 \text{ kJ}}{0.23901 \text{ kcal}} = 11.62 \frac{\text{kJ}}{\text{s}}$$

$$\Delta \dot{H} + \frac{1}{2}\left(10 \frac{\text{kg}}{\text{h}} \frac{\text{h}}{3600 \text{ s}}\right)\left(360^2 \left[\frac{\text{m}}{\text{s}}\right]^2 - 60^2 \left[\frac{\text{m}}{\text{s}}\right]^2\right) + 10 \frac{\text{kg}}{\text{h}} \frac{\text{h}}{3600 \text{ s}} 9.8 \frac{\text{m}}{\text{s}^2}$$
$$\times [0 - (-5)] = \dot{Q} - \dot{W}_s$$

$$\Delta \dot{H} + \frac{1}{2}\left(10 \frac{\text{kg}}{\text{h}} \frac{\text{h}}{3600 \text{ s}}\right)\left(360^2 \left[\frac{\text{m}}{\text{s}}\right]^2 - 60^2 \left[\frac{\text{m}}{\text{s}}\right]^2\right) + 10 \frac{\text{kg}}{\text{h}} \frac{\text{h}}{3600 \text{ s}} 9.8 \frac{\text{m}}{\text{s}^2}$$
$$\times [0 - (-5)\text{m}] = \dot{Q} - \dot{W}_s$$

$$\Delta \dot{H} + 630{,}000 \frac{\text{kg m}^2}{\text{s}^2} \frac{1 \text{ N}}{1 \frac{\text{kg m}}{\text{s}^2}} \frac{1 \text{ N}}{\text{N m}} \frac{1 \text{ kJ}}{1000 \text{ J}} + 0.13625 \frac{\text{kg m}}{\text{s}^2} \frac{1 \text{ N}}{1 \frac{\text{kg m}}{\text{s}^2}} \frac{1 \text{ N}}{\text{N m}}$$
$$\times \frac{1 \text{ kJ}}{1000 \text{ J}} = \dot{Q} - \dot{W}$$

$$\Delta \dot{H} + 630 \frac{\text{kJ}}{\text{s}} + 0.13625 \frac{\text{kJ}}{\text{s}} = 70 \frac{\text{kJ}}{\text{s}} - 11.6 \frac{\text{kJ}}{\text{s}}$$

$$\Delta \dot{H} = -571.76 \frac{\text{kJ}}{\text{s}}$$

Example 7.21 Compressor horsepower

Calculate the horsepower needed by a large compressor to compress superheated steam for use in a new process plant. The feed steam is superheated (300°C) and is available at $P_{in} = 20$ bar absolute pressure. It enters the new compressor upstream piping (0.1 m inside diameter) at a velocity of 20 m/s. The discharge piping after the compressor has a smaller inside diameter and the discharge velocity is 169 m/s. The discharged superheated steam (350°C) leaves at a pressure of $P_{out} = 60$ bar absolute. The vendor selling this line of compressors lists its efficiency at 60%. Heat losses to the atmosphere from your new compressor total 5 kW. What is the horsepower required for the compressor? The system will be mounted horizontally to the floor. Superheated steam cannot be considered as an ideal gas. Carry out the system energy balance to determine the compressor horsepower?

Solution
The open steady-state energy balance can be used to describe the booster compressor system.

$$\Delta H + \Delta E_k + \Delta E_p = Q - W_s$$

The system is located on a horizontal plane, hence $\Delta E_p = 0$, and the energy balance becomes

$$\Delta H + \Delta E_k = Q - W_s$$

Determination of specific enthalpy and specific volume:
From the superheated steam table (Appendix C):

$$P_1 = 20 \text{ barabs}, \ T_1 = 300°C: \quad \hat{H}_1 = 3025 \text{ kJ/kg}, \ \hat{V}_1 = 0.125 \text{ m}^3/\text{kg}$$
$$P_2 = 60 \text{ barabs}, \ T_2 = 350°C: \quad \hat{H}_2 = 3046 \text{ kJ/kg}, \ \hat{V}_2 = 0.0422 \text{ m}^3/\text{kg}$$

Determination of mass flow rate of steam:

$$[20 \text{ m/s}][(0.1 \text{ m})(0.1 \text{ m})/4] \ [(1 \text{ kg}/0.125 \text{ m}^3)] = 1.257 \text{ kg/s}$$

$m = 1.257$ kg/s superheated steam
  Discharge velocity out (v_2):

$$v_2 = 169 \text{ m/s}$$

Determination of ΔH:

$$H = \dot{m}(\hat{H}_2 - \hat{H}_1) = 1.257 \text{ kg/s} [3046 - 3025 \text{ kJ/kg}] = 26.4 \text{ kJ/s} = 26.4 \text{ kW}$$

Determination of $\Delta E_k = (1/2)m[v_2^2 - v_1^2]/g_c$

$$
\begin{aligned}
\Delta E_k &= (1/2)m[(169 \text{ m/s})^2 - (20 \text{ m/s})^2]/g_c \\
&= \{(1/2)(1.257 \text{ kg/s})[(28093) \text{ m}^2/\text{s}^2]\}/\{[(1 \text{ kg m/s}^2)/\text{N}] \ [1 \text{ N}/(1 \text{ kg m/s}^2)]\} \\
&= 17657 \text{ kg m}^2/\text{s}^3 \\
&= [17657 \text{ kg m}^2/\text{s}^3] \ [1 \text{ kJ}/1000 \text{ kg m}^2/\text{s}^2] = 17.7 \text{ kJ/s} = 17.7 \text{ kW}
\end{aligned}
$$

$$\Delta E_k = 17.7 \text{ kW}$$

$$Q = -5 \text{ kW (loss by the system to the surroundings)}$$

$$\Delta H + \Delta E_k = Q - W_s$$

$$26.4 + 17.7 = -5 + (-0.6) \ W_s$$

$$W_s = -(26.4 + 17.7 + 5)/0.6 = 81.8 \text{ kW}$$

$$\text{Power} = -81.8 \text{ kW}(1000 \text{ W/kW})(1.341 \times 10^{-3} \text{ hp}/1 \text{ W}) = 109.7 \text{ hp}$$

Example 7.22 Cylinder fitted with a frictionless piston
A cylinder is fitted with a frictionless floating piston (oriented horizontally with gravity) and contains 24.5 L of air at 25°C and 1 bar. The system is then heated to 250°C. (Air is a diatomic gas: $C_v = 5R/2$)

    a. What is the change in kinetic energy?
    b. What is the change in potential energy?
    c. How much heat was added to the system?
    d. How much work was done by the system on the surroundings?
    e. What is the change in internal energy?
    f. What is the change in enthalpy?

Solution
Closed system energy balance: $\Delta E_k + \Delta E_p + \Delta U = Q - W$
Constant pressure: $\Delta E_k + \Delta E_p + \Delta H = Q - W_s$

    a. $\Delta E_k = 0$
    b. $\Delta E_p = 0$
    c. $W_S = 0, \; Q = \Delta H = n \int\limits_{25°C}^{250°C} C_P \, dT$

$$\text{Ideal gas: } n = \frac{PV}{RT} = \frac{1 \text{ bar } 24.8 \text{ L}}{83.14 \dfrac{\text{cm}^3 \text{ bar}}{\text{mol K}} \; 298.15 \text{ K}} = 1.00 \text{ mol}$$

$$C_P = C_V + R = \frac{7}{2} R$$

$$Q = (1 \text{ mol}) \frac{7}{2} \left(8.314 \frac{\text{J}}{\text{mol K}} \right) (225°C) = 6547 \text{ J}$$

    d. $W = \int P dV = P\Delta V = 1 \text{ bar} \left\{ 24.8 \, 1 \left(\dfrac{523 \text{ K}}{298 \text{ K}} - 1 \right) \right\}$

$$W = 18.71 \text{ 1 bar} \frac{\text{m}^3}{10^3 \text{ 1}} \frac{10^5 \text{ Pa}}{\text{bar}} = 1871 \text{ J}$$

    e. $\Delta U = n \int\limits_{25°C}^{250°C} C_V \, dT = (1 \text{ mol}) \frac{5}{2} \left(8.314 \frac{\text{J}}{\text{mol K}} \right) (225°C) = 4677 \text{ J}$

We could also solve for work: $\Delta H - \Delta U = W$

$$W = 6547 \text{ J} - 4677 \text{ J} = 1871 \text{ J}$$

Example 7.23   Use of coal to boil water

An 800 MW power plant burns coal to boil water producing saturated steam (100% quality) at 70 bar. This steam is expanded in a turbine to steam at 100°C and 1 bar. The steam enters the turbine at 10 m/s and exits 5 m below entrance point level at 100 m/s. The turbine is connected to an electrical generator by a shaft. The efficiency of the turbine is 60%.

    a. Draw and label the flow diagram of the process.

    b. What is the shaft work done by the turbine (MW)?

    c. How much heat is lost to the surroundings from the turbine (MW)?

    d. What is the mass flow rate of steam to the turbine (kg/h)?

Solution

    (a) The process flow sheet is shown in Figure E7.23.

    (b) Given work: $W = 800$ MW

    (c) $\eta = 0.6 \Rightarrow \eta = \dfrac{W_{out}}{E_{in} - E_{out}} \Rightarrow 1 - \eta = \dfrac{Q_{out}}{E_{in} - E_{out}} \Rightarrow \dfrac{1 - \eta}{\eta} = \dfrac{Q_{out}}{W_{out}}$

$$Q = 800 \text{ MW} \left(\frac{0.4}{0.6}\right) = 533.3 \text{ MW}$$

    (d) $\Delta\dot{E}_k + \Delta\dot{E}_p + \Delta\dot{H} = \dot{Q} - \dot{W}_s$

$$\dot{m}_{steam}\left\{\frac{1}{2}\left[(v_f)^2 - (v_i)^2 + g\Delta z + \hat{H}_f - \hat{H}_i\right]\right\} = \dot{Q} - \dot{W}_s$$

$$\dot{m}_{steam}\left\{\frac{1}{2}\left[\left(100\,\frac{m}{s}\right)^2 - \left(10\,\frac{m}{s}\right)^2 + 9.81\,\frac{m^2}{s}(0 - 5\text{ m}) + 2676\,\frac{kJ}{kg} - 2773.5\,\frac{kJ}{kg}\right]\right\}$$

$$= -533.3 \text{ MW} - 800 \text{ MW}$$

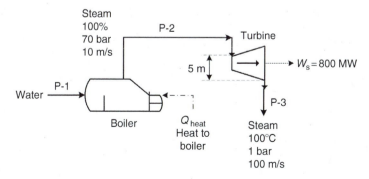

FIGURE E7.23
Steam generation process flow sheet.

Knowing that $\dfrac{m^2}{s^2} = \dfrac{J}{kg} = \dfrac{10^{-3}kJ}{kg} \Rightarrow$ since $J = N \cdot m = \dfrac{kg \times m}{s^2} \cdot m = \dfrac{kg \cdot m^2}{s^2}$

$$\dot{m}_{steam}\left(4.95\,\frac{kJ}{kg} - 0.049\,\frac{kJ}{kg} - 97.5\,\frac{kJ}{kg}\right) = 1.333 \times 10^6\,\frac{kJ}{s}$$

$$\dot{m}_{steam} = 1.4 \times 10^4\,\frac{kg}{s}\frac{3600\,s}{h} = 5.05 \times 10^7\,\frac{kg}{h}$$

7.11 Psychrometric Chart

The psychrometric chart (Figure 7.4) displays the relationship between dry-bulb, wet-bulb, and dew point temperatures, and specific and relative humidity. Given any two properties, the others can be calculated.

To use the chart, take the point of intersection of the lines of any two known factors (interpolate if necessary), and, from that intersection point, follow the lines of the unknown factors to their numbered scales to obtain the corresponding values. The thermophysical properties found on most psychrometric charts are as follows:

Dry-bulb temperature is the temperature of an air sample, as determined by an ordinary thermometer, the thermometer bulb being dry. It is typically the abscissa, or horizontal axis of the graph. The SI unit for temperature is Celsius; the other unit is Fahrenheit.

Wet-bulb temperature is the temperature of an air sample after it has passed through a constant-pressure, ideal, adiabatic saturation process, that is, after the air has passed over a large surface of liquid water in an insulated channel. In practice, this is the reading of a thermometer whose sensing bulb is covered with a wet sock evaporating into a rapid stream of the sample air. The wet-bulb temperature is the same as the dry-bulb temperature when the air sample is saturated with water.

Dew point temperature is that temperature at which a moist air sample at the same pressure would reach water vapor saturation. At this saturation point, water vapor would begin to condense into liquid water fog.

Relative humidity is the ratio of the mole fraction of water vapor to the mole fraction of saturated moist air at the same temperature and pressure. Relative humidity is dimensionless, and is usually expressed as a percentage.

Humidity ratio, also known as moisture content, mixing ratio, or specific humidity, is the proportion of mass of water vapor per unit mass of dry air at the given conditions. For a given dry-bulb temperature, there will be a particular humidity ratio for which the air sample is at 100% relative humidity. Humidity ratio is dimensionless, but is sometimes expressed as grams of water per kilogram of dry air.

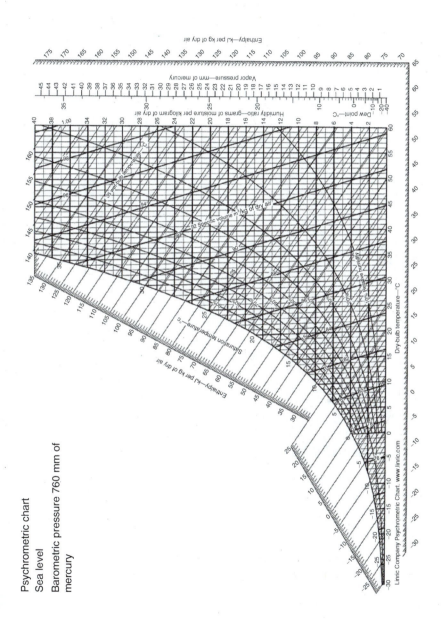

FIGURE 7.4
Psychrometric chart (reference states H_2O (L, 0°C, 1 atm), dry air (0°C, 1 atm)).

Specific enthalpy, also called heat content per unit mass, is the sum of the internal (heat) energy of the moist air in question, including the heat of the air and water vapor within. In the approximation of ideal gases, lines of constant enthalpy are parallel to lines of constant wet-bulb temperature.

Specific volume, also called inverse density, is the volume per unit mass of the air sample. The SI unit is cubic meters per kilogram of air; the other unit is cubic feet per pound of dry air.

Example 7.24   Psychrometric chart

Humid air at 28°C has a dew point of 8°C. Using the psychrometric chart provided determine the following: relative humidity, absolute humidity, wet-bulb temperature, dry-bulb temperature, humid volume, specific enthalpy, mass of air that contains 2 kg of water, and volume occupied by air that contains 2 kg of water.

Solution

Humid air at 28°C has a dew point of 8°C. Using the psychrometric chart provided, determine the following:

1. Relative humidity $= 30\%$ (Figure E7.24a)
2. Absolute humidity $= 0.007$ kg water/kg dry air (Figure E7.24b)
3. Wet-bulb temperature $= 16°C$ (Figure E7.24c). Follow constant enthalpy line from the intersection of the dry-bulb and dew point temperatures
4. Dry-bulb temperature $= 28°C$
5. Humid volume $= 0.86$ m^3/kg (Figure E7.24d)
6. Specific enthalpy $= 46$ kJ/kg $- 0.3$ kJ/kg $= 45.7$ kJ/kg (Figure E7.24e)
7. Mass of air that contains 2 kg of water

$$2\text{kg H}_2\text{O}\frac{\text{kg dry air}}{0.007 \text{ kg H}_2\text{O}} = 285.7 \text{ kg dry air}$$

8. Volume occupied by air that contains 2 kg of water

$$\frac{0.86 \text{ m}^3}{\text{kg dry air}}\frac{\text{kg dry air}}{0.007 \text{ kg H}_2\text{O}}2 \text{ kg H}_2\text{O} = 245.7 \text{ m}^3$$

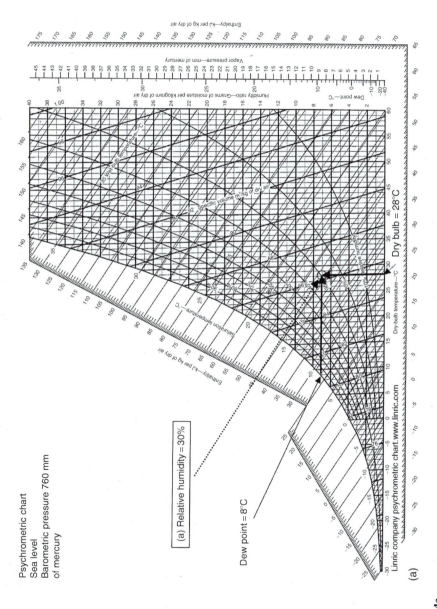

FIGURE E7.24a
Calculation of relative humidity.

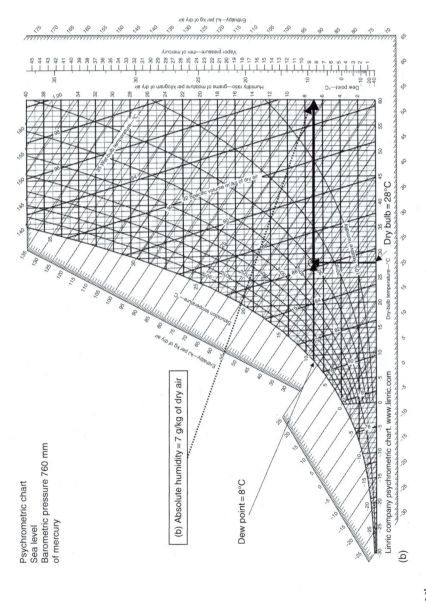

FIGURE E7.24b

Calculation of absolute humidity.

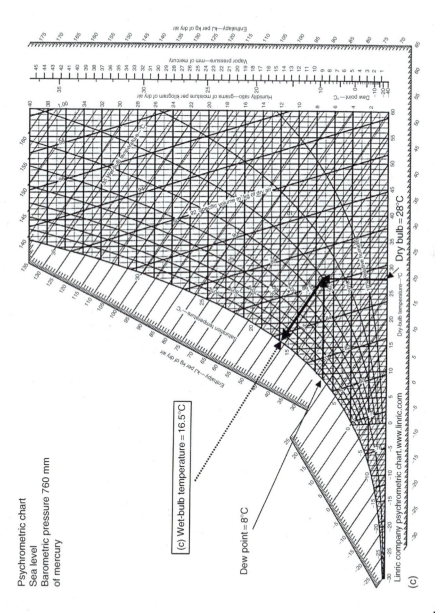

FIGURE E7.24c

Calculation of wet-bulb temperature.

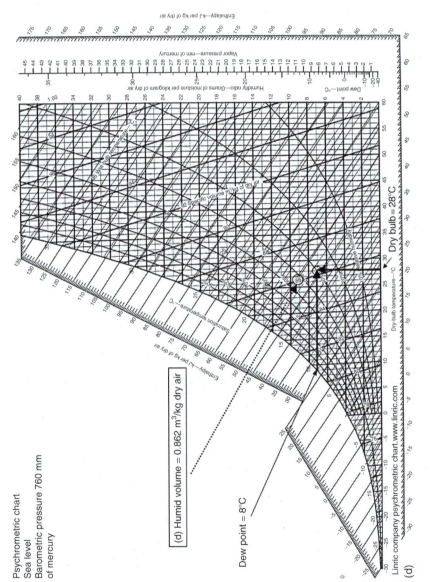

FIGURE E7.24d
Calculation of wet volume.

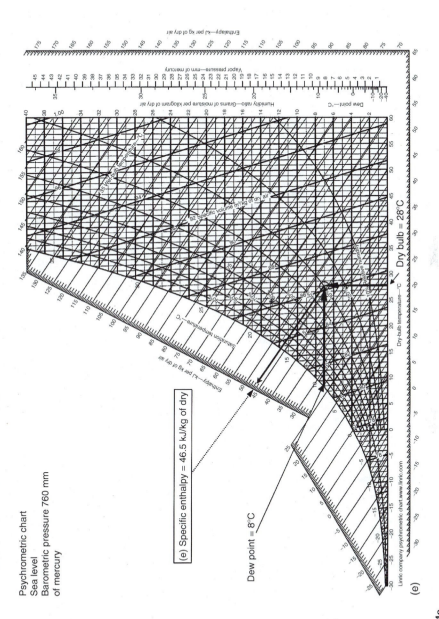

Psychrometric chart
Sea level
Barometric pressure 760 mm
of mercury

(e) Specific enthalpy = 46.5 kJ/kg of dry

Dew point = 8°C

Dry bulb = 28°C

Linric company psychrometric chart.www.linric.com

(e)

FIGURE E7.24e
Calculation of specific enthalpy.

7.12 Summary on Energy Balance without Reactions

a. The first law of thermodynamics for a closed system is

$$\Delta U + \Delta E_k + \Delta E_p = Q - W$$

b. The first law of thermodynamics for an open system at steady state (i.e., continuous) is

$$\Delta \dot{H} + \Delta \dot{E}_k + \Delta \dot{E}_p = \dot{Q} - \dot{W}_s$$

c. Procedure for energy balance calculations
   1. Draw and completely label a process flowchart.
   2. Perform all material balance calculations.
   3. Write the appropriate form of the energy balance equation and remove any negligible terms.
   4. Choose a reference state (phase, T, P) for each species involved. If using enthalpy tables, use reference state to generate table. If no tables are available, choose one inlet or outlet condition as the reference state for the species.
   5. Construct an inlet–outlet enthalpy table.
   6. Calculate all required values of $\hat{U}_i$ or $\hat{H}_i$ and insert values into table.
   7. Calculate ΔU or ΔH (e.g., $\Delta H = \Sigma m_i \hat{H}_i - \Sigma m_i \hat{H}_i$).
   8. Calculate any other terms in the energy balance equation (i.e., W, ΔE_k, ΔE_p).
   9. Solve for the unknown quantity in the energy balance equation.

7.13 Problems

7.13.1 Vaporization of Liquid Methanol

Liquid methanol at 25°C is heated and vaporized for use in a chemical reaction. How much heat is required to heat and vaporize 2 mol of methanol to 600°C. (Answer: $Q = 149,400$ J.)

7.13.2 Heating of Propane

Propane gas at 40°C and 250 kPa enters a continuous adiabatic heat exchanger (no heat is lost from the outside of the unit—that is, it is insulated on the outside but heat transfers between the two fluids inside the heat exchanger) and exits the heat exchanger at 240°C. The flow rate of propane is 100 mol/min. Saturated steam at 5 bar absolute pressure enters the heat exchanger with a flow rate of 6 kg/min. Calculate the exit temperature of steam. (Answer: $\hat{H}_{out} = 2742.3$ kJ/kg, $T_f = 133$°C.)

7.13.3 Expansion of Wet Steam

Wet steam at 20 bar with 97% quality leaks through a defective steam trap and expands to atmospheric pressure. The process can be considered to take place in two stages: A rapid adiabatic expansion to 1 atm accompanied by complete evaporation of the liquid droplets in the wet stream, followed by cooling at 1 atm to ambient temperature. Calculate the temperature of the superheated steam immediately following the rapid adiabatic expansion. (Answer: $T = 132$°C.)

7.13.4 Open System Energy Balance (Heating of Methanol)

Methanol is heated by condensing steam in a concentric cylinder heat exchanger as depicted below. Methanol, flowing through the inner pipe at 64.0 kg/s, enters at 20°C and exits at 60°C. Steam enters the outer pipe at 350°C and 80 bar (absolute) and leaves the heat exchanger as saturated water at 1.0 atm. Assume that the outer pipe is well insulated and no heat is lost to the surroundings (Figure P7.13.4).

   a. Determine the mass flow rate of the steam. (Answer: 57 kg/s.)
   b. Calculate the degrees of superheat of the entering steam. (Answer: 55°C.)

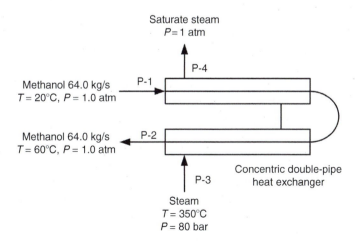

FIGURE P7.13.4
Flow sheet of double-pipe heat exchanger.

7.13.5 Open System Energy Balance (Heating of Liquid Methanol)

Calculate heat rate required to heat 32.04 kg/s of liquid methyl alcohol (CH_3OH) at 5°C and 44 atm to vapor at 500°C and 2 atm. The heat of vaporization of methanol at 64.7°C and 1 atm is 35.27 kJ/mol. (Answer: 68.0 MW.)

7.13.6 Vaporization of Liquid *n*-Hexane

Determine the total amount of heat required to convert 2.00 mol of liquid *n*-hexane (C_6H_{14}) at 10°C to vapor at 55°C in a closed container. Assume that hexane vapor behaves as an ideal gas at the system pressure. Neglect any effect of a change in pressure on the liquid enthalpy. The heat of vaporization of hexane at 68.74°C and 1 atm is 28.85 kJ/mol. (Answer: 77.9 kJ.)

7.13.7 Closed System Energy Balance (Heating of Acetone)

A volume of 734 cm^3 of liquid acetone is contained in a closed (vertical) cylinder fitted with a moveable frictionless piston at 10°C. The acetone is heated to vapor at 500°C. The piston area is 50.0 cm^2, and the piston weighs 200 kg. The heat of vaporization of acetone at its normal boiling point (56.0°C) is 30.2 kJ/mol.

a. Calculate the heat (kJ) required to convert acetone from 10°C to 500°C. Assume the cylinder is perfectly insulated, and no heat is lost to the surroundings. (Answer: 881 kJ.)
b. If the heat is provided by superheated steam at 530°C and 1.0 bar, how much steam (kg) is needed? The final condition of the steam is saturated at 100°C. (Answer: 1.00 kg.)

7.13.8 Open System Energy Balance (Power Output of Turbine)

A hydroelectric project has a volumetric throughput of 1.2 m^3/s. The water flowing in the river at atmospheric pressure and 20°C falls vertically for 300 m and passes through a turbine. The water exits the turbine at atmospheric pressure and 20.5°C. What is the power output of the turbine? (Answer: 1.01 MW.)

7.13.9 Open System Energy Balance (Power Requirement of Compressor)

Air at 100 kPa and 10°C enters a compressor and is compressed to 1000 kPa and 50°C. The constant pressure heat capacity of air is 1.01 kJ/kg K. If 15 kg of air are to be compressed every minute, determine the power requirement of the compressor. State your assumptions. (Answer: −10.1 kW.)

Further Readings

1. Reklaitis, G.V. (1983) *Introduction to Material and Energy Balances*, John Wiley & Sons, New York.
2. Felder, R.M. and R.W. Rousseau (1999) *Elementary Principles of Chemical Processes*, 3rd edn, John Wiley, New York.
3. Himmelblau, D.M. (1974) *Basic Principles and Calculations in Chemical Engineering*, 3rd edn, Prentice-Hall, New Jersey.
4. Whirwell, J.C. and R.K. Toner (1969) *Conservation of Mass and Energy*, Blaisdell, Waltham, MA.
5. Cordier, J.-L., B.M. Butsch, B. Birou, and U. von Stockar (1987) The relationship between elemental composition and heat of combustion of microbial biomass. *Appl. Microbiol. Biotechnol.* 25, 305–312.
6. Atkinson, B. and F. Mavituna (1991) *Biochemical Engineering and Biotechnology Handbook*, 2nd edn, Macmillan, Basingstoke.
7. Doran, P. (1995) *Bioprocess Engineering Principles*, Academic Press, San Diago, CA.
8. Bailey, J.E. and D.F. Ollis (1986) *Biochemical Engineering Fundamentals*, 2nd edn, McGraw-Hill, Boston, MA.
9. Shuler, M.L. and F. Kargi (2002) *Bioprocess Engineering—Basic Concepts*, 2nd edn, Prentice-Hall International Series, New Jersey.

8

Energy Balance with Reaction

At the End of This Chapter You Should Be Able to

1. Compute standard heat of reaction from heat of formation for a given reaction.
2. Calculate the heat of reaction at any reaction temperature.
3. Determine exit stream temperatures when inlet stream temperature is given, or vice versa.
4. Calculate the adiabatic reaction temperature.

8.1 Introduction

- In any chemical reaction, energy is required to break the bonds of the reacting chemical species and energy is released when the product bonds are formed.
- The large changes in enthalpy and internal energy during a chemical reaction necessitate substantial heat transfer (heating or cooling) from the reactor in order to eventually maintain the reactor at its desired operating temperature.
- The net change of enthalpy is called the heat of reaction, which is the energy that must either be transferred to or from the reactor to maintain the desired reactor temperature.

8.2 Heats of Reaction

By definition, the heat of reaction, $\Delta \hat{H}_r$ (T, P), is the enthalpy change for a process in which stoichiometric quantities of reactants at temperature T and pressure P react completely to form products at the same temperature and pressure.

Consider the following reaction:

$$aA + bB \rightarrow cC + dD$$

$\Delta \widehat{H}_r$ is calculated as the difference between the product and reactant enthalpies (at constant T, P), which are weighted by their stoichiometric coefficients. Therefore,

$$
\begin{aligned}
\Delta \widehat{H}_r(T, P) \text{ [kJ/mol]} &= H_{\text{products}} - H_{\text{reactants}} \\
&= c\widehat{H}_C(T, P) + d\widehat{H}_D(T, P) - a\widehat{H}_A(T, P) - b\widehat{H}_B(T, P) \\
&= \sum v_i \widehat{H}_i(T, P)
\end{aligned}
$$

The units of $\Delta \widehat{H}_r$ are kilojoules per mole—but, per mole of what? Recall that the reported $\Delta \widehat{H}_r$ applies to stoichiometric quantities of each species. For example,

$$2A + B \rightarrow 3C, \quad \Delta \widehat{H}_r(100°\text{C}, 1 \text{ atm}) = -50 \text{ kJ/mol}$$

The enthalpy change for the given reaction is

$$\frac{-50 \text{ kJ}}{2 \text{ mol } A \text{ consumed}} = \frac{-50 \text{ kJ}}{1 \text{ mol } B \text{ consumed}} = \frac{-50 \text{ kJ}}{3 \text{ mol } C \text{ generated}}$$

If 150 mol of C/s was generated at 100°C and 1 atm,

$$\Delta \dot{H} = \left(\frac{-50 \text{ kJ}}{3 \text{ mol } C \text{ generated}} \right) \left(\frac{150 \text{ mol } C \text{ generated}}{\text{s}} \right) = -2500 \text{ kJ/s}$$

8.3 Heats of Reaction Using the Extent of Reaction

In general, if n_{Ar} moles of A is generated or consumed by a reaction at a temperature T and pressure P, and v_A is the stoichiometric coefficient of the reactant or product, the associated enthalpy change is

$$\Delta \dot{H} = \dot{\xi} \Delta \widehat{H}_r(T, P)$$

Recall that the extent of reaction, ξ, is a measure of how far a reaction has proceeded:

$$(\dot{n}_i)_{\text{out}} = (\dot{n}_i)_{\text{in}} + v_i \dot{\xi} \Rightarrow \dot{\xi} = \frac{(\dot{n}_i)_{\text{out}} - (\dot{n}_i)_{\text{in}}}{v_i}$$

8.3.1 Notes on Heats of Reaction

1. If $\Delta\widehat{H}_r(T)$ is negative, the reaction is exothermic, that is, energy must be removed from the reactor to prevent the temperature from increasing.

2. If $\Delta\widehat{H}_r(T)$ is positive, the reaction is endothermic, that is, energy must be added to the reactor to prevent the temperature from decreasing.

3. At low and moderate pressure, $\Delta\widehat{H}_r(T, P)$ is nearly independent of pressure. Therefore, $\Delta\widehat{H}_r(T, P) \approx \Delta\widehat{H}_r(T)$.

4. The value of the heat of reaction depends on how the stoichiometric equation is written. For example,

$$CH_4 \text{ (g)} + 2O_2 \text{ (g)} \rightarrow CO_2 \text{ (g)} + 2H_2O \text{ (l)}: \Delta H_{r1} \text{ (25°C)} = -890.3 \text{ kJ/mol}$$

$$2CH_4 \text{ (g)} + 4O_2 \text{ (g)} \rightarrow 2CO_2 \text{ (g)} + 4H_2O \text{ (l)}: \Delta H_{r2} \text{ (25°C)} = -1780.6 \text{ kJ/mol}$$

5. The value of the heat of reaction depends on the phase of the reactants and products. For example,

$$CH_4 \text{ (g)} + 2O_2 \text{ (g)} \rightarrow CO_2 \text{ (g)} + 2H_2O \text{ (l)}: \Delta H_{r1} \text{ (25°C)} = -890.3 \text{ kJ/mol}$$

$$CH_4 \text{ (g)} + 2O_2 \text{ (g)} \rightarrow CO_2 \text{ (g)} + 2H_2O \text{ (g)}: \Delta H_{r2} \text{ (25°C)} = -802.3 \text{ kJ/mol}$$

6. The standard heat of reaction, ΔH_r^0, is the heat of reaction when both reactants and products are at standard conditions, that is, at 25°C and 1 atm. The symbol 0 denotes standard conditions.

Example 8.1 Calculation of enthalpy change due to butane combustion

The standard heat of the combustion on n-butane vapor is

$$C_4H_{10}(g) + \frac{13}{2}O_2 \text{ (g)} \rightarrow 4CO_2 \text{ (g)} + 5H_2O \text{ (l)}, \quad \Delta\widehat{H}_r^0 = -2878 \text{ kJ/mol}$$

Calculate the rate of enthalpy change, $\Delta\dot{H}$ (kJ/s), if 2400 mol/s of CO_2 is produced in this reaction and the reactants and products are all at 25°C.

Solution

$$\Delta\dot{H} = \dot{\xi}\Delta\widehat{H}_r(T, P)$$

Enthalpy change:

$$\Delta\dot{H} = \left(\frac{-2878 \text{ kJ}}{4 \text{ mol } CO_2 \text{ generated}}\right)\left(\frac{2400 \text{ mol } CO_2 \text{ produced}}{s}\right)$$

$$= 1{,}726{,}800 \text{ kJ/s}$$

Example 8.2 Standard heat of reaction

What is the standard heat of the reaction below?

$$2C_4H_{10} \text{ (g)} + 13O_2 \text{ (g)} \rightarrow 8CO_2 \text{ (g)} + 10H_2O \text{ (l)}$$

Calculate $\Delta \dot{H}$, if 2400 mol/s of CO_2 is produced in this reaction and the reactants and products are all at 25°C.

Solution

The standard heat of reaction from the heat of formation is as follows:

$$\Delta \widehat{H}_r^0 \text{ (25°C, 1 atm)} = \sum v_i \Delta H_f^0$$

$$\Delta \widehat{H}_r^0 \text{ (25°C, 1 atm)} = 10\Delta \widehat{H}_f^0, \ H_2O \text{ (l)} + 8\Delta \widehat{H}_f^0, \ CO_2 - 2\Delta \widehat{H}_f^0, \ C_4H_{10} - 13\Delta \widehat{H}_f^0, O_2$$

$$\Delta \widehat{H}_r^0 \text{ (25°C, 1 atm)} = 10(-285.84) + 8(-393.5) - 2(-124.7) - 13(0)$$

$$\Delta \widehat{H}_r^0 \text{ (25°C, 1 atm)} = -5757 \text{ kJ mol}$$

$$\Delta \dot{H} = \frac{-5757 \text{ kJ}}{8 \text{ mol } CO_2 \text{ generated}} (2400 \text{ mol/s}) = -1{,}727{,}100 \text{ kJ/s}$$

8.4 Reactions in Closed Processes

What if the reaction takes place in a closed system of constant volume?

$$\text{Energy balance: } \Delta U + \Delta E_k + \Delta E_p = Q - W$$

The internal energy, $\Delta \widehat{U}_r(T)$, is calculated as the difference between the product and reactant internal energies if stoichiometric quantities of reactants react completely at temperature T.

$$\Delta \widehat{U}_r(T) = U_{products} - U_{reactants}$$

Assuming ideal gas behavior, the internal energy is related to the heat of reaction by

$$\widehat{U}_r(T) = \widehat{H}_r(T) - PV$$
$$\Delta \widehat{U}_r(T) = \Delta \widehat{H}_r(T) - \Delta(PV)$$
$$\Delta \widehat{U}_r(T) = \Delta \widehat{H}_r(T) - \Delta(nRT)$$
$$\Delta \widehat{U}_r(T) = \Delta \widehat{H}_r(T) - RT(\Delta n)$$
$$\Delta \widehat{U}_r(T) = \Delta \widehat{H}_r(T) - RT \sum v_i$$

where v_i is the stoichiometric coefficient of the ith gaseous reactant or product (+ for product, − for reactant).

Example 8.3   Internal energy

The standard heat of the reaction

$$C_2H_4 \text{ (g)} + 2Cl_2 \text{ (g)} \rightarrow C_2HCl_3 \text{ (l)} + H_2 \text{ (g)} + HCl \text{ (g)}$$

is $\Delta \widehat{H}_r^0 = -420.8$ kJ/mol. Calculate $\Delta \widehat{U}_r^0$ for this reaction.

Solution

$$\Delta \widehat{U}_r(T) = \Delta \widehat{H}_r(T) - RT \sum v_i$$

$$\Delta \widehat{U}_r(T) = -420.8 \text{ kJ/mol} - 8.314 \text{ kJ/(mol K)}(25 + 273.15 \text{ K})(1 + 1 + 1 - 2 - 1)$$

$$\Delta \widehat{U}_r(T) = -420.8 \text{ kJ/mol}$$

8.5   Measurement of Heats of Reaction

Heats of reaction are measured in a calorimeter. A calorimeter is a closed reactor that is submersed in a fluid and enclosed in an insulated vessel. The increase or decrease in fluid temperature determines the amount of energy released or absorbed and using the heat capacities of the reactants and products, $\Delta \widehat{H}_r^0$ can be determined. However, this measurement technique will not work for every reaction. For example, consider the following reaction:

$$C \text{ (s)} + \frac{1}{2}O_2 \text{ (g)} \rightarrow CO \text{ (g)}, \quad \Delta \widehat{H}_r^0 \text{ (25°C, 1 atm)} = \, ?$$

- Only minimal amounts of CO would form since the rate of reaction at 25°C is too low.
- Higher reaction temperatures (higher reaction rates) would not lead to the formation of pure CO but rather a mixture of CO and CO_2.

What if we cannot measure experimentally $\Delta \widehat{H}_r^0$ for our desired reaction?

$$C + \frac{1}{2}O_2 \rightarrow CO, \quad \Delta \widehat{H}_r^0 = \, ?$$

However, we can measure the following heats of reaction.

$$C + O_2 \rightarrow CO_2, \ \Delta \widehat{H}^0_{r2} = -393.51 \ \text{kJ/mol}$$

$$CO + \frac{1}{2}O_2 \rightarrow CO_2, \ \Delta \widehat{H}^0_{r3} = -282.99 \ \text{kJ/mol}$$

8.6 Hess' Law

The previous result could be more readily obtained, if we treated the stoichiometric equations as algebraic equations. That is,

$$C + O_2 - CO - \frac{1}{2}O_2 \rightarrow CO_2 - CO_2 \ (\text{reaction 2} - \text{reaction 3})$$

$$\Downarrow$$

$$C + \frac{1}{2}O_2 \rightarrow CO \ (\text{desired reaction 1})$$

$$\Delta \widehat{H}^0_{r1} = \Delta \widehat{H}^0_{r2} - \Delta \widehat{H}^0_{r3} = -393.51 - (-282.99) = -110.52 \ \text{kJ/mol}$$

Hess' Law: If a set of reactions can be manipulated through a series of algebraic operations to yield the desired reaction, then the desired heat of reaction can be obtained by performing the same algebraic operations on the heats of reaction of the manipulated set of reactions.

Example 8.4   Using Hess' law

(a) The standard heats of the following combustion reactions have been determined experimentally:

$$C_2H_6 + \frac{7}{2}O_2 \rightarrow 2CO_2 + 3H_2O, \ \Delta \widehat{H}^0_{r1} = -1559.8 \ \text{kJ/mol}$$

$$C + O_2 \rightarrow CO_2, \ \Delta \widehat{H}^0_{r2} = -393.5 \ \text{kJ/mol}$$

$$H_2 + \frac{1}{2}O_2 \rightarrow H_2O, \ \Delta \widehat{H}^0_{r3} = -285.8 \ \text{kJ/mol}$$

Use Hess' law and the given heats of reaction to determine the standard heat of the reaction.

$$2C + 3H_2 \rightarrow C_2H_6, \ \Delta H^0_{r4} = ?$$

Solution
Using Hess' law

$$\Delta \widehat{H}^0_{r4} = 3(\Delta \widehat{H}^0_{r3}) + 2(\Delta \widehat{H}^0_{r2}) - (\Delta \widehat{H}^0_{r1})$$

$$\Delta \widehat{H}^0_{r4} = 3(-285.8) + 2(-393.5) - (-1559.8) = 201.2 \ \text{kJ/mol}$$

(b) Calculate the heat of combustion for C_2H_6 from the following information:

$$C_2H_4 + 3O_2 \rightarrow 2CO_2 + 2H_2O, \ -1409.5 \text{ kJ}$$

$$C_2H_4 + H_2 \rightarrow C_2H_6, \ -136.7 \text{ kJ}$$

$$H_2 + \frac{1}{2}O_2 \rightarrow H_2O, \ -285.5 \text{ kJ}$$

Solution

$$C_2H_6 \rightarrow C_2H_4 + H_2, \ +136.7 \text{ kJ}$$
$$+ \ C_2H_4 + 3O_2 \rightarrow 2CO_2 + 2H_2O, \ -1409.5 \text{ kJ}$$
$$= \ C_2H_6 + 3O_2 \rightarrow 2CO_2 + 2H_2O + H_2, \ -1272.8 \text{ kJ}$$
$$+ \ H_2 + \frac{1}{2}O_2 \rightarrow H_2O, \ -285.5 \text{ kJ}$$
$$= \ C_2H_6 + 3.5O_2 \rightarrow 2CO_2 + 3H_2O, \ -1558.3 \text{ kJ}$$

So, -1558.3 kJ is the final answer.

8.7 Calculating Heat of Reaction ($\Delta \widehat{H}_r^0$) from Heats of Formation

$\Delta \widehat{H}_r^0$ can be calculated using standard heats of formation. A formation reaction of a compound is the reaction in which the compound is formed from its elemental constituents as they would occur in nature (e.g., O_2 rather than O). The enthalpy change associated with the formation of 1 mol of the compound at 25°C and 1 atm is the standard heat of formation $\Delta \widehat{H}_f^0$ of the compound (available in Appendix A, Table A.2). The values of $\Delta \widehat{H}_f^0$ for many compounds can be obtained from tabulated data. For example, the $\Delta \widehat{H}_f^0$ of ammonium nitrate (NH_4NO_3 (s)) is -365.14 kJ/mol. This signifies that

$$N_2 \ (g) + 2H_2 \ (g) + \frac{3}{2}O_2 \ (g) \rightarrow NH_4NO_3 \ (s), \ \Delta \widehat{H}_r^0 = -365.14 \text{ kJ/mol}$$

A consequence of Hess' law is that $\Delta \widehat{H}_r^0$ of any reaction can be calculated as

$$\Delta \widehat{H}_r^0 = \sum_i \nu_i \Delta \widehat{H}_{f,i}^0$$

where ν_i is the stoichiometric coefficient of reactant or product species i. $(\Delta \widehat{H}_f^0)_i$ is the standard heat of formation of species i. The standard heats of formation of all elemental species (e.g., O_2, N_2, Zn, etc.) are zero.

Example 8.5 Heat of reaction for combustible liquids

Determine the standard heat of reaction for the combustion of liquid n-pentane.

$$C_5H_{12} \text{ (l)} + 8O_2 \text{ (g)} \rightarrow 5CO_2 \text{ (g)} + 6H_2O \text{ (l)}$$

The standard heat of combustion of a species, $\Delta \widehat{H}_c^0$, is the enthalpy change associated with the complete combustion of moles of a species with oxygen at 25°C and 1 atm such that

- All the carbon forms CO_2 (g)
- All the hydrogen forms H_2O (l)
- All the sulfur forms SO_2 (g)
- All the nitrogen forms N_2 (g)

$\Delta \widehat{H}_c^0$: Values for combustible species are available in Table A.2. For example, $\Delta \widehat{H}_c^0$ of ethanol is -1366.9 kJ/mol. This signifies that

$$C_2H_5OH \text{ (l)} + 3O_2 \text{ (g)} \rightarrow 2CO_2 \text{ (g)} + 3H_2O \text{ (l)}, \ \Delta \widehat{H}_r^0 = -1366.9 \text{ kJ/mol}$$

Solution

Heat of reaction from heat of formation

$$\Delta \widehat{H}_r^0 = 3(\Delta \widehat{H}_{fH_2O \text{ (l)}}^0) + 2(\Delta \widehat{H}_{fCO_2 \text{ (g)}}^0) - (\Delta \widehat{H}_{fC_2H_5OH \text{ (l)}}^0)$$

Heat of reaction from heat of combustion

$$\Delta \widehat{H}_r^0 = (\Delta \widehat{H}_{c, \, C_2H_5OH \text{ (l)}}^0) + 3(\Delta \widehat{H}_{c, \, O_2}^0) - 3(\Delta \widehat{H}_{c, \, H_2O \text{ (l)}}^0) - 2(\Delta \widehat{H}_{c, \, CO_2 \text{ (g)}}^0)$$

8.8 Calculating $\Delta \widehat{H}_r$ from Heats of Combustion

A consequence of Hess' law is that $\Delta \widehat{H}_r^0$ of any reaction involving only oxygen and a combustible species can be calculated as

$$\Delta \widehat{H}_r^0 = -\sum_i \nu_i (\Delta \widehat{H}_c^0)_i = \sum_{\text{reactants}} |\nu_i| (\Delta \widehat{H}_c^0)_i - \sum_{\text{products}} |\nu_i| (\Delta \widehat{H}_c^0)_i$$

Note: This is the reverse of determining the heat of reaction from heats of formation, where v_i is the stoichiometric coefficient of reactant or product species i. $(\Delta \widehat{H}_c^0)_i$ is the standard heat of formation of species i. If any reactants or products are combustion products (i.e., CO_2, $H_2O(l)$, SO_2), their heats of combustion are zero.

Example 8.6 Dehydrogenation of ethane

Calculate the standard heat of reaction from the dehydrogenation of ethane:

$$C_2H_6 \rightarrow C_2H_4 + H_2$$

Answer:

$$\Delta H_r^0 = \Delta \widehat{H}_{c,\ C_2H_6}^0 - \Delta \widehat{H}_{c,\ C_2H_4}^0 - \Delta \widehat{H}_{c,\ H_2}^0$$

8.9 Determining $\Delta \widehat{H}_f^0$ from $\Delta \widehat{H}_c^0$

For many substances, it is much easier to measure $\Delta \widehat{H}_c^0$ than $\Delta \widehat{H}_f^0$.
  For example, consider the formation of pentane:

$$5C\ (s) + 6H_2\ (g) \rightarrow C_5H_{12}\ (l),\ \Delta \widehat{H}_f^0 = ?$$

Carbon, hydrogen, and pentane can all be burned and their standard heats of combustion can be determined experimentally. Therefore,

$$(\Delta \widehat{H}_f^0)_{C_5H_{12}\ (l)} = 5(\Delta \widehat{H}_c^0)_{C\ (s)} + 6(\Delta \widehat{H}_c^0)_{H_2\ (g)} - (\Delta \widehat{H}_c^0)_{C_5H_{12}\ (l)}$$

8.10 Energy Balance on Reactive Processes

Material balances could be done by either writing balances on either compounds (which requires the extent of reaction) or elements (which requires only balances without generation terms for each element). We can also do energy balances using either compounds or elements.
  The methods differ in the reference state (and thus calculation of $\Delta \dot{H}$). From material balances with reaction that we had discussed, there are two methods of analyzing these types of reactive processes:

- Atomic species balances
- Extents of reaction

For energy balances with reaction, we also have two methods for solving these types of problems:

- Heat of reaction method
- Heat of formation method

These two methods differ in the choice of the reference state.

8.10.1 Heat of Reaction Method

The heat of reaction method is ideal when there is a single reaction for which $\Delta \widehat{H}_r^0$ is known. This method requires calculation of the extent of reaction, $\dot{\xi}$, on any reactant or product for which the feed and product flow rates are known. Reference state is such that all reactant and product species are at 25°C and 1 atm in the states for which the heat of reaction is known (Figure 8.1).

$\widehat{H}_i$ accounts for change in enthalpy with T and phase (if necessary). From the input–output enthalpy table, for a single reaction we obtain

$$\Delta \dot{H} = \dot{\xi}\Delta \widehat{H}_r^0 + \sum_{\text{out}} \dot{n}_i \widehat{H}_i - \sum_{\text{in}} \dot{n}_i \widehat{H}_i$$

For multiple reactions, we obtain

$$\Delta \dot{H} = \sum_{\text{reactions}} \dot{\xi}_j \Delta \widehat{H}_{rj}^0 + \sum_{\text{out}} \dot{n}_i \widehat{H}_i - \sum_{\text{in}} \dot{n}_i \widehat{H}_i$$

The reference state is such that the reactants and products are at 25°C and 1 atm.

$$\Delta \dot{H} = \sum_{\text{reactions}} \dot{\xi}_j \Delta \widehat{H}_{rj}^0 + \sum_{\text{out}} \dot{n}_i \widehat{H}_i - \sum_{\text{in}} \dot{n}_i^0 \widehat{H}_i$$

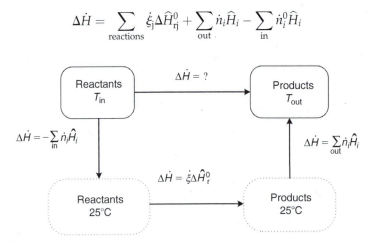

FIGURE 8.1
Heat of reaction method.

A reference temperature other than 25°C can be considered in this case, and the heat of reaction should be calculated at the new reference state.

$$\Delta \dot{H} = \sum_{\text{reactions}} \dot{\xi}_j \Delta \widehat{H}_{rj} \ (@T_{\text{ref}}) + \sum_{\text{out}} \dot{n}_i \widehat{H}_i - \sum_{\text{in}} \dot{n}_i^0 \widehat{H}_i$$

$$\Delta H_r(@T_{\text{ref}}) = \Delta H_r^0 \ (25°C) + \int_{25°C}^{T_{\text{ref}}} \Delta C_p dT$$

Example 8.7   Oxidation of ammonia

The standard heat of reaction for the oxidation of ammonia is given below:

$$4NH_3 \ (g) + 5O_2 \ (g) \rightarrow 4NO \ (g) + 6H_2O \ (v), \ \Delta \widehat{H}_r^0 = -904.7 \ \text{kJ/mol}$$

One hundred moles of NH_3/s and 200 mol O_2/s at 25°C are fed into a reactor in which ammonia is completely consumed. The product gas emerges at 300°C. Calculate the rate at which heat must be transferred to or from the reactor, assuming operation at approximately 1 atm. The heat of formation method is ideal for multiple reactions and single reactions for which $\Delta \widehat{H}_r^0$ is not known.

Solution

The reference state is such that all elemental species (e.g., C (s), O_2 (g), H_2 (g)) that constitute the reactant and product species at 25°C and 1 atm. Nonreacting species can be evaluated at any convenient temperature or the reference conditions used for the species in an available enthalpy table (e.g., N_2 (g), at 25°C at 1 atm)

$$\Delta \dot{H} = \dot{\xi} \Delta \widehat{H}_r^0 + \sum_{\text{out}} \dot{n}_i \widehat{H}_i - \sum_{\text{in}} \dot{n}_i \widehat{H}_i$$

Material balance (extent of reaction):
Basis: 100 mol of NH_3/s

$$n_1 = 100 - 4\xi, \ n_2 = 200 - 5\xi$$
$$n_3 = 0.0 + 4\xi, \ n_4 = 0.0 + 6\xi$$

Complete conversion of ammonia ($x_c = 1$)

$$1 = \frac{100 - 4\xi}{100} \Rightarrow \xi = 25 \ \text{mol}$$

Answer: $n_1 = 0.0$ mol, $n_2 = 75$ mol, $n_3 = 100$ mol, $n_4 = 150$ mol
Energy balance (reference temperature 25°C):

$$\Delta \dot{H} = \dot{\xi} \Delta \widehat{H}_r^0 + \sum_{\text{out}} \dot{n}_i \widehat{H}_i - \sum_{\text{in}} \dot{n}_i \widehat{H}_i$$

$$\Delta \dot{H} = 25 \text{ mol}(-904.7 \text{ kJ/mol}) + \left[75 \int_{25}^{300} C_{p\,O_2} dT + 100 \int_{25}^{300} C_{p\,NO} dT + 150 \int_{25}^{300} C_{p\,H_2O} dT \right]_{\text{out}}$$

$$- \left[200 \int_{25}^{25} C_{p\,NH_3} dT + 200 \int_{25}^{25} C_{p\,O_2} dT \right]_{\text{in}}$$

$$\Delta \dot{H} = 25 \text{ mol} \left[\frac{-904.7 \text{ kJ}}{\text{mol}} \right] + [75(8.47) + 100(8.45) + 150(9.57)] - [0 + 0]$$

$$= -19{,}700 \text{ kJ/s}$$

8.10.2 Heat of Formation Method: Process Path

In the heat of formation method, the heat of reaction terms ($\Delta \widehat{H}_r^0$) are not required as they are implicitly included when heats of formation of the reactants are subtracted from the products. For single and multiple reactions,

$$\Delta \dot{H} = \sum_{\text{out}} \dot{n}_i \widehat{H}_i - \sum_{\text{in}} \dot{n}_i \widehat{H}_i$$

$\widehat{H}_i$ accounts for change in enthalpy with T and phase (if necessary) $+ \Delta \widehat{H}_f^0$ (Figure 8.2).

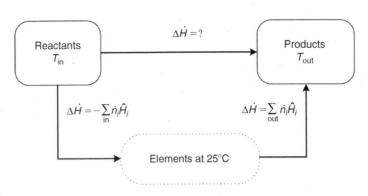

FIGURE 8.2
Energy balance using heat of formation method.

Example 8.8 Oxidation of methane
Methane is oxidized by air to produce formaldehyde in a continuous reactor. A competing reaction is the combustion of methane to form CO_2.

$$CH_4 \text{ (g)} + O_2 \rightarrow HCHO \text{ (g)} + H_2O \text{ (v)} \; \xi_1, \; \Delta\widehat{H}_{r1}^0 = ?$$

$$CH_4 \text{ (g)} + 2O_2 \rightarrow CO_2 + 2H_2O \text{ (v)} \; \xi_2, \; \Delta\widehat{H}_r^0 = ?$$

A flowchart of the process for an assumed basis of 100 mol of methane fed to the reactor is shown in Figure E8.8.

Solution (Extent of reaction method)
Balance on HCHO in order to find the extent of reaction.

$$30 = 0 + \xi_1 \Rightarrow \xi_1 = 30 \text{ mol}$$

Balance on CO_2 so as to calculate the extent of the second reaction.

$$10 = 0 + \xi_2 \Rightarrow \xi_2 = 10$$

$$\Delta\widehat{H}_{r1}^0 = \Delta\widehat{H}_f^0, \; H_2O \text{ (v)} + \Delta\widehat{H}_f^0, \; HCHO \text{ (g)} - \Delta\widehat{H}_f^0, \; CH_4 \text{ (g)}$$

$$= (-241.83) + (-115.9) - (-74.85)$$

$$= -282.88 \text{ kJ/mol}$$

$$\Delta\widehat{H}_{r2}^0 = 2\Delta\widehat{H}_f^0, \; H_2O \text{ (v)} + \Delta\widehat{H}_f^0, \; CO_2 \text{ (g)} - \Delta\widehat{H}_f^0, \; CH_4 \text{ (g)}$$

$$= 2(-241.83) + (-393.5) - (-74.85)$$

$$= -802.31 \text{ kJ/mol}$$

$$\Delta\dot{H} = \sum_{\text{reactions}} \dot{\xi}_j \Delta\widehat{H}_{rj}^0 + \sum_{\text{out}} \dot{n}_i \widehat{H}_i - \sum_{\text{in}} \dot{n}_i^0 \widehat{H}_i^0$$

Basis: 100 mol of CH_4
Reference: 25°C

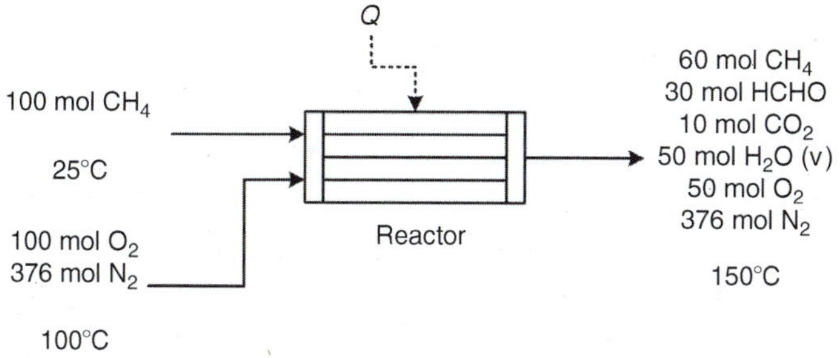

FIGURE E8.8
Oxidation of ethane.

$$\Delta\dot{H} = \dot{\xi}_1\Delta\hat{H}^0_{r1} + \dot{\xi}_2\Delta\hat{H}^0_{r2} + 60\int_{25}^{150}C_{p\,CH_4}\,dT + 30\int_{25}^{150}C_{p\,HCHO_{(g)}}\,dT + 10\int_{25}^{150}C_{p\,CO_2}\,dT$$

$$+ 50\int_{25}^{150}C_{p\,H_2O}\,dT + 50\int_{25}^{150}C_{p\,O_2}\,dT + 376\int_{25}^{150}C_{p\,N_2}\,dT$$

$$- 100\int_{25}^{25}C_{p\,CH_4}\,dT - 100\int_{25}^{100}C_{p\,O_2}\,dT - 376\int_{25}^{100}C_{p\,N_2}\,dT$$

$$\Delta\dot{H} = 30(-282.88) + 10(-802.31) + 60(4.9) + 30(4.75) + 10(4.75)$$
$$+ 50(4.27) + 50(3.758) + 376(3.655)$$
$$- 100(0) - 100(2.235) - 376(2.187)$$

$$\Delta\dot{H} = 15{,}300 \text{ kJ}$$

Example 8.9 Energy balance on elements

To set up an energy balance using the elements as a reference, consider the oxidation reaction of methane: $CH_4 + 2O_2 \rightarrow CO_2 + 2H_2O$ (Figure E8.9).

The feed to the reactor will be 1 mol of CH_4 at T_1 and 3 mol of O_2 at T_2. The following reaction occurs in the reactor, and the effluent stream exits at a temperature of T_3. Let us set up the energy balance so as to calculate the heat load, Q, on the reactor for this process. We will reference all enthalpies to the elements at 25°C.

Solution

The main difference between using a component balance and an element balance is that we must calculate the heat of reaction when doing a

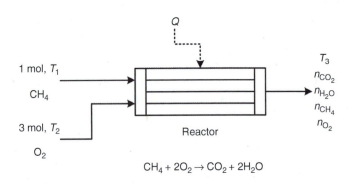

FIGURE E8.9
Flow sheet of methane oxidation process.

component balance, but we simply use heats of formation when writing down a balance based on the elements as the reference. We proceed by calculating the enthalpy at the stream condition for each individual compound in each stream relative to the elements at 25°C. We then plug these enthalpies into the normal energy balance expression. For the process that we are considering here, that is, the combustion of methane with oxygen, we can prepare the following table of enthalpies for each component in each stream. How to do these calculations is shown in the following table:

Species	n_{in}	H_{in}	n_{out}	H_{out}
CH_4	1	$\widehat{H}_{CH_4}(T_1)$	0	—
O_2	3	$\widehat{H}_{O_2}(T_2)$	1	$\widehat{H}_{O_2}(T_3)$
CO_2	0	—	1	$\widehat{H}_{CO_2}(T_3)$
H_2O	0	—	2	$\widehat{H}_{H_2O}(T_3)$

A. Enthalpy of CH_4 at T_1: Form CH_4 from the elements at 25°C, then raise the temperature to T_1:

$$\widehat{H}_{CH_4}(T_1) = \Delta \widehat{H}^0_{f,CH_4}(25°C) + \int_{25}^{T_1} C_{p\,CH_4}\,dT$$

B. Enthalpy of O_2 at T_2: Oxygen is already in the reference state at 25°C, so simply raise the temperature to T_2:

$$\widehat{H}_{O_2}(T_2) = 0 + \int_{25}^{T_2} C_{p\,O_2}\,dT$$

C. Enthalpy of O_2 at T_3: Oxygen is already in reference state at 25°C, so simply raise the temperature to T_3:

$$\widehat{H}_{O_2}(T_3) = 0 + \int_{25}^{T_3} C_{p\,O_2}\,dT$$

D. Enthalpy of CO_2 at T_3: Form CO_2 from the elements at 25°C, then raise the temperature to T_3:

$$\widehat{H}_{CO_2}(T_3) = \Delta \widehat{H}^0_{f,CO_2}(25°C) + \int_{25}^{T_3} C_{p\,CO_2}\,dT$$

E. Enthalpy of H_2O at T_3: Form H_2O from the elements at 25°C, then raise the temperature to T_3:

$$\widehat{H}_{H_2O}(T_3) = \Delta \widehat{H}^0_{f,H_2O}(25°C) + \int_{25}^{T_3} C_{p\,H_2O}\,dT$$

Now, having filled in the above table with the values that we calculate by the procedures shown, we simply plug these values of enthalpy into the general energy balance, namely,

$$Q - W_s = \left\{ n_{CO_2}\widehat{H}_{CO_2}(T_3) + n_{O_2}\widehat{H}_{O_2}(T_3) + n_{H_2O}\widehat{H}_{H_2O}(T_3) \right\}_{out}$$
$$- \left\{ n_{O_2}\widehat{H}_{O_2}(T_2) + n_{CH_2}\widehat{H}_{CH_4}(T_1) \right\}_{in}$$

This will be illustrated in the following examples.

8.11   General Procedure for Energy Balance with Reaction

1. Construct the flow diagram.
2. Complete the material balance calculations on the reactor to the greatest extent possible (using either extent of reaction or atomic species balances).
3. Prepare the inlet–outlet enthalpy table, inserting known molar amounts (or flow rates) for each stream component (and phase).
4. Choose your reference state for specific enthalpy calculations.
5. Calculate each unknown stream component enthalpy, $\widehat{H}_i$.
6. Calculate $\Delta\dot{H}$ for the reactor.
7. Using the general energy balance equation, solve for the unknown quantity.

Steps 4–6 depend on the energy balance method used (i.e., heat or reaction or heat of formation).

8.12   Processes with Unknown Outlet Conditions

In the previous problems, the inlet and outlet conditions were supplied or could be determined and the heat input or output could be calculated from the general energy balance equation. Another set of problems involves calculating the outlet temperature when the inlet conditions and heat input (or output) are specified. These types of problems require that the enthalpies (relative to the chosen reference temperature) be evaluated in terms of the unknown outlet temperature. The resultant enthalpy expressions are then substituted into the general energy balance equation and solved for the outlet temperature.

Example 8.10 Dehydrogenation of ethanol

The dehydrogenation of ethanol to form acetaldehyde

$$C_2H_5OH \ (v) \rightarrow CH_3CHO \ (v) + H_2 \ (g)$$

is carried out in an adiabatic reactor. Ethanol vapor is fed to the reactor at 400°C, and conversion of 30% is obtained. Calculate the product temperature (Figure E8.10).

Solution

Basis: 1 mol of ethanol
Reference: 25°C

The exit temperature is unknown, so a simultaneous material and energy balance is required.
Material balance (extent of reaction method):

$$n_1 = 1 - \xi, \ n_2 = 0 + \xi, \ n_3 = 0 + \xi$$

$$\text{Conversion} = 0.3 = \frac{1 - n_1}{1} \Rightarrow n_1 = 0.7$$

$$0.7 = 1 - \xi \Rightarrow \xi = 0.3 \text{ mol}$$

$$n_1 = 0.7 \text{ mol}, \ n_2 = 0.3 \text{ mol}, \ n_3 = 0.3 \text{ mol}$$

Energy balance: (heat of reaction method)
The general energy balance equation

$$\Delta \dot{H} = \xi \Delta \widehat{H}_r^0 \ (25°C, \ 1 \text{ atm}) + \sum_{\text{out}} n_i H_i - \sum_{\text{in}} n_i H_i$$

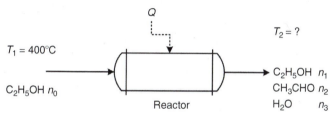

$$C_2H_5OH \ (V) \rightarrow CH_3CHO \ (V) + H_2 \ (g)$$

FIGURE E8.10
Flow sheet of ethanol dehydrogenation.

$$\Delta \widehat{H}_r^0 \,(25°C, 1\ atm) = \Delta H_{f,H_2}^0 + \Delta H_{f,CH_3CHO}^0 - \Delta H_{f,C_2H_5OH}^0$$

$$\Delta \widehat{H}_r^0 \,(25°C, 1\ atm) = 0 + (-166.2) - (-235.31) = 69.11\ kJ/mol$$

Open system and adiabatic process

$$Q = \Delta \dot{H} = 0$$

Since the reaction is adiabatic and endothermic the exit temperature should be less than the inlet temperature.

$$\Delta \dot{H} = \xi\, \Delta \widehat{H}_r^0 + \left\{ n_1 \int_{25}^{T_2} C_{PC_2H_5OH}\ dT + n_2 \int_{25}^{T2} C_{PCH_3CHO}\ dT + n_3 \int_{25}^{T_2} C_{PH_2O}\ dT \right\}$$

$$- \left\{ n_0 \int_{25}^{400} C_{PH_2O}\ dT \right\}$$

Solve for the exit temperature: $T_2 = 185°C$.

Example 8.11   Methanol dehydrogenation

Methanol at 675°C and 1 bar is fed at a rate of 100 mol/h to an adiabatic reactor where 25% of it is dehydrogenated to formaldehyde according to the reaction

$$CH_3OH\ (g) \rightarrow HCHO\ (g) + H_2\ (g)$$

Calculate the temperature of the gases leaving the reactor, assuming that constant average heat capacity of 17, 12, and 7 cal/mol°C for CH_3OH, $HCHO$, and H_2, respectively, are acceptable over the temperature range.

Solution

The process flow sheet is shown in Figure E8.11.

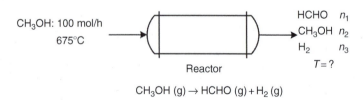

FIGURE E8.11
Methanol dehydrogenation process.

Basis: 100 mol/h of methanol

Material balance

Extent of reaction method is used to calculate exit number of moles:

$$n_1 = 0 + \xi, \ n_2 = 100 - \xi, \ n_3 = 0 + \xi$$

$$x = 0.25 = \frac{100 - n_2}{100}$$

Solving the above set of the equations leads

$$\xi = 25 \text{ mol/h}, \ n_1 = 25 \text{ mol/h}, \ n_2 = 75 \text{ mol/h}, \ n_3 = 25 \text{ mol/h}$$

Energy balance

The system is open so the first law of thermodynamics for open systems is used as follows:

$$Q - W_s = \Delta \dot{H} + \Delta \dot{E}_k + \Delta \dot{E}_p$$

Neglecting kinetic and potential energies, no shaft work and the reactor are adiabatic. The energy balance equation is simplified to the following form:

$$0 = \Delta \dot{H}$$

The enthalpy consists of enthalpy of reaction and the sensible heat:

$$\Delta H = \Delta H_r(@T_{ref}) + \sum_{product} n_i \int_{T_{ref}}^{T_{out}} C_{pi} \, dT - \sum_{reactant} n_i \int_{T_{ref}}^{T_{in}} C_{pi} \, dT$$

Taking the reference temperature at 675°C

$$\Delta \dot{H} = \Delta H_r(@675°C) + \sum_{product} n_i \int_{675°C}^{T} C_{pi} \, dT - \sum_{reactant} n_i \int_{675°C}^{675°C} C_{pi} \, dT$$

The heat capacity of this problem is constant and the effect of temperature on the heat capacity is negligible:

$$\Delta H_r(@675°C) = \Delta H_r^0(25°C) + \Delta C_p(675 - 25)$$

$$\Delta H_r^0(25°C) = \Delta H_{f,HCHO}^0 - \Delta H_{f,CH_3OH}^0$$
$$= -27.7 - (-48.08) = 20.38 \text{ kcal/gmol}$$

$$\Delta C_p = C_{p\,H_2} + C_{p\,HCHO} - C_{p\,CH_3OH}$$

ΔH_r (@ 675°C) = 20.38 kcal/mol

$$+ (7 + 12 - 17)\left(\frac{\text{cal}}{\text{mol°C}} \middle| \frac{1\,\text{kcal}}{1000\,\text{cal}}\right)(675 - 25)°C = 21.68\,\text{kcal/mol}$$

$$\Delta \dot{H} = \Delta H_r \text{ (@ 675°C)} + \left(n_1 C_{p_1} + n_2 C_{p_2} + n_3 C_{p_3}\right)(T - 675) - 0$$

$$\Delta \dot{H} = 0 = 21.68\,\text{kcal/mol} + (25 \times 12 + 75 \times 17 + 25 \times 7)\left(\frac{1\,\text{kcal}}{1000\,\text{cal}}\right)(T - 675)$$

Solving for the exit temperature, T, leads to $T = 365.3°C$

Example 8.12   Burning of carbon monoxide

Carbon monoxide (CO) at 10°C is completely burnt at 1 atm pressure with 50% excess air that is fed to a burner at a temperature of 540°C. The combustion products leave the burner chamber at a temperature of 425°C. Calculate the heat involved, Q, from the burner (Figure E8.12).

Solution

(a) Identify the reaction and calculate the heat of reaction:

$$\text{Reaction: CO (g)} + \frac{1}{2}O_2\,(g) \rightarrow CO_2\,(g)$$

$$\Delta \hat{H}_r^0 \text{ (@ 25°C)} = \Delta \hat{H}_f^0 CO_2 \text{ (@ 25°C)} - \Delta \hat{H}_f^0 CO \text{ (@ 25°C)}$$
$$= -393.5\,\text{kJ/mol} - (-110.52\,\text{kJ/mol})$$
$$= -282.98\,\text{kJ/mol}$$

(b) Material balance:

Theoretical O_2: 1 mol CO (0.5 mol O_2/1 mol CO) = 0.5 mol O_2
For 50% excess air, O_2 in = 1.5(0.5 mol O_2) = 0.75 mol O_2 in
Then, N_2 in = 3.76 (O_2 in) = (3.76)(0.75) = 2.82 mol N_2 in

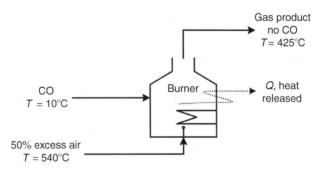

FIGURE E8.12
Flow sheet of carbon monoxide burner.

CO out $= 0$ (complete combustion)

CO_2 out $= 1$ mol CO_2 out

N_2 out $= N_2$ in $= 2.82$ mol

O_2 out $= 0.75 - 0.5 = 0.25$ mol O_2 out

(c) Energy balance ($T_{ref} = 25°C$):

$$\hat{H}_{in}(O_2) = \int_{25}^{540} C_{pO_2}\, dT = 16.38 \text{ kJ/mol}$$

$$\hat{H}_{in}(N_2) = \int_{25}^{540} C_{pN_2}\, dT = 15.49 \text{ kJ/mol}$$

$$\hat{H}_{in}(CO) = \int_{25}^{10} C_{pCO}\, dT = -0.4353 \text{ kJ/mol}$$

$$\hat{H}_{out}(O_2) = \int_{25}^{425} C_{pO_2}\, dT = 12.54 \text{ kJ/mol}$$

$$\hat{H}_{out}(N_2) = \int_{25}^{425} C_{pN_2}\, dT = 11.92 \text{ kJ/mol}$$

$$\hat{H}_{out}(CO_2) = \int_{25}^{425} C_{pCO_2}\, dT = 17.58 \text{ kJ/mol}$$

Summary of the calculated enthalpies is shown in the following table:

Substance	n_{in} (mol)	$\hat{H}_{in}$ (kJ/mol)	n_{out} (mol)	$\hat{H}_{out}$ (kJ/mol)
O_2	0.75	16.38	0.25	12.54
N_2	2.82	15.49	2.82	11.92
CO	1.00	−0.44	0	—
CO_2	0	—	1	17.58

(d) Calculate Q:

$$Q = \Delta H = \xi \Delta \hat{H}_r \ (@\ 25°C) + \sum_{out} n_i \hat{H}_i - \sum_{in} n_i \hat{H}_i$$

$$= -282.98 \text{ kJ/mol} + [(0.25)(12.54) + (2.82)(11.92) + (1)(17.58)]$$
$$- [0.75(16.38) + 2.82(15.49) + 1(-0.44)]$$

$$Q = -284.177 \text{ kJ/mol}$$

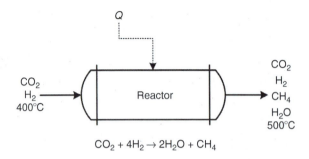

FIGURE E8.13
Flow sheet of gas phase reaction.

Example 8.13 Gas phase reaction
The gas phase reaction

$$CO_2 + 4H_2 \rightarrow 2H_2O + CH_4$$

proceeds with 80% conversion. Estimate the heat that must be provided or removed, if the gases enter at 400°C and leave at 500°C (Figure E8.13).

Solution
First, prepare the following data table using the polynomial form of the specific heat capacity, $\hat{C}_p = a + bT + cT^2 + dT^3$, and the heats of formation at 25°C. The ΔC_p for the reaction is immediately obtained from the Δ row for use in integration from one temperature to another. The Δ in the last column (Table E8.13) is the Δ of the heats of formation, that is, it is the heat of reaction at 25°C.

Now let us use these data that we have generated to evaluate Q in three different ways.

Method #1
In this approach, we use a component energy balance and the reference temperature of 500°C:

$$T_{ref} = 500°C$$

TABLE E8.13

Heat Capacity of Inlet and Exit Gases

Components	a	$b \times 100$	$c \times 10^5$	$d \times 10^9$	ΔH_f^0 (25°C)
CO_2	36.11	4.233	−2.887	7.464	−393.5
H_2	28.84	0.00765	0.3288	−0.8698	0
H_2O	33.46	0.6880	0.7604	−3.593	−241.83
CH_4	34.31	5.469	0.3661	−11.00	−74.85
Δ	−50.24	2.581	3.46	−22.2	−165

Note: $\Delta = CH_4 + 2H_2O - 4H_2 - CO_2$.

This requires that we know the heat of reaction at 500°C. The energy balance for this process includes the sensible heat to change the temperature of everything in the inlet stream from 400°C to 500°C and the heat of reaction at 500°C times the extent of reaction. Recall that the heat of reaction was expressed in kilojoules per mole, where the mole basis was per molar extent of reaction. Thus,

$$Q = \sum_{out} n_i \hat{H}_i \Big|_{@500°C} - \sum_{in} n_i^0 \hat{H}_i \Big|_{@400°C} + \xi \Delta \hat{H}_r \ (@T_{ref} = 500°C)$$

$$= \int_{500°C}^{500°C} \sum_i n_i C_{p_i} \ dT - \int_{500°C}^{400°C} \sum_i n_i^0 C_{p_i} \ dT + \xi \Delta \hat{H}_r \ (500°C)$$

$$Q = 0 + \int_{400°C}^{500°C} \sum_i n_i^0 C_{p_i} dT + \xi, \ \Delta \hat{H}_r \ (500°C)$$

But the heat of reaction at 500°C must be calculated. This is done, as we have seen, by using the heat of reaction at 25°C and the Δ heat capacity to change the reaction temperature to 500°C, i.e.,

$$\Delta \hat{H}_r (500°C) = \Delta \hat{H}_r (25°C) + \int_{25}^{500} \Delta C_p \ dT$$

$$\Delta \hat{H}_r \ (500°C) = -165 \ \text{kJ/mol} + [(-50.24)(475) + (1/2)(2.581 \times 10^{-2})(500^2 - 25^2)$$
$$+ (1/3)(3.46 \times 10^{-5})(500^3 - 25^3) - (1/4)(22.2 \times 10^{-9})(500^4 - 25^4)] \ \text{J/mol}$$
$$= -184.8 \ \text{kJ/mol}$$

Now the sensible heat term (first terms shown in the equation above) includes only the moles of those compounds in the inlet stream. This gives

$$\int_{400°C}^{500°C} \sum_i n_i^0 C_{p_i} \ dT = (1 \ \text{mol}) \int_{400}^{500} C_{p \ CO_2} + (400) \int_{400}^{500} C_{p \ H_2} \ dT = 4.998 \ \text{kJ} + 9.98 \ \text{kJ}$$

$$= 15.00 \ \text{kJ}$$

The extent of reaction is given by

$$x = 0.8 = \frac{(n_{CO_2,in} - n_{CO_2,out})}{n_{CO_2,in}} = \frac{\xi}{1 \ \text{mol}} \Rightarrow \xi = 0.8 \ \text{mol}$$

So finally, we obtain

$$Q = 15.00 \ \text{kJ} + (0.8 \ \text{mol})(-184.6 \ \text{kJ/mol}) = -131 \ \text{kJ}$$

Method #2

In this approach, we use a component energy balance and reference temperature of 25°C. This requires that we know the heat of reaction at 25°C, which we obtain from the heats of formation. The energy balance for this process includes the sensible heat to change the temperature of everything in the inlet stream from 400°C to 25°C and also to change everything in the product stream from 25°C to 500°C. Again, the heat of reaction at 25°C must be multiplied by the extent of reaction. Thus,

$$Q = \sum_{\text{out}} n_i \widehat{H}_i \bigg|_{@500°C} - \sum_{\text{in}} n_i^0 \widehat{H}_i \bigg|_{@400°C} + \xi \Delta \widehat{H}_r \ (@T_{\text{ref}} = 500°C)$$

$$= \int_{25°C}^{500°C} \sum_{\text{out}} n_i C_{p_i} \, dT - \int_{25°C}^{400°C} \sum_{\text{in}} n_i^0 C_{p_i} \, dT + \xi \Delta \widehat{H}_r(25°C)$$

The first term in this equation, the cooling of everything in the inlet stream to 25°C, is simply

$$\int_{25°C}^{400°C} \sum_i n_i^0 C_{p_i} \, dT = (1 \text{ mol}) \int_{25}^{400} C_{p\,CO_2} \, dT + (4 \text{ mol}) \int_{25}^{400} C_{p\,H_2} \, dT$$

$$= 16.35 \text{ kJ} + 43.54 \text{ kJ} = 59.89 \text{ kJ}$$

The heat of reaction term at 25°C is found from the Δ term in the table for the heats of formation. Likewise, we have already found in method 1 that $\xi = 0.8$ mol. Thus, the reaction term is

$$\xi \, \Delta H_r = (0.8 \text{mol})(-165 \text{ kJ/mol}) = -132 \text{ kJ}$$

Then we must calculate the sensible heat term for heating everything in the outlet stream from 25°C to 500°C. Notice that we still have some unused reactants, CO_2 and H_2, that end up in the outlet stream. Do not forget to include the sensible heat of everything that appears in the outlet stream:

$$\int_{25°C}^{500°C} \sum_i n_i C_{p_i} \, dT = (0.2 \text{ mol}) \int_{25}^{500} C_{p\,CO_2} \, dT + (0.8 \text{ mol}) \int_{25}^{500} C_{p\,H_2} \, dT$$

$$+ (0.8 \text{ mol}) \int_{25}^{500} C_{p\,CH_4} \, dT + (1.6 \text{ mol}) \int_{25}^{500} C_{p\,H_2O} \, dT$$

$$\int_{25°C}^{500°C} \sum_i n_i C_{p_i} \, dT = (4.27 + 11.06 + 18.48 + 27.22) \text{ kJ} = 61.03 \text{ kJ}$$

So finally, we obtain $Q = \sum_{\text{out}} n_i \widehat{H}_i - \sum_{\text{in}} n_i^0 \widehat{H}_i^0 + \xi \Delta H_r^0$

$$Q = 61.03 \text{ kJ} - 59.89 \text{ kJ} - 132 \text{ kJ} = -131 \text{ kJ}$$

Method #3

In this approach, we will use an element energy balance so that we find the enthalpy of all of the compounds relative to the elements at 25°C. We then plug these enthalpies directly into the energy balance expression. In this case, no heat of reaction is needed to be calculated at all. We write the energy balance as

$$Q = \sum_{\text{out}} n_i \widehat{H}_i - \sum_{\text{in}} n_i^0 \widehat{H}_i^0$$

Now we compute the enthalpy of each component in each stream relative to their elements. That is, we form the compound from its elements at 25°C (this is the heat of formation) and then we raise the temperature of the compound up to the temperature of the stream. Thus,

Outlet terms

$$\widehat{H}_{CH_4} = \Delta \widehat{H}_f^0(25) + \int_{25}^{500} C_{p\,CH_4} \, dT = -74.85 + 23.10 = -51.75 \, \frac{\text{kJ}}{\text{mol}}$$

$$\widehat{H}_{H_2O} = \Delta \widehat{H}_f^0(25) + \int_{25}^{500} C_{p\,H_2O} \, dT = -241.83 + 17.01 = -224.8 \, \frac{\text{kJ}}{\text{mol}}$$

$$\widehat{H}_{CO_2} = \Delta \widehat{H}_f^0(25) + \int_{25}^{500} C_{p\,CO_2} \, dT = -393.5 + 21.34 = -372.2 \, \frac{\text{kJ}}{\text{mol}}$$

$$\widehat{H}_{H_2} = 0 + \int_{25}^{500} C_{p\,H_2} \, dT = -13.83 \, \frac{\text{kJ}}{\text{mol}}$$

Thus, the sum of all of the outlet enthalpies is

$$\sum H_{\text{out}} = (0.2)(-372.2) + (0.8)(13.83) + (0.8)(-51.75) + (1.6)(-224.8)$$
$$= -464.5 \text{ kJ}$$

Inlet terms

$$\widehat{H}^0_{CO_2} = \Delta \widehat{H}^0_f(25) + \int_{25}^{400} C_{p\,CO_2}\, dT = -393.5 + 16.35 = -377.2 \frac{kJ}{mol}$$

$$\widehat{H}^0_{H_2} = 0 + \int_{25}^{400} C_{p\,H_2}\, dT = 10.89 \frac{kJ}{mol}$$

Thus, the sum of all of the inlet enthalpies is

$$\sum H_{in} = (1)(-377.2) + (4)(10.89) = -333.6 \text{ kJ}$$

Finally, from the energy balance we obtain

$$Q = \sum H_{out} - \sum H_{in} = -464.5 \text{ kJ} + 333.6 \text{ kJ} = -131 \text{ kJ}$$

Conclusions

Obviously all three methods are equivalent. There are times when you know a heat of reaction at some particular temperature and the extent of reaction, making the component energy balance convenient, and there are times when you have only a few different components, but multiple equations, making the element balances more convenient. It is really a matter of convenience, but you will understand the concepts much better if you see how all three of these approaches are really equivalent, and are comfortable with using both component and element balances.

8.13 Energy Balance in Bioprocesses

Bioprocesses contributions to sensible heat amount are insignificant compared with the total magnitude of ΔH^0_r and can be ignored without much loss of accuracy. This situation is typical of most reactions in bioprocessing where the actual temperature of reaction is not sufficiently different from 25°C.

Example 8.14 Fermentation and citric acid production

Citric acid is manufactured using a submerged culture of *Aspergillus niger* in a batch reactor operated at 30°C. Over a period of 2 days, 2500 kg glucose and 860 kg oxygen are consumed to produce 1500 kg citric acid, 500 kg biomass, and other products. Ammonia is used as a nitrogen source. Power input to the system by mechanical agitation of the broth is about 15 kW; approximately 100 kg water is evaporated during the culture period. Estimate the cooling requirements. The latent heat of evaporation of water at 30°C is 2430.7 kJ/kg. The heat of reaction at 30°C is -460 kJ/(g mol) O_2 consumed.

Solution

The reaction that takes place in the current fermentation process (Figure E8.14) follows the reaction:

$$\text{Glucose} + O_2 + NH_3 \rightarrow \text{biomass} + CO_2 + H_2O + \text{citric acid}$$

All components are involved in the reaction; the mass balance need not be completed as the sensible energy associated with inlet and outlet streams is negligible. The integral energy balance used for batch culture is used to calculate the amount of heat that must be removed to produce zero accumulation of energy in the system.

Basis: 1500 kg citric acid produced

$$\xi\Delta H_{reaction} + m_v\Delta H_v = Q - W_s$$

$$\xi\Delta H_{reaction} = 860 \text{ kg } O_2 \text{ consumed} \left|\frac{1000 \text{ g}}{1 \text{ kg}}\right|\frac{1 \text{ mol}}{32 \text{ g}} = -1.24 \times 10^7 \text{ kJ}$$

$$m_v\Delta H_v = (100 \text{ kg})(2430.7 \text{ kJ/kg}) = 2.43 \times 10^5 \text{ kJ}$$

$$W_s = (-15 \text{ kJ/s})(2 \text{ days of operation})\frac{3600 \text{ s}}{1 \text{ h}}\left|\frac{24 \text{ h}}{1 \text{ day}}\right| = -2.59 \times 10^6 \text{ kJ}$$

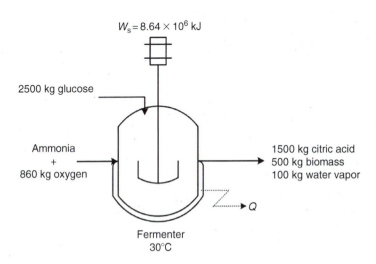

$W_s = 8.64 \times 10^6$ kJ

2500 kg glucose

Ammonia
+
860 kg oxygen

1500 kg citric acid
500 kg biomass
100 kg water vapor

Q

Fermenter
30°C

FIGURE E8.14
Flow sheet for microbial production of citric acid.

$$\xi\Delta H_{rxn} + m_v \Delta H_v = Q - W_s$$

$$-1.24 \times 10^7 \text{ kJ} + 2.43 \times 10^5 \text{ kJ} = Q - (-2.59 \times 10^6 \text{ kJ})$$

$$Q = -1.47 \times 10^7 \text{ kJ}$$

Q is negative, indicating that heat is removed from the system.

8.14 Problems

8.14.1 Estimation of Heat of Reaction

Given the following heats of formation (ΔH_f^0, 25°C) and heats of combustion (ΔH_c^0, 25°C) data, determine the heat of reaction for the liquid-phase esterification of lactic acid ($C_3H_6O_3$) with ethanol (C_2H_5OH) to form ethyl lactate ($C_5H_{10}O_3$) and liquid water at 25°C. The following table shows the standard heat of combustion and heat of reaction:

Compound	ΔH_c^0 (25°C, 1 atm) (kJ/mol)	ΔH_c^0 (25°C) (kJ/mol)
$C_3H_6O_3$	—	−687.0
C_2H_5OH	—	−277.6
$C_5H_{10}O_3$	−2685	—
H_2O (liquid)		−285.8
CO_2 (gas)		−393.5
O_2 (gas)	—	0

The standard states of products are CO_2 (g) and H_2O (l).
   (Answer: heat of reaction $= -32.7$ KJ/mol)

8.14.2 Production of Superheated Steam

Superheated steam (40 bar, 350°C) is produced from liquid water (40 bar, 50°C) in a methane-fired boiler. To insure complete combustion of the methane 1% excess air is provided. Both methane and combustion air enter the boiler at 25°C. Determine the outlet temperature of the flue gas from the boiler, if 19.85 kg/min of superheated steam is produced from the combustion of 1.4 kg/min of methane. Assume the boiler is perfectly insulated.

   (Answer: The outlet temperature $= 360.158$°C)

8.14.3 Ammonia Synthesis Process

Ammonia is synthesized through the reaction of nitrogen with hydrogen as follows:

$$N_2 + 3H_2 \rightarrow 2NH_3$$

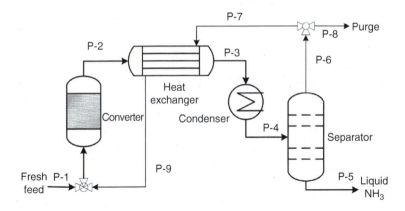

FIGURE P8.14.3
Process flow sheet for the ammonia synthesis process.

Figure P8.14.3 shows a process flow sheet for the ammonia synthesis process. In this process, the fresh feed consists of argon (1 mol%) and stoichiometric amounts of N_2 and H_2. The reactor feed has a molar flow rate of 100 mol/min and a composition of 15 mol% argon, 21.25 mol% N_2, and 63.75 mol% H_2. The reactor feed temperature is 400°C. The fractional conversion of N_2 to NH_3 in the reactor is 0.15 mol N_2 reacted/mol N_2 feed to the reactor. The hot reactor effluent gas is used to heat the recycle gas from the separator in a combined reactor effluent/recycle heat exchanger. After passing through this heat exchanger, the reactor effluent gas passes through a condenser where the NH_3 product is condensed. The liquid NH_3 is separated from the noncondensable recycle gases. A purge stream is taken off the separator off-gas to maintain the level of argon at 15 mol% feed to the reactor. In this particular process, the converter is operated adiabatically and the heat of reaction at 400°C was found to be -53.109 kJ/mol at the pressure of the reactor. The following table gives the heat capacities at the pressure of the reactor. Note that the given heat capacities are assumed to be constant over the temperature range found in the reactor:

Compound	C_p (J/mol°C)
NH_3	49.4
H_2	29.5
N_2	31.0
Argon	20.8

Determine the flow rates in moles per minute and compositions in mole percent of

(a) Fresh feed stream. (Answer: flow rate = 12.5 mol/min)
(b) Separator purge gas stream. (Answer: Purge = 0.74 mol/min)

(c) Recycle gas stream. (Answer: R = 87.5 mol/min)

(d) Estimate the temperature of the effluent gases from the converter. (Answer: T = 460.776 K)

(e) If the recycle gas stream enters the heat exchanger at 50°C and leaves the exchanger at 400°C, determine the outlet temperature of the reactor effluent stream from the heat exchanger. Assume no condensation of ammonia in the heat exchanger. (Answer: T = 242.24°C)

8.14.4 Catalytic Transalkylation of Toluene to Benzene

A process flow sheet for the catalytic transalkylation of toluene to benzene and xylene is shown in Figure P8.14.4. Toluene transalkylates in the presence of hydrogen to form benzene and xylene.

$$2C_7H_8 \rightarrow C_6H_6 + C_8H_{10}$$

Toluene also may dealkylate in the reactor to form benzene and methane.

$$C_7H_8 + H_2 \rightarrow C_6H_6 + CH_4$$

In this process, toluene reacts with a fractional conversion of 0.80 (moles of toluene reacted/moles of toluene fed to the unit) resulting in benzene and xylene yields of 0.505 and 0.495, respectively. Yields are defined as moles of product/moles of toluene reacted. In this process, the reactor effluent is condensed and separated. The separator liquid is sent to a benzene distillation column where benzene with a purity of 99.5 mol% is removed as the overhead product. The bottom products contain benzene, unreacted toluene, and xylene. A purge stream containing hydrogen and methane is taken off the separator gas stream. The rest of this separator gas stream is combined with a fresh hydrogen stream containing hydrogen and methane with concentrations of 95.0 and 5.0 mol%, respectively. These combined hydrogen streams are added to the fresh toluene feed and sent to the reactor heater.

(a) If 225 kmol/h of toluene is fed to the reactor, and the benzene column recovers 99.0 of the benzene at a purity of 99.5 mol% (balance of overhead is toluene), determine the molar flow rates of each component in the benzene column overhead and bottom product streams (Figure P8.14.4).

(b) Determine the molar flow rates of hydrogen and methane in the fresh hydrogen stream and molar flow rates of hydrogen and methane in the separator off-gas purge stream to maintain a hydrogen purity of 90.0 mol% in the separator off-gas stream.

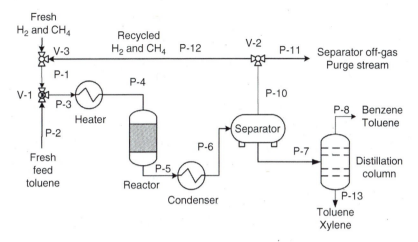

FIGURE P8.14.4
Process flow sheet of Problem 8.5.

The fresh hydrogen stream contains hydrogen and methane with concentrations of 95.0 and 5.0 mol%, respectively.

(c) If the liquid toluene fresh feed stream, fresh hydrogen stream, and recycled streams are all at 25°C and 15 bar absolute, determine the heat requirements for the reactor heater to provide the reactor with a combined feed at 400°C and 15 bar absolute in the vapor phase. It is desired to have a hydrogen to toluene molar ratio of 3 to the reactor. Assume ideal gases and no pressure effects.

8.14.5 Combustion of Methane

Methane at 25°C is burned in a boiler furnace with 10.0% excess air preheated to 100°C. Ninety percent of the methane fed is consumed, the product gas contains 10.0 mol CO_2/mol CO, and the combustion products leave the furnace at 400°C. Calculate the heat transferred from the furnace for a basis of 100 mol CH_4 fed per second (Figure P8.14.5). (Answer: $Q = -204$ kW.)

8.14.6 Anaerobic Yeast Fermentation

Saccharomyces cerevisiae is grown anaerobically in continuous culture at 30°C. Glucose is used as carbon source and ammonia is the nitrogen source. A mixture of glycerol and ethanol is produced. Mass flow streams to and from the reactor at steady state are shown in Figure P8.14.6. Estimate the cooling requirement. (Answer: $Q = -1.392 \times 10^4$ kJ.)

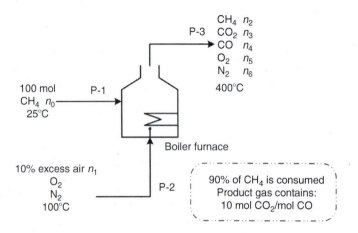

FIGURE P8.14.5
Methane burned in a boiler furnace.

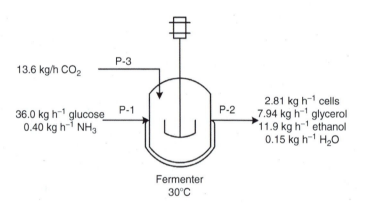

FIGURE P8.14.6
Flow sheet for anaerobic yeast fermentation.

Further Readings

1. Felder, R.M. and R.W. Rousseau (1999) *Elementary Principles of Chemical Processes*, 3rd edn, John Wiley, New York.
2. Reklaitis, G.V. (1983) *Introduction to Material and Energy Balances*, John Wiley & Sons, New York.
3. Himmelblau, D.M. (1996) *Basic Principles and Calculations in Chemical Engineering*, 6th edn, Prentice-Hall, New Jersey.
4. Whirwell, J.C. and R.K. Toner (1969) *Conservation of Mass and Energy*, Blaisdell, Waltham, MA.

5. Atkinson, B. and F. Mavituna (1991) *Biochemical Engineering and Biotechnology Handbook*, 2nd edn, Macmillan, Basingstoke.
6. Doran, P. (1995) *Bioprocess Engineering Principles*, Academic Press, San Diago, CA.
7. Bailey, J.E. and D.F. Ollis (1986) *Biochemical Engineering Fundamentals*, 2nd edn, McGraw-Hill, Boston, MA.

9

Combined Material and Energy Balances

At the End of This Chapter You Should Be Able to

1. Understand the basic definitions needed in solving material and energy balance problems, for example, conversion, yield, extent of reaction, and standard heat of reaction.
2. Develop steady-state material and energy balance equations.
3. Solve simultaneously the developed material and energy balance equations.
4. Explain the rational change in concentration versus time.

9.1 Material Balances

The general material balance equation takes the form

$$\text{Accumulation} = \text{in} - \text{out} + \text{generation} - \text{consumption}$$

There are frequently used definitions which should be known when solving material and energy balance problems involving chemical reactions. As the definitions include conversion, yield, selectivity, and extent of reaction these are briefly explained in subsequent sections.

9.1.1 Conversion

Generally, syntheses of chemical products do not involve a single reaction but rather multiple reactions. The purpose, in this case, is to maximize the production of the desirable product and minimize the production of unwanted by-products.

Conversion is the ratio of the moles that react to the moles that are fed to a reactor. Relative to species (i), the fractional conversion can be calculated using the following equation:

{Fractional conversion of component i}

$$= \frac{\{\text{moles of component } i\}_{\text{in}} - \{\text{moles of component } i\}_{\text{out}}}{\text{moles of component } i \text{ in}} \qquad (9.1)$$

$$x_i = \frac{n_{i0} - n_i}{n_{i0}}$$

9.1.2   Yield

The yield of a reaction is the ratio of the desired product formed (in moles) to the total amount that could have been produced if conversion of the limiting reactant was completed (i.e., 100%) and no side reactions occurred.

$$\text{Yield} = \frac{\text{moles of desired product formed}}{\text{moles formed if there were no side reactions and limiting reactant reacted completely}} \qquad (9.2)$$

9.1.3   Selectivity

The selectivity of a reaction is the ratio of the desired product formed (in moles) to the undesired product formed (in moles):

$$\text{Selectivity} = \frac{\text{moles of desired product formed}}{\text{moles of undesired product formed}} \qquad (9.3)$$

9.1.4   Extent of Reaction (ξ)

The concept of extent of reaction can also be applied to multiple reactions, with each reaction having its own extent. The extent of reaction is the amount in moles (or molar flow rate) that is converted in a given reaction (we used ξ in Chapter 5). If a set of reactions take place in a batch or continuous steady-state reactor, we can write

$$n_i = n_{i0} + \sum_j \nu_{ij}\xi_j \qquad (9.4)$$

where
  ν_{ij} is the stoichiometric coefficient of substance i in reaction j
  ξ_j is the extent of reaction for reaction j

For a single reaction, the above equation reduces to the following equation:

$$\dot{n}_i = \dot{n}_{i0} + \nu_i \dot{\xi} \qquad (9.5)$$

9.2 Energy Balances

The general energy balance equation for an open system at steady state is as follows:

$$\dot{Q} - \dot{W}_s = \Delta \dot{H} + \Delta \dot{E}_{KE} + \Delta \dot{E}_{PE} \tag{9.6}$$

$$\Delta \dot{E}_{KE} = \tfrac{1}{2}\dot{m}(v_2^2 - v_1^2) \tag{9.7}$$

$$\Delta \dot{E}_{PE} = \dot{m}g(z_2 - z_1) \tag{9.8}$$

Methods differ in reference state (and thus in the calculation of $\Delta\dot{H}$).

9.2.1 Heat of Reaction Method

In this method the reference state is such that the reactants and products are at 25°C and 1 atm.

$$\Delta \dot{H} = \sum_{\text{reactions}} \dot{\xi}_j \Delta \widehat{H}_{rj}^0 + \sum_{\text{out}} \dot{n}_i \widehat{H}_i - \sum_{\text{in}} \dot{n}_i \widehat{H}_i \tag{9.9}$$

$$\widehat{H}_i\big|_{@T} = \int_{T_{\text{ref}}}^{T} \widehat{C}_{p_i}\, dT \tag{9.10}$$

$$\Delta H_{rj}^0 = \sum_{i}^{n} v_i \Delta H_f^0 \tag{9.11}$$

9.2.2 Heat of Formation Method

In this method the reference state is the elemental species that constitutes the reactants and products in the states they occur in nature at 25°C and 1 atm.

$$\Delta \dot{H} = \sum_{\text{out}} \dot{n}_i \widehat{H}_i - \sum_{\text{in}} \dot{n}_i \widehat{H}_i \tag{9.12}$$

The enthalpy in this case includes the sensible heat and the enthalpy of formation.

$$\widehat{H}_i\big|_{@T} = \Delta H_f^0 + \int_{T_{\text{ref}}}^{T} \widehat{C}_{p_i}\, dT \tag{9.13}$$

9.2.3 Concept of Atomic Balances

Consider the reaction of hydrogen with oxygen to form water.

$$H_2 + O_2 \rightarrow H_2O$$

We may attempt to do our calculations with this reaction, but there is something seriously wrong with this equation. It is not balanced; as written, it implies that an atom of oxygen is somehow "lost" in the reaction, but this is in general impossible. Therefore, we must compensate by writing:

$$H_2 + \tfrac{1}{2}O_2 \rightarrow H_2O$$

or some multiple thereof. Notice that in so doing we have made use of the conservation law, which is actually the basis of the conservation of mass: The number of atoms of any given element does not change in any reaction (assuming that it is not a nuclear reaction).

9.2.4 Mathematical Formulation of the Atom Balance

Now recall the general balance equation.

$$\text{In} - \text{out} + \text{generation} - \text{consumption} = \text{accumulation}$$

Moles of atoms of any element are conserved, therefore generation $= 0$. So we have the following balance on a given element A:

$$\sum \dot{n}_{A,\text{in}} - \sum \dot{n}_{A,\text{out}} = 0 \qquad (9.14)$$

Note: When analyzing a reacting system you must choose either an atom balance or a molecular species balance but not both. An atomic balance often yields simpler algebra, but also will not directly tell you the extents of reaction, and will not tell you whether the system specifications are actually impossible to achieve for a given set of equilibrium reactions.

9.2.5 Degree of Freedom Analysis for the Atom Balance

As before, to do a degree of freedom analysis, it is necessary to count the number of unknowns and the number of equations one can write, and then subtract them. However, there are a couple of important things to be aware of with these balances:

When doing atom balances, the extent of reaction does not count as an unknown, while with a molecular species balance it does. This is the primary advantage of this method. The extent of reaction does not matter since atoms of elements are conserved regardless of how far the reaction has proceeded.

When doing an atom balance, only reactive species are included, and not inerts.

Example 9.1   Natural gas burner

Suppose a mixture of nitrous oxide (N_2O) and oxygen is used in a natural gas burner. The following chemical reaction occurs in it. How many atomic balance equations can you write?

$$CH_4 + 2O_2 \rightarrow 2H_2O + CO_2$$

Solution

There would be four equations that you can write: three atom balances (C, H, and O) and a molecular balance on nitrous oxide. You would not include the moles of nitrous oxide in the atom balance on oxygen.

Example 9.2   Equilibrium reactions

Suppose that you are working in an organic chemistry laboratory in which 10 kg of compound A is added to 100 kg of 16% aqueous solution of B (which has a density of 57 lb/ft^3). The following reaction occurs:

$$A + 2B \rightleftharpoons 3C + D$$

A has a molar mass of 25 g/mol and B has a molar mass of 47 g/mol. If the equilibrium constant, K, for this reaction is 187 at 298 K, how much of compound C could you obtain from this reaction? Assume that all products and reactants are soluble in water at the design conditions. Adding 10 kg of A to the solution causes the volume to increase by 5 L. Assume that the volume does not change over the course of the reaction.

Solution

The process flowchart is shown in Figure E9.2.

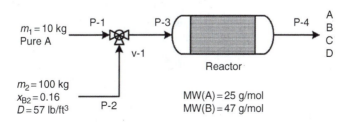

FIGURE E9.2
Equilibrium reaction block diagram.

Since all of the species are dissolved in water, we should write the equilibrium constant in terms of molarities:

$$K = 187 = \frac{[C]^3[D]}{[A][B]^2}$$

First find the number of moles of A and B we have initially.

$$n_{A0} = 10 \text{ kg A} \times \frac{1 \text{ mol A}}{25 \text{ g A}} \left| \frac{1000 \text{ g}}{1 \text{ kg}} \right. = 400 \text{ mol A}$$

$$n_{B0} = 100 \text{ kg solution} \times \frac{0.16 \text{ kg B}}{\text{kg s ln}}$$

$$= 16 \text{ kg B} \times \frac{1 \text{ mol B}}{47 \text{ g B}} \left| \frac{1000 \text{ g}}{1 \text{ kg}} \right. = 340.43 \text{ mol B}$$

Now, the volume contributed by 100 kg of 16% B solution is

$$V = \frac{m}{\rho} = \frac{100 \text{ kg}}{57 \frac{\text{lb}}{\text{ft}^3} \times \frac{1 \text{ kg}}{2.2 \text{ lb}} \times \frac{1 \text{ ft}^3}{28.317 \text{ L}}} = 109.3 \text{ L}$$

Since adding A contributes 5 L to the volume, the volume after the two is mixed is 109.3 L.

$$L + 5 \text{ L} = 114.3 \text{ L}$$

By definition then, the molarities of A and B before the reaction occurs are

$$[A]_0 = \frac{400 \text{ mol A}}{114.3 \text{ L}} = 3.500 \text{ M}$$

$$[B]_0 = \frac{340.42 \text{ mol B}}{114.3 \text{ L}} = 2.978 \text{ M}$$

In addition, there is no C or D in the solution initially.

$$[C]_0 = [D]_0 = 0$$

According to the stoichiometry of the reaction, $a = 1$, $b = 2$, $c = 3$, and $d = 1$. Therefore, we now have enough information to solve for the conversion. Plugging all the known values into the equilibrium equation for liquids, the following equation is obtained:

$$187 = \frac{\left(\dfrac{3\xi}{114.3}\right)^3 \left(\dfrac{\xi}{114.3}\right)}{\left(3.5 - \dfrac{\xi}{114.3}\right)\left(2.978 - \dfrac{2\xi}{114.3}\right)^2}$$

This equation can be solved using E–Z solve or any other of the numerical methods available to yield

$$\xi = 146.31 \text{ mol}$$

Since we seek the amount of compound C that is produced, we have $\xi = \dfrac{\Delta n_C}{c}$.

Since $c = 3$, $n_{C0} = 0$, and $\xi = 146.31$, this yields $n_C = 3 \times 146.31 = 438.93$ mol C. A total of 438.93 mol of C can be produced by this reaction.

9.2.6 Implementing Recycle on the Separation Process

Recycle may improve reaction conversion enough to eliminate the need for a second reactor to achieve an economical conversion. Recycle reduces the amount of waste that a company generates. Not only it is the most environmentally sound way to go about but it also saves the company money in disposal costs. Using recycle, it is possible to recover expensive catalysts and reagents. Catalysts are not cheap, and if we do not try to recycle them into the reactor, they may be lost in the product stream. This not only gives us a contaminated product but also wastes a lot of catalyst.

Example 9.3 Separation of binary mixture with recycle
Consider the process flow sheet (see Figure E9.3), where a recycle system is set up in which half of stream number three is split and recombined with the feed (recycle stream is at the same composition as stream number three).

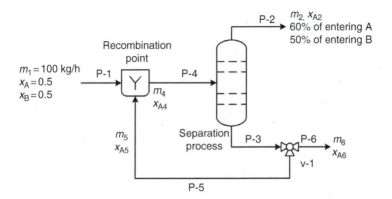

FIGURE E9.3
Flow sheet of a separation process with recycle.

Calculate the concentrations of A in streams two and three. Is separation more or less effective than without recycle? Can you see a major limitation of this method? How might this be overcome? This is a rather involved problem, and must be taken one step at a time. The analyses of the cases for recycling each stream are similar, so the first case will be considered in detail and the second will be left for the reader. You must be careful when drawing the flowchart because the separator separates 60% of all A that enters it into stream, 2, not 60% of the fresh feed stream, and 50% of all A entering the separation process goes into stream two (P-2).

Solution
The process flow sheet is shown in Figure E9.3.

Degree of freedom (DOF) analysis
Recombination point: 4 unknowns − 2 mass balances = 2 DOF.

Separator: 6 unknowns (nothing is specified) − 2 independent pieces of information − 2 mass balances = 2 DOF.

Splitting point: 6 unknowns (again, nothing is specified) − 2 mass balances − 1 assumption that concentration remains constant − 1 splitting ratio = 2 DOF.

Total = 2 + 2 + 2 − 6 = 0. Thus, the problem is completely specified.

Basis: 100 kg/h of fresh feed.

First, look at the entire system, since none of the original processes individually had DOF of 0.

Overall balance: 4 unknowns −2 equations = 2 DOF

Overall mass balance on A: $0.5 \times 100 \frac{\text{kg}}{\text{h}} = \dot{m}_2 \times x_{A2} + \dot{m}_6 \times x_{A6}$

Overall mass balance on B: $50 \frac{\text{kg}}{\text{h}} = \dot{m}_2 \times (1 - x_{A2}) + \dot{m}_6 \times (1 - x_{A6})$

We have four unknowns and two equations at this point. First, combine this information with the splitting ratio and constant concentration at the splitter.

Splitting ratio: $\dot{m}_6 = \frac{\dot{m}_3}{2}$

Constant concentration: $x_{A6} = x_{A3}$

Plugging these into the overall balances, we have

$$A: 50 = \dot{m}_2 \times x_{A2} + \frac{\dot{m}_3}{2} \times x_{A3}$$

$$B: 50 = \dot{m}_2 \times (1 - x_{A2}) + \frac{\dot{m}_3}{2} \times (1 - x_{A3})$$

Relations:

If 60% of A entering the separator goes into stream 2, then

$$\dot{m}_2 \times x_{A2} = 0.6 \times x_{A4} \times \dot{m}_4$$

If 40% of A entering the separator goes into stream 3, then

$$\dot{m}_3 \times x_{A3} = 0.4 \times x_{A4} \times \dot{m}_4$$

If 50% of B entering the separator goes into stream 2, then

$$\dot{m}_2 \times (1 - x_{A2}) = 0.5 \times (1 - x_{A4}) \times \dot{m}_4$$

If 50% of B entering the separator goes into stream 3, then

$$\dot{m}_3 \times (1 - x_{A3}) = 0.5 \times (1 - x_{A4}) \times \dot{m}_4$$

Translating words in the relations into algebraic equations and plugging in all of these into the existing balances, we finally obtain two equations in two unknowns:

$$\text{A: } 50 = 0.6 \times \dot{m}_4 \times x_{A4} + \frac{0.4}{2} \dot{m}_4 \times x_{A4}$$

$$\text{B: } 50 = 0.5 \dot{m}_4 \times (1 - x_{A4}) + \frac{0.5}{2} \dot{m}_4 \times (1 - x_{A4})$$

Solving these equations gives

$$\dot{m}_4 = 129.17 \frac{kg}{h}, \quad x_{A4} = 0.484$$

The mass balances on the separator can be solved using the same method as that without a recycle system. The results are

$$\dot{m}_2 = 70.83 \frac{kg}{h}, \quad x_{A2} = 0.530, \quad \dot{m}_3 = 58.33 \frac{kg}{h}, \quad x_{A3} = 0.429$$

Now since we know the flow rate of stream 3 and the splitting ratio we can find the rate of stream 6:

$$\dot{m}_6 = \frac{\dot{m}_3}{2} = 29.165 \frac{kg}{h}, \quad x_{A6} = x_{A3} = 0.429$$

Example 9.4   Methane oxidization

Methane and oxygen react in the presence of a catalyst to form formaldehyde. In a parallel reaction, methane is oxidized to carbon dioxide and water.

$$CH_4(g) + O_2(g) \rightarrow HCHO(g) + H_2O(g) \quad (1)$$
$$CH_4(g) + 2O_2(g) \rightarrow CO_2(g) + 2H_2O(g) \quad (2)$$

Methane and oxygen at 25°C are fed in stoichiometric amounts (according to the first reaction) to the reactor:

1. Calculate the standard heat of reaction for both reactions at 25°C.
2. Would you expect the final temperature to be higher or lower than 25°C, if the reaction took place in an insulated tank, and why?
3. If the product gases emerge at 400°C, and the mole fraction of CO_2 in the effluent gases is 0.15, and there is no remaining O_2, determine the composition of effluent gas per mole of CH_4 fed to the reactor.
4. Determine the amount of heat removed from the reactor per mole of CH_4 fed to the reactor.

Solution

The process flow sheet is shown in Figure E9.4.

1. Standard heat of reaction for both reactions at 25°C:

$$CH_4 (g) + O_2 (g) \rightarrow HCHO (g) + H_2O (g) \quad (1)$$
$$CH_4 (g) + 2O_2 (g) \rightarrow CO_2 (g) + 2H_2O (g) \quad (2)$$

$$\Delta H_{fCH_4} = -74.85 \text{ kJ/mol}$$

$$\Delta H_{fO_2} = 0$$

$$\Delta H_{fHCHO} = -115.9 \text{ kJ/mol}$$

$$\Delta H_{fH_2O} = 241.83 \text{ kJ/mol}$$

$$\Delta H_{fCO_2} = -393.5 \text{ kJ/mol}$$

$$\Delta H_{r_1}^0 = \Delta H_{f,H_2O}^0 (g) + \Delta H_{f,HCHO}^0 (g) - \Delta H_{f,CH_4}^0 (g) - \Delta H_{f,O_2}^0 (g)$$

$$\Delta H_{r_2}^0 = 2\Delta H_{f,H_2O}^0 (g) + \Delta H_{f,CO_2}^0 (g) - \Delta H_{f,CH_4}^0 (g) - 2^\Delta H_{f,O_2}^0 (g)$$

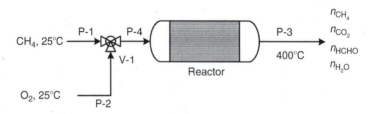

FIGURE E9.4
Block diagram of methane oxidation process.

$$\Delta H_{r_1}^0 = -115.9 - 241.83 - (-74.85) - (0) = -282.88 \ \frac{kJ}{mol}$$

$$\Delta H_{r_2}^0 = 2(-241.83) + (-393.5) - (-74.85) - 2(0) = -802.31 \frac{kJ}{mol}$$

2. Final temperature is higher than 25°C.

$$T_{final} > 25°C$$

Because the reaction is exothermic, it releases heat that will increase T of the system.

3. The composition of the exit gases is calculated using the extent of reaction method. There is enough information to complete the material balance.

Basis: 1 mol of CH_4 fed to the reactor

$$n_{CH_4} = 1 - \xi_1 - \xi_2, \ n_{O_2} = 1 - \xi_1 - 2\xi_2 = 0$$

$$n_{HCHO} = \xi_1, \ n_{CO_2} = \xi_2, \ n_{H_2O} = \xi_1 + 2\xi_2$$

Since $1 - \dot{\xi}_1 - 2\dot{\xi}_2 = 0 \Rightarrow n_{CO_2} = \dot{\xi}_2 = 0.15 \Rightarrow \dot{\xi}_1 = 0.7 \ mol/s$

$\dot{n}_{CH_4} = 0.15 \ mol/s, \ \dot{n}_{HCHO} = 0.7 \ mol/s, \ \dot{n}_{CO_2} = 0.15 \ mol/s$

$$\dot{n}_{H_2O} = 1 \ mol/s, \ total \ moles = 2$$

4. The amount of heat removed from the reactor per mole of CH_4 fed to the reactor. The heat capacities are taken from tabulated values as a function of temperature.

Reference temperature $= 25°C$

$$C_p \left(\frac{J}{mol°C} \right) = a + bT + cT^2 + dT^3$$

From the specific heat table (Table B.1)

HCHO: $a = 34.28, \ b = 4.268 \times 10^{-2}, \ c = 0.0, \ d = -8694 \times 10^{-9}$

CH$_4$: $a = 34.31, \ b = 5.469 \times 10^{-2}, \ c = 0.3661 \times 10^{-5}, \ d = -11 \times 10^{-9}$

CO$_2$: $a = 36.11, \ b = 4.233 \times 10^{-2}, \ c = -2.887 \times 10^{-5}, \ d = 7.464 \times 10^{-9}$

H$_2$O: $a = 33.46, \ b = 0.688 \times 10^{-2}, \ c = 0.7604 \times 10^{-5}, \ d = -3.593 \times 10^{-9}$

Calculation of enthalpies of outlet and inlet streams relative to the reference temperature (i.e., 25°C) is shown below.

Enthalpy of outlet stream components:

$$\widehat{H}_{HCHO} = \int_{25}^{400} C_p \, dT = \int (34.28 + 4.268 \times 10^{-2} T - 8694 \times 10^{-9} T^3) dT$$

$$\widehat{H}_{HCHO} = \int_{25}^{400} C_p \, dT = \left[34.28T + 4.268 \times 10^{-2} \frac{T^2}{2} - 3(8694 \times 10^{-9}) \frac{T^3}{4} \right]_{25}^{400}$$

$$\widehat{H}_{HCHO} = \int_{25}^{400} C_p \, dT$$

$$= 34.28(400 - 25) + 4.268 \times 10^{-2} \frac{(400^2 - 25^2)}{2} - 8694 \times 10^{-9}$$

$$\times \frac{(400^4 - 25^4)}{4}$$

Be careful: $(400^2 - 25^2) \neq (400 - 25)^2$

The results are

$$\widehat{H}_{HCHO} = 16.2 \frac{kJ}{mol}$$

$$\widehat{H}_{CH_4} = 17.23 \frac{kJ}{mol}$$

$$\widehat{H}_{CO_2} = 16.35 \frac{kJ}{mol}$$

$$\widehat{H}_{H_2O} = 1323 \frac{kJ}{mol}$$

$$Q = \sum_{out} \dot{n}_i \widehat{H}_i - \sum_{in} \dot{n}_i \widehat{H}_i + \xi_1 \Delta H^0_{rxn1} + \xi_2 \Delta H^0_{rxn2}$$

$$\dot{Q} = \left\{ \dot{n}_{CH_4}(\widehat{H}_{CH_4}) + \dot{n}_{CO_2}(\widehat{H}_{CO_2}) + \dot{n}_{HCHO}(\widehat{H}_{HCHO}) + \dot{n}_{H_2O}(\widehat{H}_{H_2O}) \right\}_{OUT@400°C}$$

$$- \left\{ \dot{n}_{CH_4}(\widehat{H}_{CH_4}) + \dot{n}_{CO_2}(\widehat{H}_{CO_2}) \right\}_{in@25°C} + \dot{\xi}_1 \Delta \widehat{H}^0_{rxn1} + \dot{\xi}_2 \Delta \widehat{H}^0_{rxn2}$$

The enthalpy of inlet stream components is zero because inlet temperature is at reference temperature; therefore, enthalpy of inlet components relative to reference temperature of 25°C is zero.

$$\dot{Q} = \{0.15(17.23) + 0.15(16.35) + 0.7(16.2) + 1(13.23)\}_{out} - \{0 + 0\}_{in}$$

$$+ 0.7(-282.88) + 0.15(-802.31)$$

$$= -288.68 \text{ kJ/s}$$

Example 9.5 Adiabatic saturation temperature
Air at 50°C (dry bulb) and 10% relative humidity is to be humidified adiabatically to 40% relative humidity. Use the psychrometric chart to estimate the adiabatic saturation temperature of the air.

Solution

$$T_{db} = 50°C, \ h_r = 10\%, \ T_{as} = T_{wb} = 23.6°C$$

$$h_a = 0.0077 \frac{\text{kg H}_2\text{O}}{\text{kg DA}}$$

Estimate the final temperature of the air and the rate at which water must be added to humidify 15 kg/min of the entering air, $T_{final} = 35°C$:

$$ha = 0.014 \text{ kg H}_2\text{O/kg DA}$$

H_2O added

$$\frac{15 \text{ kg air}}{\text{min}} \left| \frac{1 \text{ kg DA}}{1.0077 \text{ kg air}} \right| \frac{0.014 - 0.0077}{\text{kg DA}} \right| = 0.0938 \frac{\text{kg H}_2\text{O}}{\text{min}}$$

Example 9.6 Condensation of cyclopentane
A stream of pure cyclopentane vapor flowing at a rate of 1550 L/s at 150°C and 1 atm enters a cooler, also at 1 atm. Seventy-five percent of the feed is condensed and exits the cooler at 1 atm.

1. What is the temperature of the exiting streams from the cooler? (Hint: There are both vapor and liquid present at 1 atm.)
2. Prepare and fill in an inlet–outlet enthalpy table and calculate the required cooling rate in kilowatts.

Solution
The process flow diagram is shown in Figure E9.6.

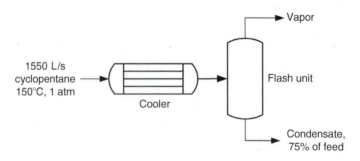

FIGURE E9.6
Condensation of cyclopentane.

1. The exit stream temperature must be 49.3°C for cyclopentane, because the boiling point of cyclopentane at 1 atm pressure is 49.3°C.

2. Taking the reference temperature at 49.3°C and cyclopentane in liquid state (i.e., $T_{ref} = 49.3$°C, liquid state), the number of moles of inlet cyclopentane in vapor phase, using the ideal gas law, is

$$n = \frac{PV}{RT} = \frac{1 \text{ atm} \times 1550 \frac{L}{s}}{0.08206 \frac{L \times atm}{mol\,K} \times (150 + 273.15)K} = 44.6 \; \frac{mol}{s}$$

The enthalpy of vapor cyclopentane at 150°C relative to reference temperature (49.3°C, liquid) is shown in the following equation:

$$\Delta \widehat{H}_v = \int_{49.3°C}^{150°C} C_p \, dT$$

Heat capacity of cyclopentane vapor (C_{pv}) as a function of temperature is shown below:

The constants are:

$$a = 73.39 \times 10^{-3},\ b = 39.28 \times 10^{-5},\ c = -25.54 \times 10^{-8},\ d = 68.66 \times 10^{-12}$$

$$C_p \left(\frac{kJ}{mol\,K} \right) = a + bT + cT^2 + dT^3$$

$$C_{p_v} \left(\frac{kJ}{mol\,K} \right) = 73.39 \times 10^{-3} + 39.28 \times 10^{-5}T - 25.54 \times 10^{-8}T^2$$
$$+ 68.66 \times 10^{-12}T^3$$

TABLE E9.6

Enthalpies Relative to Reference Conditions ($T_{ref} = 49.3°C$, liquid state, 1 atm)

Substance	$\dot{n}_{in}(mol/s)$	$\widehat{H}_{in}(kJ/mol)$	$\dot{n}_{out}(mol/s)$	$\widehat{H}_{out}(kJ/mol)$
Cyclopentane (V)	44.6	38.36	11.1	27.3
Cyclopentane (L)	—	—	33.5	0

$$\Delta\widehat{H}_v = \int_{49.3°C}^{150°C} C_p \, dT = \left[aT + \frac{b}{2}T^2 + \frac{c}{3}T^3 + \frac{d}{4}T^4\right]_{49.3°C}^{150} = 11.06\frac{kJ}{mol}$$

The heat of evaporation of cyclopentane and at 1 atm at its boiling point 49.3°C is

$$\Delta\widehat{H}_{vap} = 27.30\frac{kJ}{mol}$$

Since cyclopentane enters the condenser at 150°C in vapor phase, the enthalpy at this temperature relative to reference temperature (49.3°C) and liquid state is the summation of the latent heat of vaporization required to transfer the liquid phase to vapor phase and the heat required to raise the temperature of the vapor phase from reference temperature to 150°C.

$$\widehat{H}_{in} = \Delta\widehat{H}_v + \int_{49.3}^{150} C_{p_v} dT$$
$$= 27.3 + 11.06 = 38.36 \text{ kJ/mol}$$

The resultant enthalpies can be summarized in Table E9.6.
   Total energy balance: $Q = \sum n_{out}\widehat{H}_{out} - \sum n_{in}\widehat{H}_{in}$

$$= \left(11.1 \frac{mol}{s}\right)\left(27.30 \frac{kJ}{mol}\right) - 44.6 \frac{mol}{s}(38.36) \frac{kJ}{mol}$$

$$= -1407\frac{kJ}{s} = -1.4 \times 10^3 \text{ kW}$$

Example 9.7 Heating of liquid methanol (enthalpy of phase change)

Liquid methanol (CH_3OH, specific gravity $= 0.792$) is fed to a space heater at a rate of 12.0 L/h and burned with excess air. The product gas is analyzed and the following dry-basis mole percentages are determined: $CH_3OH = 0.45\%$, $CO_2 = 9.03\%$, and $CO = 1.81\%$. (Hint: You have here the exit mole fractions containing all the reacted carbon.)

(a) Write the equations of the chemical reactions taking place in the heater.

(b) Draw and label a flowchart and verify that the system has zero DOF.

(c) Calculate the fractional conversion of methanol, the percentage excess air fed, and the mole fraction of water in the product gas.

Solution

(a) The chemical reactions taking place are

$$CH_3OH + \tfrac{3}{2}O_2 \rightarrow CO_2 + 2H_2O \quad (1)$$
$$CH_3OH + O_2 \rightarrow CO + 3H_2O \quad (2)$$

(b) The labeled flowchart is shown in Figure E9.7a.
   Degree of Freedom Analysis (Atomic Balance)

Number of unknowns	4
Atomic balance equations (C, H, O)	3
Number of relations	1
DF	0

(c) The fractional conversion of methanol, the percentage excess air fed, and the mole fraction of water in the product gas are calculated by converting volumetric flow rate to molar flow rate.

$$\dot{m} = \dot{V} \times \rho$$

$$\dot{m} = 12\frac{lCH_3OH}{h} \times \left(0.79 \times 1000\ \frac{g}{L}\right) = 9504\ \frac{g}{h}\ \frac{1}{32.04\ g/mol} = 297\ \frac{mol}{s}$$

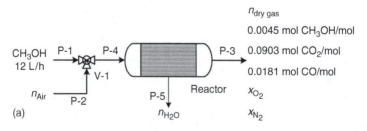

(a)

Number of moles of methanol (MW = 32.04 g/mol)

$$\dot{n} = 9504 \frac{g}{h} \left| \frac{1}{32.04 \text{ g/mol}} \right. = 297 \frac{\text{mol}}{\text{s}}$$

Atomic balance
Basis: 297 mol/s of inlet methanol
C balance:

$$n_{\text{dry gas}} = \frac{297 \text{ mol/h}}{0.1129} = 2627 \text{ mol/h}$$

$$297 \frac{\text{mol}}{\text{h}} = (0.0045 + 0.0903 + 0.0131)n_{\text{dry gas}}$$

H balance:

$$197 \text{ mol} \frac{CH_3OH}{h} \times \frac{4 \text{ mol H}}{\text{mol } CH_3OH}$$

$$= n_{H_2O} \times 2 + 0.0045 \text{ mol } CH_3OH \times \frac{4 \text{ mol H}}{\text{mol } CH_3OH} \times n_{\text{dry gas}} \ n_{H_2O}$$

$$= 569.2 \text{ mol/s}$$

Total moles out = 569.2 + 2627 = 3196 mol $\quad$ Fraction of water in the product stream = $\frac{569.2}{3196} = 0.178$

Fractional conversion of methanol = $x = \frac{297 - 0.0045 \times 2627}{297} = 0.96$

In calculating excess air
O balance:

$$2(0.21 \times n_{\text{Air}}) + 297 = (0.0045 + 2 \times 0.0903 + 0.0181 + 2 \times y_{O_2}) \times 2627 + 569.2$$

N balance:

$$2(0.79 \times n_{\text{Air}}) = 2 \times (1 - 0.0045 - 0.0903 - 0.0181 - y_{O_2}) \times 2627$$

The results using *E–Z* solve are shown in Figure E9.7b.

$$n_{\text{Air}} = 2733.41 \text{ mol/s}, \ y_{N_2} = 0.822, \ y_{O_2} = 0.065$$

To calculate the percent excess air, first calculate the theoretical oxygen using the complete combustion.

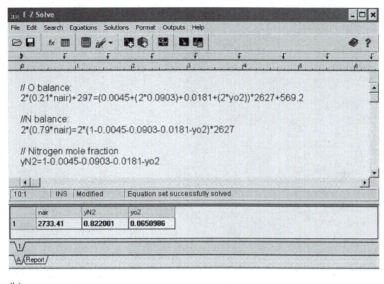

(b)

FIGURE E9.7b
Results using *E–Z* solve.

$$\text{Theoretical oxygen} = \frac{297 \text{ mol CH}_3\text{OH}}{s} \left| \frac{\frac{3}{2}\text{mol O}_2 \text{ conusmed}}{1 \text{ mol CH}_3\text{OH}} = 445.5\frac{\text{mol}}{s} \right.$$

$$\text{Theoretical air} = \frac{n_{O_2}}{0.21} = \frac{445.5 \text{ mol O}_2}{0.21} = 2121.43 \text{ mol air}$$

$$\% \text{ Excess air} = \frac{\text{total inlet air} - \text{theoretical air}}{\text{theoretical air}} = \frac{2733.41 - 2121.43}{2121.43} \times 100\%$$

$$= 28.85$$

Example 9.8 Methanol synthesis

Methanol (CH$_3$OH) is synthesized from carbon monoxide (CO) and hydrogen (H$_2$) in a catalytic reactor. The fresh feed to the process contains 31 mol% CO, 66 mol% H$_2$, and 3 mol% N$_2$. This stream is mixed with a recycle stream in a ratio of 4 mol recycle to 1 mol of fresh feed to enter the reactor (hint: $n_4 = 4\,\dot{n}_1$), and the stream entering the reactor contains 11 mol% N$_2$. The reactor effluent goes to a condenser from which two streams emerge: a liquid stream containing pure liquid CH$_3$OH and a gas stream containing all the CO, H$_2$, and N$_2$. The gas stream from the condenser is split into a purge stream and the remainder is recycled to mix with the fresh feed to enter the reactor. The process flowchart is shown in Figure E9.8.

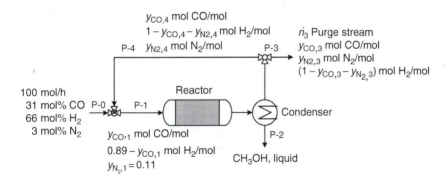

FIGURE E9.8
Flow sheet of methanol synthesis process.

1. Assuming a basis of 100 mol/h fresh feed, write the overall process degree of freedom analysis and the degree of freedom analysis around the mixing point.
2. Calculate the production rate of methanol (mol/h), the molar flow rate and composition of the purge gas, and the overall conversion of CO.

Solution
Basis: 100 mol/h of fresh feed
Degree of freedom analysis

Degree of Freedom Analysis	Overall Process (Component Balance)	Mixing Point (Atomic Balance)
Number of unknowns	4 ($\dot{n}_2, \dot{n}_3, y_{CO}, y_{N_2}$)	5 ($\dot{n}_1, y_{CO,1}, \dot{n}_4, y_{N_2,4}, y_{CO,4}$)
Number of reactions	1	0
Number of equations	4	4
Number of relations	—	1
DF	1	0

System: Mixing point balances

From the relation: recycle stream = 4 times of the fresh feed stream, $n_4 = 400$, $n_1 = 500$
N_2 balance:

$$3 + y_{N_2,4} \times 400 = 0.11 \times 500 \Rightarrow y_{N_2,4} = 0.13$$

System: Overall balance
N_2 balance: $3\ \text{mol} = y_{N_2}, 3 \times n_3$

Since the composition of recycle stream and purge stream are the same

$$y_{N_2,3} = y_{N_2,4} = 0.13$$

The flow rate of purge: $n_3 = 23$ mol/h
Atomic C balance:

$$31 \text{ mol CO} \times 1 = n_2 \text{ mol CH}_3\text{OH} + y_{CO,3} \times 23$$

$$n_2 = 31 - 23 \times y_{CO,3}$$

Atomic H balance:

$$(66 \text{ mol H}_2 \times 2)_{\text{fresh feed}}$$
$$= \left(n_2 \text{ mol CH}_3\text{OH} \times \frac{4 \text{ mol H}}{\text{mol CH}_3\text{OH}} \right)_{\text{product}}$$
$$+ \left((1 - 0.13 - y_{CO,3}) \times \frac{2 \text{ mol H}}{\text{mol H}_2\text{O}} \times 23 \text{ mol purge} \right)_{\text{purge}}$$

Substituting $n_2 = 31 - 23 y_{CO}$ in the above equation yields

$$\Rightarrow 132 = (31 - 23 y_{CO,3}) \times 4 + 22.62 - 46 y_{CO,3}$$

$$-14.6 = -138 \times y_{CO,3}$$

$$y_{CO} = 0.106$$

$$y_{H_2,3} = 0.87 - 0.106 = 0.76 \frac{\text{mol H}_2}{\text{mol}}$$

$$n_2 = 28.6 \text{ mol CH}_3\text{OH}$$

$$\text{Overall conversion} = \frac{31 \text{ mol} - 0.106 \times 23}{31} = 0.92 \Rightarrow 92\% \text{ conversion of CO}$$

Example 9.9   Heating of propane gas

Propane gas at 40°C and 250 kPa enters a continuous adiabatic heat exchanger (no heat is lost from the outside of the unit—i.e., it is insulated on the outside). The stream exits at 240°C. The flow rate of propane is 100 mol/min, and superheated steam at 5 bar absolute pressure and 300°C enters the heat exchanger with a flow rate of 6 kg/min. The steam exits the heat exchanger at 1 bar absolute pressure. Calculate the temperature of the exit steam.

Solution

The process flow sheet is shown in Figure E9.9.
For a heat capacity of propane:

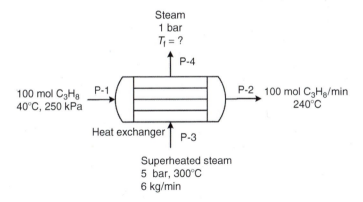

FIGURE E9.9
Block diagram of a heat exchanger process.

$$a = 68.032 \times 10^{-3}, \quad b = 22.59 \times 10^{-5}$$
$$c = -13.11 \times 10^{-8}, \quad d = 31.71 \times 10^{-12}$$

$$\Delta \widehat{H}_{C3H8} = \int\limits_{40°C}^{240°C} C_p \, dT = 68.032 \times 10^{-3}[240 - 40] + \frac{22.59 \times 10^{-5}}{2}[240^2 - 40^2]$$

$$+ \frac{-13.11 \times 10^{-8}}{3}[240^3 - 40^3] + \frac{31.71 \times 10^{-12}}{4}[240^4 - 40^4]$$

$$= 13.6064 + 6.3252 - 0.013 + 0.0263 = 19.36 \frac{kJ}{mol}$$

The amount of heat transferred from the steam to heat propane is

$$\dot{n}\Delta\widehat{H} = 100\frac{mol}{min} \times 19.36\frac{kJ}{mol} = 1936\frac{kJ}{min}$$

$$\dot{m}\Delta\widehat{H} = 6\frac{kg}{min}\left[\widehat{H}_{out} - 3065\right]\frac{kJ}{kg}$$

$$\dot{Q} = 0 = 1936\frac{kJ}{min} + 6\frac{kg}{min}\left[\widehat{H}_{out} - 3065\right]\frac{kJ}{kg}$$

$$= \frac{1936}{6} + 3065 = \widehat{H}_{out,H_2O}, \quad \widehat{H}_{out} = 2742.3\frac{kJ}{kg}$$

Using the superheated steam table calculate the temperature at 1 bar absolute pressure and enthalpy, $\widehat{H} = 2742.3$ kJ/mol. Since the value of the calculated enthalpy at 1 bar does not exist in the steam table, interpolation is required to get the value of the exit temperature.

$$\frac{2776 - 2676}{150 - 100} = \frac{2776 - 2742.3}{150 - T_f} \Rightarrow T_f = 133°C$$

Example 9.10   Heating of liquid methanol

Liquid methanol at 25°C is heated and vaporized for use in a chemical reaction. How much heat is required to heat and vaporize 2 kmol of methanol to 600°C.

Solution

The inlet and exit conditions of the heater are shown in Figure E9.10. Consider $T_{ref} = 25°C$ and methanol in liquid phase as reference conditions. The normal boiling point of methanol is 64.7°C.
The exit stream enthalpy is

$$\Delta \widehat{H}_{C_3H_8} = \int_{25°C}^{64.7°C} C_{pl} \, dT = 75.86 \times 10^{-3}[64.7 - 25] + \frac{16.83 \times 10^{-5}}{2}[64.7^2 - 25^2]$$

$$= 3012 + 0.3$$

The enthalpy change for ethanol vapor is

$$\Delta \widehat{H}_{C_3H_8} = \int_{64.7°C}^{600°C} C_{pv} \, dT = 42.93 \times 10^{-3}[600 - 64.7] + \frac{8.30 \times 10^{-5}}{2}[600^2 - 64.7^2]$$

$$= \frac{-1.87 \times 10^{-8}}{3}[600^3 - 64.7^3] - \frac{8.03 \times 10^{-12}}{4}[600^4 - 64.7^4]$$

$$= 22.98 + 14.77 - 1.34 - 0.26$$

Latent heat of vaporization is

$$\Delta \widehat{H}_{vapor} = 36.14 \frac{kJ}{mol}$$

$$Q = 2000 \text{ mol}[3.312 + 35.27 + 36.14] \frac{kJ}{mol} = 149,400 \text{ kJ}$$

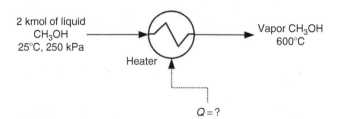

FIGURE E9.10
Flow sheet of methanol vaporization process.

Example 9.11   Turbine power plant

A gas turbine power plant receives a shipment of hydrocarbon fuel whose composition is uncertain but may be represented by the expression C_xH_y. The fuel is burned with excess air. An analysis of the product gas gives the following results (these values are in mol%) on a moisture-free basis: 9.5% CO_2, 1.0% CO, 5.3% O_2, and 84.2% N_2.

(a) Identify the number of independent chemical reactions in the process but do not try to balance them since the composition of the fuel is unknown.

(b) Write the reactions (unbalanced), identifying the reactants and products in the reaction.

(c) Do the degree of freedom analysis for this process and write the material balance equation for this process.

(d) Determine the molar ratio of hydrogen to carbon in the fuel, r, where $r=y/x$, and the percentage of excess air used in the combustion.

Solution

(a) The number of independent chemical reactions in the process is two since the flue gases contain CO, which means that there is a side reaction and the combustion is not complete.

(b) The two chemical reactions are shown below:

$$C_xH_y + \nu_{O_2,1}O_2 \rightarrow xCO_2 + \frac{y}{2}H_2O, \quad \nu_{O_2,1} = \frac{y}{4} + x$$

$$C_xH_y + \nu_{O_2,2}O_2 \rightarrow xCO + \frac{y}{2}H_2O, \quad \nu_{O_2,2} = \frac{y}{4} + \frac{x}{2}$$

The flowchart of this process is shown in Figure E9.11.

(c) Degree of freedom analysis

Degree of Freedom Analysis	Overall Process (Atomic Balance)		
Number of unknowns	5 $(n_{C_xH_y}, n_{H_2O}, x, y, n_{air})$		
Number of independent equations	4		
Number of relations	—		
DF	1		

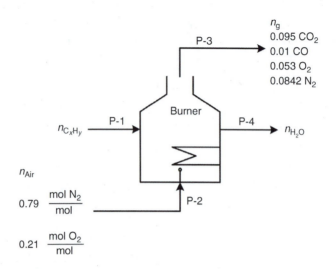

FIGURE E9.11
Methanol combustion process.

Atomic balance:
   Basis: 100 mol of flue gases (dry basis)
   C atomic balance: $x \times n_{C_xH_y} = 9.5 + 1.0$
   H atomic balance: $y \times n_{C_xH_y} = 2 \times n_{H_2O}$
   N atomic balance: $2(0.79\, n_{Air}) = 2(84.2\ \text{mol})$

$$n_{air} = 106.6\ \text{mol}$$

O atomic balance:
   $2 \times 106.6 \times 0.21 = 2 \times 9.5 + 1.0 + 2 \times 5.3 + n_{H_2O} \Rightarrow n_{H_2O} = 14.17$

(d) The molar ratio of hydrogen to carbon in the fuel $r = y/x$, and the percentage of excess air used in the combustion can thus be calculated.

From the following equations:
   C atomic balance: $x \times n_{C_xH_y} = 9.5 + 1.0$
   H atomic balance: $y \times n_{C_xH_y} = 2 \times 14.17$

$$r = \frac{y \times C_xH_y}{x \times C_xH_y} = \frac{2 \times 14.17}{10.5} \Rightarrow r = \frac{y}{x} = \frac{28.34}{10.5}$$

$$C_xH_y + \left(\frac{y}{4} + x\right)O_2 \rightarrow xCO_2 + \frac{y}{2}H_2O$$

For 106.6 mol air fed, 10.5 mol of CO_2 produced $x = 10.5$, $y = 28.34$
The percentage of excess air used in the combustion:
Moles of C_xH_y?
From the equation of C atomic balance: $10.5 \times n_{C_xH_y} = 9.5 + 1.0$
One mole of C_xH_y is fed to the burner.

$$\text{Theoretical } O_2 = \frac{28.34}{4} + 10.5 = 17.59 \text{ mol } O_2$$

$$17.59 \text{ mol } O_2 \times \frac{0.79 \text{ mol } N_2}{0.21 \text{ mol } O_2} = 66.15 \text{ mol } N_2$$

Theoretical air: $17.59 + 66.15 = 83.74$ mol air
Excess air: $106.6 - 83.74 = 27.86$ mol excess air

$$22.8/83.7 = 27\% \text{ excess air}$$

Example 9.12   Ethanol production
Ethanol (C_2H_5OH) is produced by steam (H_2O) reformation with ethylene
(C_2H_4). In an undesirable side reaction, ethylene oxide is formed.

$$C_2H_4 + H_2O \rightarrow C_2H_5OH$$
$$2C_2H_5OH \rightarrow (C_2H_5)_2O + H_2O$$

Fresh feed contains 20.0% C_2H_4 and 80.0% H_2O. The reaction products are
fed to a condenser that has two product streams: a vapor stream that
contains C_2H_4, C_2H_5OH, and water vapor; and a liquid stream that contains
the remaining ethanol product, ethylene oxide, and water. The vapor stream
from the condenser is mixed with the fresh feed (i.e., it is recycled) to be fed
to the reactor. The overall process yield is 80% of ethanol produced.

(a) Draw a flowchart of the process.
(b) Perform the degree of freedom analysis around the mixing point
and the condenser.
(c) If the ratio of water in the recycle stream to water in the product
stream is 1:10, and the ratio of the ethylene in the recycle to
ethylene in the fresh feed is 4:1, determine the recycle flow rate
and the single-pass conversion of ethylene in the reactor.

Solution

(a) The flowchart of the process is shown in Figure E9.12.

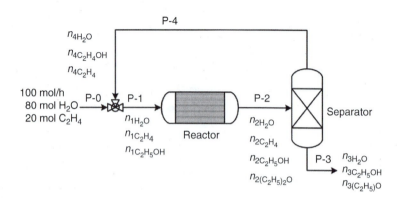

FIGURE E9.12
Process flow sheet of ethanol synthesis.

System: Overall process
(b) Degree of freedom analysis

Degree of Freedom Analysis	Overall Process
Number of unknowns	3 ($n_{3H_2O}, n_{3C_2H_4OH}, n_{3(C_2H_5)_2O}$)
Number of reactions	1
Number of atomic balances	3
Number of relations	1 (overall conversion)
DF	0

(c) Overall yield $0.8 = \dfrac{n_{C_2H_4OH}}{20} \Rightarrow n_{C_2H_4OH} = 16$ mol

C atomic balance: $20 \times 2 = n_{3(C_2H_4)_2O} \times 4 + 16 \times 2 \Rightarrow n_{3(C_2H_4)_2O} = 2$ mol
O atomic balance: $80 = 2 + 16 + n_{3H_2O}$

$$n_{3H_2O} = 62 \text{ mol } H_2O \text{ in product ratio of } H_2O \text{ in recycle}$$

$$\frac{n_{4H_2O}}{n_{3H_2O} = 62} = \frac{1}{10}$$

$$n_{4H_2O} = 6.2 \text{ mol } H_2O \text{ in recycle}$$

Mixing point balance: $n_{2H_2O} = 86.2$ mol
Water balance around the condenser:

$$n_{4H_2O} = 62 + 6.2 = 68.2 \text{ mol}$$

$$\text{Ratio of } \frac{\text{ethylene in recycle} = n_{4,C_2H_4}}{\text{ethylene in fresh feed} = 20 \text{ mol}} = \frac{4}{1}$$

$$80 \text{ mol } C_2H_4 \text{ in recycle} = n_{C_2H_4}$$

Also 80 mol C_2H_4 enters the condenser.
The exit of the reactor contains the following: 80 mol C_2H_4
2 mol$(C_2H_5)_2O$, 68.2 mol H_2O, $y_{C_2H_5OH} = 0.157$

$$\text{Single-pass conversion} = \frac{\text{reactant into reactor} - \text{reactant out from reactor}}{\text{reactant into reactor}}$$

$$x = \frac{100 - 80}{100} = 20\%$$

$$\frac{150.2}{0.843} = 178.10 = \text{total moles in reactor effluent}$$

A total of 27.9 mol of ethanol is present in the reactor effluent.
   The flow rate of recycle stream is $27.9 - 16 = 11.9$ mol of ethanol.

Recycle $= 80$ mol $C_2H_4 + 6.2$ mol $H_2O + 11.9$ mol ethanol $= 98.1$ mol

Example 9.13   Methanol combustion
Methanol (CH_3OH) and oxygen (O_2) are fed to an isothermal (25°C) reactor
at 1:1 molar ratio. The feed rate to the reactor is 480 gmol/min, and the
reactor operates at steady state. Two reactions take place:

$$CH_3OH + O_2 \rightarrow HCOOH + H_2O \quad (1)$$
$$CH_3OH + 3/2\, O_2 \rightarrow CO_2 + 2H_2O \quad (2)$$

The flow rate out of the reactor is 520 gmol/min. The reactor effluent
contains methanol, formic acid ($HCOOH$), water (H_2O) and CO_2, and no O_2.

(i) Draw and label the process flowchart.
(ii) Determine the heat that must be withdrawn to keep the reactor at
    constant temperature.
(iii) Determine the fractional conversion of methanol.
(iv) Determine the selectivity for the conversion of methanol to
    formic acid.

Solution

(i) The process flowchart is shown in Figure E9.13.
    Using the extent of reaction method:

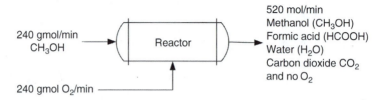

FIGURE E9.13
Flow sheet of methanol combustion.

$$CH_3OH: n_{CH_3OH} = n^0_{CH_3OH} - \xi_1 - \xi_2$$

$$HCOOH: n_{HCOOH} = 0 + \xi_1$$

$$H_2O: n_{H_2O} = 0 + \xi_1 + 2\xi_2$$

$$CO_2: n_{CO_2} = 0 + \xi_2$$

$$O_2: n_{O_2} = n^0_{O_2} - \xi_1 - \frac{3}{2}\xi_2$$

$$\text{Total: } n_{out} = n_{in} + \frac{1}{2}\xi_2$$

Total molar flow rate out: $\dot{n}_{out} = \dot{n}_{in} + \sum_k \sum_i v_{ik}\dot{\xi}_k = 480 + \frac{1}{2}\dot{\xi}_2 = 520$

Solving: $\dot{\xi}_2 = 80$ gmol/min

O_2 molar flow rate out:

$$\dot{n}_{O_2,out} = \dot{n}_{O_2,in} + \sum_k v_{O_2,k}\dot{\xi}_k = 240 - \dot{\xi}_1 - \frac{3}{2}\dot{\xi}_2 = 240 - \dot{\xi}_1 - 120 = 0$$

Solving:

$$\dot{\xi}_1 = 120 \text{ gmol/min}$$

(ii) Since the reactor is isothermal, the heat released is the heat of reaction.

$$Q = \Delta H^0_{rxn}$$

$$\Delta H^0_{rxn} = \xi_1 \Delta H^0_{rxn1} + \xi_2 \Delta H^0_{rxn2}$$

$$CH_3OH + O_2 \rightarrow HCOOH + H_2O \quad (1)$$

$$CH_3OH + 3/2 O_2 \rightarrow CO_2 + 2H_2O \quad (2)$$

$$\Delta \widehat{H}^0_{rxn1} = \Delta \widehat{H}^0_{f,H_2O} + \Delta \widehat{H}^0_{f,HCOOH} - \Delta \widehat{H}^0_{f,O_2} - \Delta \widehat{H}^0_{f,CH_3OH}$$

$$\Delta \widehat{H}^0_{rxn1} = -241.83 + (-115.9) - 0 - (-238.6) = -119.13 \text{ kJ/mol}$$

$$\Delta \widehat{H}^0_{rxn2} = 2\Delta \widehat{H}^0_{f,H_2O} + \Delta \widehat{H}^0_{f,CO_2} - \frac{3}{2}\Delta \widehat{H}^0_{f,O_2} - \Delta \widehat{H}^0_{f,CH_3OH}$$

$$\Delta \widehat{H}^0_{rxn2} = 2(-241.83) + (-393.5) - \frac{3}{2}(0) - (-238.6) = -638.56 \text{ kJ/mol}$$

$$Q = \Delta H^0_{rxn} = \xi_1 \Delta H^0_{rxn1} + \xi_2 \Delta H^0_{rxn2}$$

$$Q = 120(-119.13) + 80(-638.56) = -65380.4 \text{ kJ/min}$$

(iii) Fractional conversion of methanol: $\dfrac{\dot{\xi}_1 + \dot{\xi}_2}{\dot{n}_{MeOH,in}} = \dfrac{120 + 80}{240} = 0.833$

(iv) Fractional selectivity to formic acid: $\dfrac{\dot{\xi}_1}{\dot{\xi}_1 + \dot{\xi}_2} = \dfrac{120}{120 + 80} = 0.60$

Example 9.14   Ethane combustion

Ethane (C_2H_6) (750 gmol/h) is mixed with 20% excess air (21 mol% O_2) and fed to a burner where the mixture is completely combusted isothermally at 25°C, using cooling water. What is the air flow rate to the burner (gmol/h) and what is the amount of heat released to the cooling water?

Solution

The process flowchart is shown in Figure E9.14.

Balanced reaction is $C_2H_6 + \dfrac{7}{2}O_2 \rightarrow 2CO_2 + 3H_2O$

Air flow at 20% above the stoichiometric requirement is calculated as

$$\text{Theoretical air} = \left(\frac{750 \text{ gmol ethane}}{h}\right)\left(\frac{3.5 \text{ gmol } O_2}{\text{gmol ethane}}\right)\left(\frac{1 \text{ gmol air}}{0.21 \text{ gmol } O_2}\right)$$

$$= 1500 \text{ gmol/h}$$

Excess air:

The amount of excess air $= 0.2$ (theoretical air)

$$= 0.2 (1500 \text{ gmol/h}) = 300 \text{ gmol/h}$$

Total air fed to the burner $=$ theoretical $+$ excess

$$= 1500 + 300 = 1800 \text{ gmol/h}$$

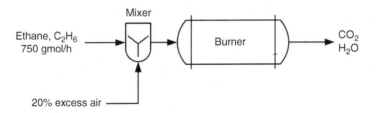

FIGURE E9.14
Ethane combustion process flow sheet.

The heat released to the coolant $(Q) = \dot{\xi} \times \Delta \hat{H}^0_{rxn}$

$$\Delta H^0_{rxn} = 3\Delta H^0_{f,H_2O} + 3\Delta H^0_{f,CO_2} - 3\Delta H^0_{f,C_2H_6} - \tfrac{7}{2}3\Delta H^0_{f,O_2}$$

$$\Delta H^0_{rxn} = 3(-241.83) + 3(-393.5) - 3(-84.67) - 0 = -1651.98 \text{ kJ/gmol}$$

$$Q = 750 \text{ gmol/h} (-1651.98 \text{ kJ/gmol}) = -1.24 \times 10^6 \text{ kJ/h}$$

Example 9.15 Chemical reactors with recycle

Butanol (C_4H_8O), used to make laundry detergents, is made by the reaction of propylene (C_3H_6) with CO and H_2:

$$C_3H_6 + CO + H_2 \rightarrow C_4H_8O$$

In an existing process, 180 kgmol/h of C_3H_6 (P) is mixed with 420 kgmol/h of a mixture containing 50% CO and 50% H_2 and with a recycle stream containing propylene and then fed to a reactor, where a single-pass conversion of propylene of 30% is achieved. The desired product butanol (B) is removed in one stream, unreacted CO and H_2 are removed in a second stream, and unreacted C_3H_6 is recovered and recycled.

Calculate:

(i) Production rate of butanol (kgmol/h).
(ii) Flow rate of the recycle stream (kgmol/h).

Solution

The process flow sheet is shown in Figure E9.15.
Basis: 180 kgmol/h of propylene feed

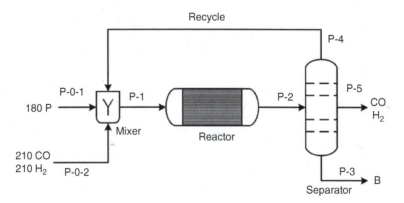

FIGURE E9.15
Process flow sheet of butanol production.

(i) First choose the entire process as the system. Since no propylene leaves the process, the overall conversion is 100% (fractional conversion $= 1.0$).

Therefore,

$$1 = \frac{\dot{\xi}}{\dot{n}_{P,fed}} = \frac{\dot{\xi}}{180} \quad \text{or} \quad \dot{\xi} = 180 \text{ kgmol/h}$$

The balance on butanol is simply $\dot{n}_{B,out} = \dot{\xi} = 180$ kg mol/h. Butanol production rate is 180 kgmol/h.

(ii) Now choose the reactor as the system. The fractional single-pass conversion $= 0.3$.

The extent of reaction is the same as in (i), because the reactor is the only unit where a reaction takes place. Therefore,

$$0.3 = \frac{\dot{\xi}}{\dot{n}_{P,fed \text{ to reactor}}} = \frac{180}{\dot{n}_{P,fed \text{ to reactor}}}$$

$$\dot{n}_{P,fed \text{ to reactor}} = 600 \text{ kgmol/h}$$

From a balance around the mixer, we can find that the recycle rate must be

$$600 - 180 = 420 \text{ kgmol/h}$$

Example 9.16 Economical selection of propylene for butanol production

A company contacts you and claims that they can supply you with a cheaper source of propylene (P) that could replace your current supply. Unfortunately, the cheaper source is contaminated with propane at a ratio of 5:95 (propane: propylene). The cheaper source is economically attractive if the butanol production rate is maintained at the current rate and if an overall conversion of 0.90 can be achieved. At your reactor conditions, propane (I) is an inert, and it is too expensive to separate propane from propylene so you decide to install a purge stream. Assume the $CO + H_2$ stream remains the same (420 kgmol/h, 50 mol% CO) as does the single-pass conversion of propylene in the reactor (0.3). The production rate of butanol is 180 kgmol/h; the same as in Example 9.15. Given this, show how the block flow diagram must be modified to accommodate the cheaper source of propylene and calculate the following

(i) Flow rate of the contaminated propylene stream to the process (kgmol/h).

(ii) mol% inert and the total flow rate (kgmol/h) of the purge stream.

(iii) mol% inert and the total flow rate (kgmol/h) of the feed to the reactor.

Solution

A splitter and purge are now essential, because of the contaminated feed and the existence of an inert. The purge is necessary to avoid inert accumulation in the process. We no longer know the incoming flow rate but we fix butanol production rate at 180 kgmol/h. The block flow diagram must be modified to accommodate the cheaper source of propylene as shown in Figure E9.16.

(i) We first group the entire process into a single system. The overall fractional conversion for the process is fixed at 0.9. Therefore,

$$0.9 = \frac{\dot{\xi}}{\dot{n}_{P,fed}} = \frac{180}{\dot{n}_{P,fed}}$$

$$\dot{n}_{P,fed} = 200 \text{ kgmol/h}$$

This feed stream is 5% inert and 95% propylene. The flow rate of the inert to the process: $(5/95) \times 200 = 10.5$ kgmol/h. The total flow of the contaminated propylene stream is 210.5 kgmol/h.

(ii) Unreacted propylene exits the system in the purge stream.

System: overall process

From a material balance on propylene with the entire process as the system:

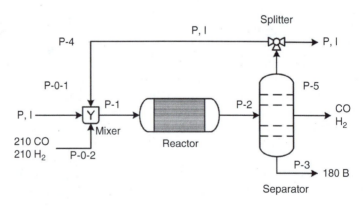

FIGURE E9.16
Flow sheet of butanol production.

- The propylene flow rate in the purge stream $= 200 - 180 = 20$ kgmol/h.
- All the inert flowing into the system needs to leave in the purge, so the inert flow in the purge $= 10.5$ kgmol/h.
- Total purge flow is therefore $10.5 + 20 = 30.5$ kgmol/h.
- The mol% inert is $(10.5/(10.5 + 20)) = 34.5$ mol%.

(iii) Since the extent of reaction and the single-pass conversion are the same as that in Example 9.15, then propylene fed to the reactor is the same or 600 kgmol/h.

A balance on propylene around the mixer shows that $600 - 200 = 400$ kgmol/h propylene must be in the recycle stream. Since the composition of the recycle stream equals that of the purge stream, the recycle stream must contain $(400(10.5/20)) = 210$ kgmol/h inert. Again, the balance on the mixer tells us that $210 + 10.5 = 220.5$ kgmol/h inert must flow into the reactor.

Therefore, the total flow into the reactor is $220.5 + 600 + 420$ (from the CO and the H_2) $= 1240.5$ kgmol/h. The mole percent inert in the reactor feed stream is $(220.5/1240.5) \times 100\% = 17.8$ mol% inert.

Example 9.17 Steam turbine

Steam (1500 kg/h) at 20 bar and 350°C is fed to a turbine that operates adiabatically and at steady state. The steam leaves the turbine at 1.0 bar and 150°C and is cooled in a heat exchanger to a saturated liquid:

(a) Draw and label the process flow diagram.
(b) How much work (kJ/h) is extracted in the turbine?
(c) How much heat (kJ/h) is removed in the heat exchanger?

Solution

(a) The process flow diagram is shown in Figure E9.17.
(b) The mass flow basis is 1500 kg/h steam entering the process. The material balance equations are trivial to solve. The system is open, so we would like to define the energy of the streams flowing in and out. We have no information about the velocity or relative position of any of the streams, so finding E_k and E_p is problematic. Because of the large change in temperature as well as the phase change (in the heat exchanger), the enthalpy of the streams is the only important contributor to the energy flows.
  - The specific enthalpy of steam at 20 bar and 350°C is 3137.7 kJ/kg ($\hat{H}_1 = 3137.7$ kJ/kg).

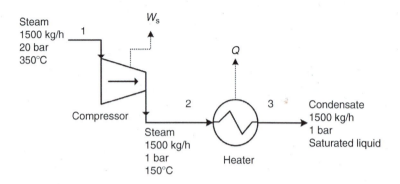

FIGURE E9.17
Steam turbine flow sheet.

- The specific enthalpy of steam at 1 bar and 150°C is 2776.6 kJ/kg ($\widehat{H}_2 = 2776.6$ kJ/kg).
- The saturated liquid at 1 bar (which is at 99.6°C) is 417.50 kJ/kg ($\widehat{H}_3 = 417.50$ kJ/kg).

System: Turbine
   With the turbine as the system, the energy balance simplifies to

$$0 = \dot{m}_1\widehat{H}_1 - \dot{m}_2\widehat{H}_2 + \dot{W}_s,$$

assuming steady state, including one inlet stream and one outlet stream, neglecting kinetic and potential energy of the inlet and outlet streams, and noting that the turbine is adiabatic and hence there is no heat term.

Rearranging:

$\dot{W}_s = 1500$ kg/h(2776.6 kJ/kg) $- 1500$ kg/h(3137.7 kJ/kg) $= -541{,}650$ kJ/h

The work term is negative because work is done on the system.

System: Heat exchanger
With the heat exchanger as the system, the energy balance simplifies to

$$0 = \dot{m}_2\widehat{H}_2 - \dot{m}_3\widehat{H}_3 + \dot{Q}$$

We plug in appropriate values to find

$\dot{Q} = 1500$ kg/h(417.50 kJ/kg) $- 1500$ kg/h(2776.6 kJ/kg) $= -3.54 \times 10^6$ kJ/h

Heat is transferred from the system to the surroundings, thus the negative sign.

9.3 Problems

9.3.1 Mixing of Hot and Cold Ethanol

Ten kilograms of hot ethanol (150°C, 1.2 atm) is cooled and condensed by mixing it with cold ethanol (5°C, 1.2 atm). If the final ethanol product is to be at 25°C and 1.2 atm, how much cold ethanol (kg) must be added?

9.3.2 Combustion of Acetylene

A quantity of 100 gmol/h acetylene (C_2H_2) is mixed with 2000 gmol/h air (79 mol% N_2, 21 mol% O_2) and the mixture (at 298 K and 1 atm) fed to a reactor, where complete combustion takes place. The reactor is equipped with cooling tubes. The combustion mixture leaving the reactor is at 1000 K and 1 atm. Draw and label the process flowchart. How much heat (kJ/h) was removed in the reactor? Suppose the coolant supply was suddenly shut off. What reactor outlet temperature would be reached? Data: $C_{p_{CO_2}} = 37$ J/mol°C, $C_{p_{H_2O}} = 33.6$ J/mol°C, $C_{p_{O_2}} = 29.3$ J/mol°C, $C_{p_{N_2}} = 29.3$ J/mol°C, assume these values to be constant and independent of temperature. (Answer: $\dot{Q} = -82370$ kJ/h.)

9.3.3 Dehydrogenation of Ethanol

Ethanol (C_2H_5OH) is dehydrogenated in a catalytic reactor to acetaldehyde (CH_3CHO), with hydrogen (H_2) as a by-product. In an existing process, 100 gmol/min liquid ethanol at 25°C and 1 atm pressure is first heated to 300°C in a heat exchanger, and then fed to the reactor. One hundred percent of the ethanol is converted to products, and the product stream leaves the reactor at 300°C and 1 atm (760 mmHg). The product stream leaving the reactor is cooled to −15°C, and sent to a flash drum, where vapor and liquid streams are separated (Figure P9.3.3).

A. How much heat must be supplied to the first heat exchanger? (Answer: $Q = 6400$ kJ/min.)

B. How much heat must be supplied to or removed from (state which one) the reactor in order to maintain a constant temperature of 300°C? (Answer: About 7100 kJ/min.)

C. What are the flow rates of the vapor and liquid streams leaving the flash drum? (Answer: The vapor flow rate is 126 gmol/min and the liquid flow rate is 74 gmol/min.)

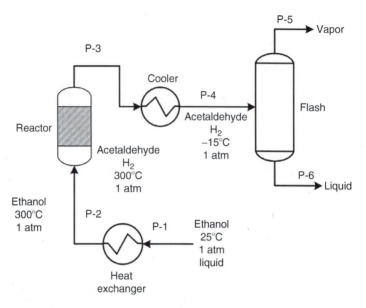

FIGURE P9.3.3
Flow sheet of ethanol dehydrogenation process.

9.3.4 Independent Chemical Reaction

(A) A gas contains the following compounds: CH_4, C_2H_6, O_2, CO_2, CO, H_2O, and H_2. What is the maximum number of independent chemical reactions that can be written involving these compounds? (You do not need to determine what the reactions are, only how many are there.) (Answer: There are seven compounds and three elements. There are 7–3 or 4 independent chemical reactions.)

(B) 12 gmol $SiCl_4$ and 20 gmol H_2 react to make solid silicon Si and HCl. Which compound ($SiCl_4$ or H_2) is the limiting reactant? What is the percent excess for the excess reactant? (Answer: 20%.)

9.3.5 Cumene Synthesis

Cumene (C_9H_{12}) is a useful intermediate in the synthesis of specialty chemicals such as pharmaceuticals and flavorings. Cumene is synthesized from propylene (C_3H_6) and benzene (C_6H_6). Unfortunately, a side reaction also occurs, in which diisopropylbenzene ($C_{12}H_{18}$) is generated by reaction of propylene with cumene. The two balanced reactions are

$$P + B \rightarrow C \quad (1)$$
$$P + C \rightarrow D \quad (2)$$

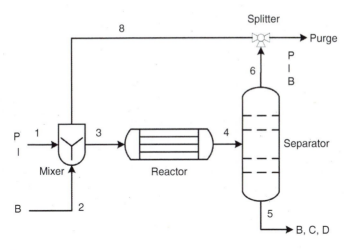

FIGURE P9.3.5
Cumene synthesis process.

where
  C is cumene
  P is propylene
  B is benzene
  D is diisopropylbenzene

A block flow diagram for a cumene manufacturing process is shown in Figure P9.3.5. A 100 kgmol/h of a gas containing 95 mol% P and 5 mol% I (stream 1) is mixed with 80 kgmol/h B (stream 2) plus a recycle stream (stream 8). The mixer outlet is fed to a reactor. The fractional conversions achieved in the reactor (based on the reactor feed stream 3) are $f_{CB}=0.9$ and $f_{CP}=0.7$. The reactor outlet is sent to a separator, where all of the P, all of the I, and 10% of the B is recovered in stream 6, and the remaining B and all of the C and D are recovered in stream 5. Stream 6 is sent to a splitter. 85% of the splitter feed is recycled to the mixer and the remainder is purged (stream 7). Calculate the following:

(1) Flow rates of P and B (kgmol/h) in stream 3. (Answer: $B_3=80.7$ kgmol/h.)

(2) Mol% I in stream 7. (Answer: 46%.)

(3) Flow rates of C and D (kgmol/h) in stream 5. (Answer: $D_5=D_4=\xi_2=16.62$ kgmol/h, $C_5=C_4=\xi_1-\xi_2=56$ kgmol/h.)

(4) Selectivity for converting B to C achieved by the overall process. (Answer: Selectivity = moles of B converted to C/moles of B consumed $=56/72.6=0.77$.)

9.3.6 Dehydrogenation of Propane

The dehydrogenation of propane is carried out in a continuous reactor. Pure propane is fed to the reactor at 1300°C and at a rate of 100 mol/h. Heat is supplied at a rate of 1.34 kW. If the product temperature is 1000°C, calculate the extent of reaction. (Answer: $\xi = 0.0236523$.)

$$C_3H_8(g) \rightarrow C_3H_6(g) + H_2(g), \quad \Delta H_r(1000°C) = 128.8 \text{ kJ/mol}$$

Further Readings

1. Reklaitis, G.V. (1983) *Introduction to Material and Energy Balances*, John Wiley & Sons, New York.
2. Himmelblau, D.M. (1996) *Basic Principles and Calculations in Chemical Engineering*, 6th edn, Prentice-Hall, New Jersey.
3. Whirwell, J.C. and R.K. Toner (1969) *Conservation of Mass and Energy*, Blaisdell, Waltham, MA.
4. Felder, R.M. and R.W. Rousseau (1999) *Elementary Principles of Chemical Processes*, 3rd edn, John Wiley, New York.

10

Unsteady-State Material and Energy Balances

At the End of This Chapter You Should Be Able to

1. Develop unsteady-state material and energy balance equations.
2. Solve simultaneously the resultant first-order ordinary differential material and energy balance equations using E–Z solve or Polymath software packages.
3. Plot temperature and concentration versus time.
4. Explain the rational changes in concentration or temperature versus time.

10.1 Unsteady-State Material Balance

Unsteady state or transient state refers to processes in which quantities or operating conditions within the system change with time. Important industrial problems that fall into this category are

- Startup of equipment.
- Batch processes.
- Change in operating conditions.
- Perturbations that develop as process conditions fluctuate.

Figure 10.1 shows an open system with multiple inputs and outputs. The general material balance equation takes the following form:

$$\frac{dm}{dt} = \dot{m}_{in} - \dot{m}_{out} + \dot{m}_{g} - \dot{m}_{c} \tag{10.1}$$

where
$\dot{m}_{in}$ is the inlet mass flow rate [mass/time]
$\dot{m}_{out}$ is the outlet mass flow rate [mass/time]

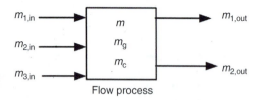

FIGURE 10.1
Material balance on an unsteady-state flow process.

$\dot{m}_g$ is the generated mass flow rate [mass/time]
$\dot{m}_c$ is the consumed mass flow rate [mass/time]
m is the mass accumulated in the system [mass]

$$\dot{m}_{in} = \sum_{in} \dot{m}_{i,in}$$

$$\dot{m}_{out} = \sum_{out} \dot{m}_{i,out}$$

Example 10.1 Filling controlled level storage tank

A storage tank that is 2.0 m in diameter is filled at the rate of 2.0 m^3/min. When the height of the liquid is 2 m in the tank, a control valve installed on the exit stream at the bottom of the tank opens and the fluid flows at a rate proportional to the head of the fluid, that is, $0.4h$ m^3/min, where h is the height of fluid in meters. Plot the height of the liquid as a function of time. What is the steady-state height of the fluid in the tank?

Solution

The tank flowchart is shown in Figure E10.1a.
Unsteady-state mass balance:
  Neither generation nor consumption occurs in the process.

$$\frac{dm}{dt} = \dot{m}_{in} - \dot{m}_{out}$$

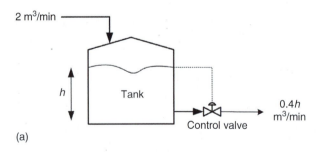

(a)

FIGURE E10.1a
Filling a water storage tank.

$$m = \rho V = \rho(Ah), \quad A = \pi D^2/4$$

where
  A is the cross-sectional area of the tank
  V is the volume of the system
  $\dot{V}$ is the volumetric flow rate

$$\dot{m} = \rho \times \dot{V}$$

$\dot{V}_{in}$ and $\dot{V}_{out}$ are the volumetric flow rates of inlet and exit streams, respectively.
  Simplifying the material balance equations in terms of one variable (h) with time gives

$$\frac{d(\rho Ah)}{dt} = \rho \dot{V}_{in} - \rho \dot{V}_{out}$$

Assume density is constant; the equation is then simplified to

$$A\frac{dh}{dt} = \dot{V}_{in} - \dot{V}_{out}$$

where
  $\dot{V}_{in} = \text{constant} = 2 \text{ m}^3/\text{min}$
  $\dot{V}_{out}$ is a function of the height of fluid in the tank $= 0.4h$

Rearrange the equation and solve using E–Z solve.

Results
The set of generated material balance equations is solved using E–Z solve as shown in Figure E10.1b. The change in height versus time is shown in

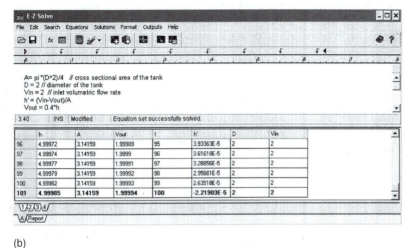

(b)

FIGURE E10.1b
E–Z solve script for the solution of the differential and algebraic equations.

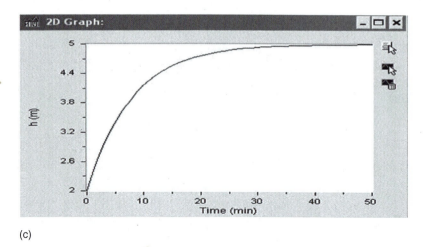

(c)

FIGURE E10.1c
Fluid height versus time using *E–Z* solve.

Figure E10.1c. It can be seen that the height of the tank increases sharply at the very beginning and slowly with time later on until it reaches the steady state, which is at approximately 5 m.

The steady-state height is calculated by setting the differential term to zero. The height is found to be approximately 5 m as shown in Figure E10.1d.

Using Polymath, the model equations at steady state are solved as in Figure E10.1e. The effect of height versus time is generated through solving

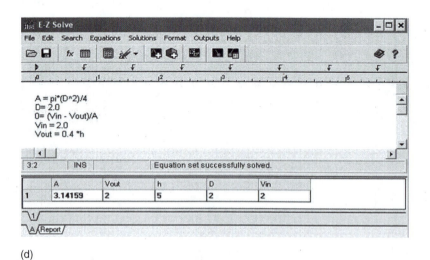

(d)

FIGURE E10.1d
Height of the tank at steady state using *E–Z* solve.

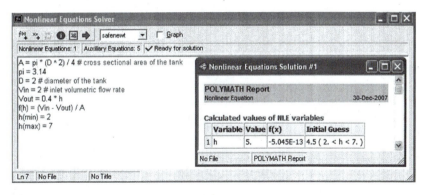

(e)

FIGURE E10.1e
Steady-state solution of the nonlinear equations using Polymath.

the differential equation along with the five auxiliary equations as shown in Figure E10.1f.

Example 10.2   Dilution of a salt solution

A tank holds 100 L of a salt water solution in which 5.0 kg of salt is dissolved. Water runs into the tank at the rate of 5 L/min and salt solution overflows at the same rate.

    a. Plot the concentration of the salt versus time.

    b. How much salt is in the tank at the end of 10 min?

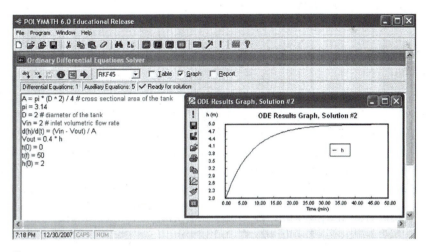

(f)

FIGURE E10.1f
Solving differential equations using Polymath.

(Assume the tank is well mixed and the density of the salt solution is essentially the same as that of pure water.)

Solution

The process flow sheet is shown in Figure E10.2a.

The general unsteady-state mass balance is

$$\frac{dm}{dt} = \dot{m}_{in} - \dot{m}_{out} + \dot{m}_{g} - \dot{m}_{c}$$

Since there are no reactions, generation and consumption terms are dropped and the general material balance equation is reduced to

$$\frac{dm}{dt} = \dot{m}_{in} - \dot{m}_{out}$$

The inlet mass flow rate as a function of salt concentration is

$$\dot{m}_{in} = \dot{V}_{in} C_{in}[=]\frac{\cancel{L}}{min}\left|\frac{kg}{\cancel{L}}\right| = \frac{kg}{min}$$

The outlet mass flow rate is

$$\dot{m}_{out} = \dot{V}_{out} C_{out}[=]\frac{\cancel{L}}{min}\left|\frac{kg}{\cancel{L}}\right| = \frac{kg}{min}$$

Accumulated mass (note that, in the accumulated mass, V is the volume of the fluid in the tank) is

$$\frac{d(m)}{dt} = \frac{d(VC)}{dt}[=]\frac{\cancel{L}\frac{kg}{\cancel{L}}|}{min} = \frac{kg}{min}$$

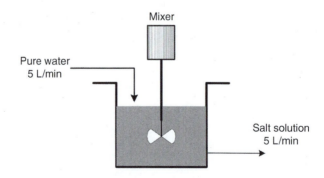

Mixer

Pure water
5 L/min

Salt solution
5 L/min

100 L of a salt-water solution in
which 5.0 kg of salt are dissolved

(a)

FIGURE E10.2a
Flow sheet of dilution of salt solution process.

Substitution of the above terms in the simplified material balance equation yields the following equation:

$$\frac{d(VC)}{dt} = (\dot{V}_{in}C_{in}) - (\dot{V}_{out}C_{out})$$

Assuming the tank is well mixed, the outlet salt concentration equals the concentration in the tank

$$\frac{d(VC)}{dt} = (\dot{V}_{in}C_{in}) - (\dot{V}_{out}C)$$

The inlet is pure water, which means inlet salt concentration is zero $\Rightarrow C_{in} = 0$. Inlet and outlet volumetric flow rates are equal $\Rightarrow \dot{V}_{in} = \dot{V}_{out}$ Volume of fluid in the tank is constant $\Rightarrow V = $ constant

$$V\frac{d(C)}{dt} = 0 - \dot{V}_{out}C$$

Solving the resultant differential equation requires an initial condition, which is the concentration of salt in the tank at time zero.

$$C|_{@t=0} = \frac{5\,kg}{100\,L} = 0.05\,\frac{kg}{L}$$

Results
Figure E10.2b reveals that the salt concentration in the tank decreases with time. At the end of 10 min the salt concentration in the tank is 0.03 kg/L (Figures E10.2b and E10.2c).

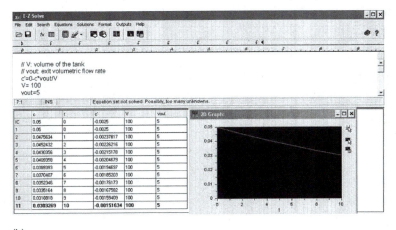

(b)

FIGURE E10.2b
Results obtained using *E–Z* solve.

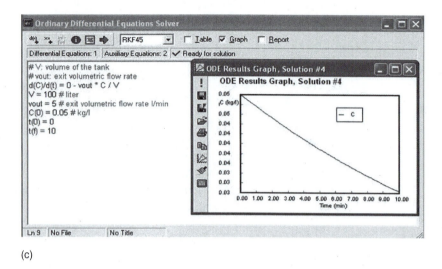

(c)

FIGURE E10.2c
Results obtained using Polymath.

Example 10.3   Dilution of salt solution

Ocean water (the average ocean salinity is 35 ppt; if we have 1 g of salt and 1000 g of water, the salinity is 1 part per thousand or 1 ppt) flows into a 100 L tank containing 1.5 kg salt at a rate of 5 L/min, and salt solution overflows out of the tank at 5 L/min. How much salt is in the tank at the end of 15 min? Assume the tank is well mixed and the density of salt solution is constant and equal to that of water.

Solution

The dilution process flow sheet is shown in Figure E10.3a.

The general unsteady-state material balance is

$$\frac{dm}{dt} = \dot{m}_{in} - \dot{m}_{out} + \dot{m}_g - \dot{m}_c$$

Since there are no reactions, generation and consumption terms are dropped and the general material balance equation is reduced to the following form:

$$\frac{dm}{dt} = \dot{m}_{in} - \dot{m}_{out}$$

Replacing mass flow rates in terms of concentration as done in Example 10.2 leads to

$$\frac{d(VC)}{dt} = (\dot{V}_{in} C_{in}) - (\dot{V}_{out} C_{out})$$

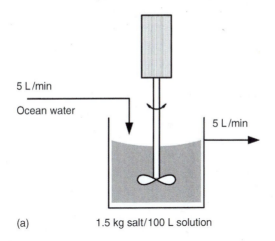

(a)   1.5 kg salt/100 L solution

FIGURE E10.3a
Schematic of dilution of salt solution.

Assuming the tank is well mixed, the outlet salt concentration equals the concentration in the tank; this assumption reduces the above equation to the following form:

$$\frac{d(VC)}{dt} = (\dot{V}_{in}\, C_{in}) - (\dot{V}_{out}\, C)$$

Inlet and outlet volumetric flow rates are equal $\Rightarrow \dot{V}_{in} = \dot{V}_{out}$. Volume of fluid in the tank is constant $\Rightarrow V = $ constant.

$$V\frac{dC}{dt} = \dot{V}_{in}\, C_{in} - \dot{V}_{out}\, C$$

Solving the resultant differential equation requires an initial condition, which is the concentration of salt in the tank at time zero.

$$C\big|_{@t=0} = \frac{1.5 \text{ kg}}{100 \text{ L}} = 0.015 \ \frac{\text{kg}}{\text{L}}$$

The ocean salt concentration in kilograms per liter is

$$C_{in} = 35 \text{ ppt} \left|\frac{\frac{1 \text{ g salt}}{1000 \text{ g water}}}{1 \text{ ppt}}\right| \frac{1000 \text{ g water}}{\text{L}} \left|\frac{1 \text{ kg}}{1000 \text{ g}}\right| = 0.035 \ \frac{\text{kg}}{\text{L}}$$

Results are obtained using E–Z solve as shown in Figure E10.3b.

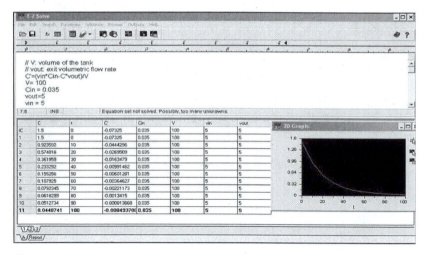

(b)

FIGURE E10.3b
Results obtained using *E–Z* solve.

Example 10.4   Sewage treatment

In a sewage treatment plant, a large concrete tank initially contains 440,000 L liquid and 10,000 kg fine suspended solids. To flush this material out of the tank, water is pumped into the vessel at a rate of 40,000 L/h, and liquid containing solids leave at the same rate. Estimate the concentration of suspended solids in the tank at the end of 4 h.

Solution

The process flow diagram is shown in Figure E10.4a.

The general unsteady-state material balance equation is

$$\frac{dm}{dt} = \dot{m}_{in} - \dot{m}_{out} + \dot{m}_g - \dot{m}_c$$

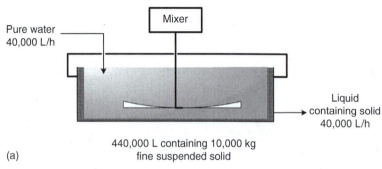

(a)

FIGURE E10.4a
Schematic diagram of sewage tank.

Since, there are no reactions, generation and consumption terms are dropped, and the general material balance equation is reduced to the following form:

$$\frac{dm}{dt} = \dot{m}_{in} - \dot{m}_{out}$$

Replacing mass flow rates in terms of concentration as done in Example 10.3 leads to

$$\frac{d(VC)}{dt} = (\dot{V}_{in}C_{in}) - (\dot{V}_{out}C_{out})$$

Assuming the tank is well mixed, the outlet suspended solid concentration equals the concentration in the tank; this assumption reduces the above equation to following form:

$$\frac{d(VC)}{dt} = (\dot{V}_{in}C_{in}) - (\dot{V}_{out}C)$$

Inlet and outlet volumetric flow rates are equal $\Rightarrow \dot{V}_{in} = \dot{V}_{out} = 40,000$ L/h. Volume of fluid in the tank is constant $\Rightarrow V = $ constant $= 440,000$ L.

Pure water is pumped into the vessel at a rate of 40,000 L/h; concentration of solids in the pure water is zero, $C_{in} = 0$:

$$V\frac{d(C)}{dt} = 0 - (\dot{V}_{out}C)$$

Solving the resultant differential equation requires an initial condition, which is the concentration of solids in the sewage tank at time zero:

$$C|_{@t=0} = \frac{10,000 \text{ kg}}{440,000 \text{ L}} = 0.023 \frac{\text{kg}}{\text{L}}$$

Results
Concentration versus time is shown in Figure E10.4b.

Example 10.5   Diffusion of a solid into a liquid
A compound dissolves in water at a rate proportional to the product of the amount of undissolved solid and the difference between the concentration in a saturated solution and the actual solution, that is, $C_{sat} - C(t)$. The dissolution rate is 7.14×10^{-5} s^{-1}. A saturated solution of this compound contains 0.4 g solid/g water. In a test run starting with 20 kg of undissolved compound in 100 kg of water, how many kilograms of compound will remain undissolved after 7 h? Assume that the system is isothermal.

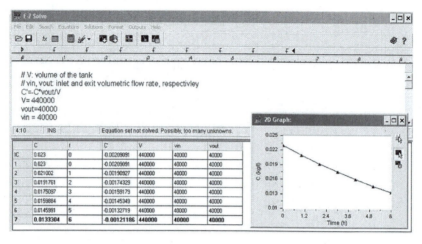

(b)

FIGURE E10.4b
Solution of the differential equation using *E*–*Z* solve.

Solution
System: solid compound

Let us assign *m* for the mass of the undissolved compound at any time, m_0 is the initial mass of the undissolved compound at time zero, and *C* is the concentration of the dissolved compound in water. Figure E10.5a is a schematic of diffusion of solids in water.

General material balance on the undissolved compound is

$$\frac{dm}{dt} = \dot{m}_{in} - \dot{m}_{out} + \dot{m}_{g} - \dot{m}_{c}$$

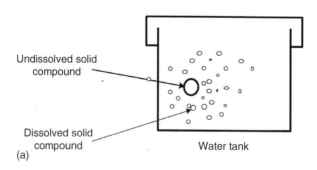

FIGURE E10.5a
Schematic diagram for dissolved solids in water.

There is no inlet or outlet mass flow rate to the tank $\dot{m}_{in} = \dot{m}_{out} = 0$.
There is no generation of the undissolved solid: $\dot{m}_g = 0$.
The rate of consumption: $\dot{m}_c = k\, m(C_{sat} - C)$
   Rearranging the general material balance equation leads to the following equation:

$$\frac{dm}{dt} = -km(C_{sat} - C)$$

Mass of solids in the tank as a function of concentration is

$$C = \frac{m_0 - m}{W} \Rightarrow m = m_0 - C \times W$$

Differentiation of the above relation leads to

$$\frac{dC}{dt} = -\frac{1}{W}\frac{dm}{dt}$$

Rearranging so as to replace mass by solid concentration results in

$$\frac{dm}{dt} = -W\frac{dC}{dt}$$
$$\frac{dC}{dt} = \frac{km}{W}(C_{sat} - C)$$

Substituting concentration instead of mass of undissolved compound

$$\frac{dC}{dt} = \frac{k(m_0 - C{\cdot}W)}{W}(C_{sat} - C)$$

With the following data, solve the above equations using E–Z solve and Polymath.

$$k = 7.14 \times 10^{-5}\ s^{-1}$$

$$W = 100\ kg\left|\frac{1000\ g}{1\ kg}\right. = 100{,}000\ g$$

$$C_{sat} = 0.4\,\frac{g\ solid}{g\ water}$$

Initial undissolved solid: $m_0 = 20\ kg\left|\frac{1000\ g}{1\ kg}\right. = 20{,}000\ g$

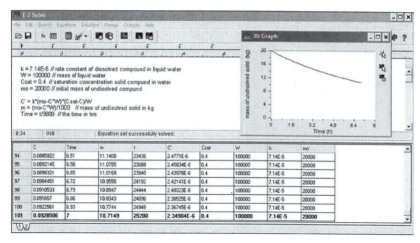

(b)

FIGURE E10.5b
Result of material balance equations using *E–Z* solve.

Figures E10.5b and E10.5c show the solution using *E–Z* solve and Polymath, respectively.

The concentration of the dissolved solid in liquid water is shown in Figure E10.5d. The diagram reveals solid concentration increasing with increasing time.

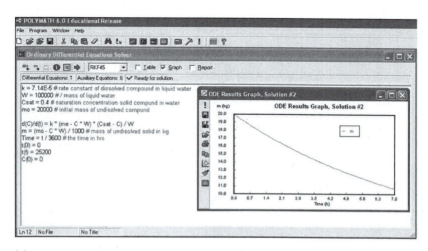

(c)

FIGURE E10.5c
Result of material balance equations using Polymath.

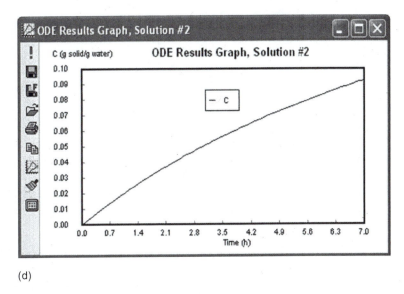

(d)

FIGURE E10.5d
Concentration of the dissolved compound in water.

10.2 Unsteady-State Energy Balance

Consider the mixing tank shown in Figure 10.2 where heat is added or removed from the tank through jacketed inlet and exit streams.

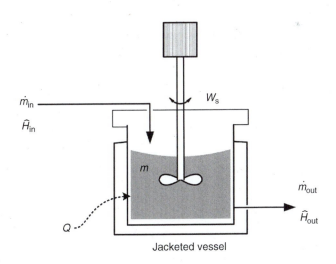

FIGURE 10.2
Energy balance on an unsteady-state process.

The general form of the energy balance under unsteady-state conditions takes the form

$$d(U)/dt = \sum \dot{m}_{in}\,\hat{H}_{in} - \sum \dot{m}_{out}\,\hat{H}_{out} + \dot{Q} - \dot{W}_s \qquad (10.2)$$

where

U is the internal energy; $U = mC_v T$

$\hat{H}_{in}$ is the specific enthalpy of the inlet stream

$\hat{H}_{out}$ is the specific enthalpy of the exit stream

$\dot{Q}$ is the heat added to the system $(+)$. If the heat is lost or transferred from the system to the surrounding then heat sign is negative.

$\dot{W}_s$ is the work done by the system on the surrounding $(+)$. If the work is done on the system then the sign is negative $(-)$.

For liquids and solids $C_v \approx C_p$ because $\frac{d(PV)}{dt} \approx 0$

$$H = U + PV$$

$$\frac{dH}{dt} = \frac{dU}{dt} + \frac{d(PV)}{dt}$$

This leads to $\dfrac{dH}{dt} = \dfrac{dU}{dt}$

Example 10.6   Heating of a closed system

A kettle used to boil water containing 1.00 L of water at 20°C is placed on an electric heater $(Q = 1950\ J/s)$. Find the time at which water begins to boil. (Hint: Normal boiling point of water is 100°C.)

Solution

The kettle heater diagram is shown in Figure E10.6a.

The general energy balance equation is

$$\frac{d(U)}{dt} = \Sigma \dot{m}_{in}\hat{H}_{in} - \Sigma \dot{m}_{out}\hat{H}_{out} + \dot{Q} - \dot{W}_s$$

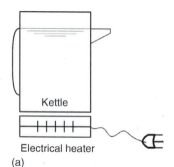

FIGURE E10.6a   Electrical heater

Schematic diagram of water kettle.   (a)

Assumptions

The kettle is batch; inlet and outlet mass flow rates are zero; $\dot{m}_{in} = \dot{m}_{out} = 0$.

No shaft work: $W_s = 0$

For liquids: $C_p \approx C_v$

Specific heat of water: $C_p = 4.18$ J/g °C

The general energy balance equation is simplified to the following equation:

$$\frac{d(mC_pT)}{dt} = \dot{Q}$$

The mass of liquid water inside the kettle is

$$m = \rho V$$

Substituting m in the simplified energy balance equation leads to

$$\frac{d[mC_pT]}{dt} = \dot{Q}$$

Rearranging

$$\frac{dT}{dt} = \frac{\dot{Q}}{\rho V C_p}$$

Solve the above ordinary differential equation using the following data:

Heat added to the kettle: $\dot{Q} = 1950$ J/s

Specific heat of water: $C_p = 4.18$ J/g °C

Volume of the kettle: $V = 1.0$ L

Density of the water in the kettle: $\rho = 1000 \frac{g}{L}$

Results obtained using E–Z solve are shown in Figure E10.6b and by Poly-math are shown in Figure E10.6c.

Example 10.7   Heating a tank with flow

Oil at 20°C is being heated in a stirred tank. Oil enters the tank at the rate of 500 kg/h at 20°C and leaves at temperature T. The tank holds 2300 kg of oil, which is initially all at 20°C. The heat is provided by steam condensing at 130°C in coils submerged in the tank. The rate of heat transfer is given by

$$Q = h(T_{steam} - T_{oil})$$

The heat capacity of the oil is given by $C_p = 2.1$ J/(g°C) and the heat transfer coefficient is $h = 115$ J/(s°C). The shaft work of the stirrer is 560 W. When the process is started, how long does it take before the oil leaving the tank is at 30°C?

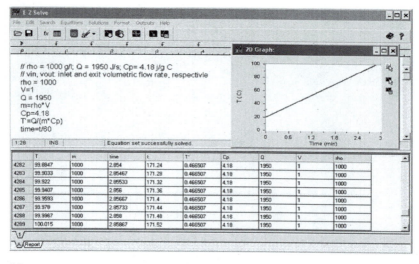

(b)

FIGURE E10.6b
Solution of the differential and algebraic equations using *E–Z* solve.

Solution
The general energy balance equation is

$$\frac{d(u)}{dt} = \sum \dot{m}_{in}\widehat{H}_{in} - \sum \dot{m}_{out}\widehat{H}_{out} + \dot{Q} - \dot{W}_s$$

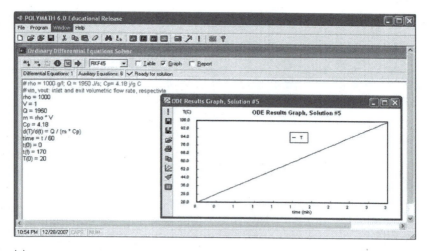

(c)

FIGURE E10.6c
Solution of the differential and algebraic equations using Polymath.

Data and assumptions

Reference: inlet oil temperature (i.e., $T_{ref} = T_0 = 20°C$).
System: oil in the tank (Figure E10.7a).
For liquids, C_p and C_v are approximately equal.
Specific heat of oil: $C_p = 2.1 \text{ J/g°C}$.
The shaft work of the stirrer is 560 W; the work is done on the system $W_s = -560 \text{ W} = -560 \text{ J/s}$.
The heat added to the oil from the steam is $Q = h \, (T_{steam} - T)$.
The heat transfer coefficient is $h = 115 \text{ J/(s°C)}$.

$$\text{Mass of oil in the tank } m = 2300 \text{ kg} \left| \frac{1000 \text{ g}}{1 \text{ kg}} \right. = 2.3 \times 10^6 \text{g}$$

$$\text{Inlet and exit mass flow rates: } \dot{m} = \frac{500 \text{ kg}}{h} \left| \frac{1000 \text{ g}}{1 \text{ kg}} \right| \frac{h}{3600 \text{ s}} = 139 \frac{g}{s}$$

The process flow diagram is shown in Figure E10.7a.
The general energy balance equation is simplified to the following equation:

$$\frac{d(mC_pT)}{dt} = \sum \dot{m}_{in}\widehat{H}_{in} - \sum \dot{m}_{out}\widehat{H}_{out} + \dot{Q} - \dot{W}_s$$

Rearranging the equation in order to collect and separate variables leads to

$$mC_p \frac{d(T)}{dt} = 0 - \dot{m}C_p(T - T_0) + h(T_{steam} - T) - \dot{W}_s$$

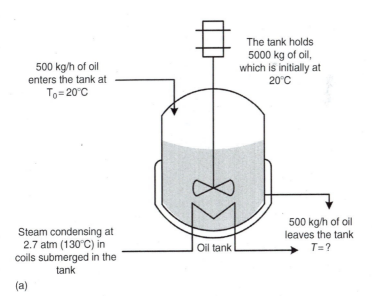

500 kg/h of oil
enters the tank at
$T_0 = 20°C$

The tank holds
5000 kg of oil,
which is initially at
20°C

Steam condensing at
2.7 atm (130°C) in
coils submerged in the
tank

Oil tank

500 kg/h of oil
leaves the tank
$T = ?$

(a)

FIGURE E10.7a
Heating oil using condensing steam.

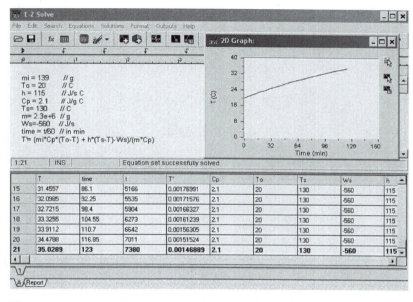

(b)

FIGURE E10.7b
Solution of the differential and algebraic equations using E–Z solve.

$$\frac{dT}{dt} = [0 - \dot{m}C_p(T - T_0) + h(T_{steam} - T) - \dot{W}_s]/mC_p$$

Solving the above differential equation using E–Z solve (Figure E10.7b) and Polymath (Figure E10.7c) reveals that the temperature of oil reaches 35°C in approximately 123 min.

Example 10.8   Quenching of an iron bar

An iron bar 2 cm × 3 cm × 10 cm at a temperature of 95°C is dropped into a barrel of water at 25°C. Density of the iron bar is 11.34 g/cm³. The barrel is large enough so that the water temperature rise is negligible as the bar cools. The rate at which heat is transferred from the bar to the water is given by the expression

$$Q(J/min) = UA(T_b - T_w)$$

where
$U = 0.050$ J/(min cm²°C)
U is the heat transfer coefficient
A is the exposed surface area of the bar
T_b (in °C) is the surface temperature of the bar
T_w (in °C) is the temperature of the water

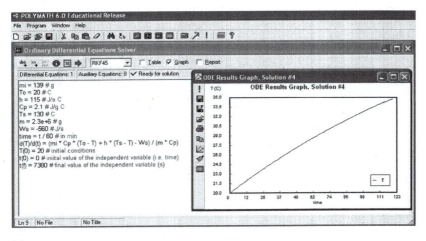

(c)

FIGURE E10.7c
Solution of differential and algebraic equations using Polymath.

Plot the temperature of the bar as a function of quench time and calculate the time for the bar to cool to 30°C. The heat capacity of the bar is 0.460 J/(g°C). Heat conduction in iron is rapid enough for the temperature of the bar to be uniform throughout. This latter concept is an important approximation called lumped capacitance, and it allows us to considerably simplify the problem because we do not have to worry about heat transfer within the solid bar itself.

Solution
System: iron block
Reference: 25°C
    Schematic of the process is shown in Figure E10.8a.

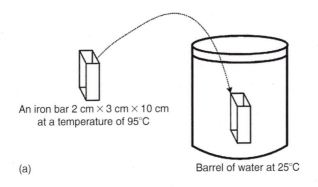

An iron bar 2 cm × 3 cm × 10 cm
at a temperature of 95°C

(a)

Barrel of water at 25°C

FIGURE E10.8a
Schematic diagram for immersion of iron hot block in a large barrel of water.

The general energy balance equation is

$$\frac{d(mC_pT)}{dt} = \sum \dot{m}_{in}\widehat{H}_{in} - \sum \dot{m}_{out}\widehat{H}_{out} + \dot{Q} - \dot{W}_s$$

Notice that for solids and liquids C_p and C_v are equal.

Data and assumptions

No inlet or exit flowing streams: $\sum \dot{m}_{in}\widehat{H}_{in} = \sum \dot{m}_{out}\widehat{H}_{out} = 0$.

The barrel is large enough so that the water temperature rise is negligible as the bar cools; this means the water temperature remains constant, T_w = constant.

No stirrer or shaft work: $W_s = 0$.

The heat transfer from the iron block: $Q(J/min) = - UA(T_b - T_w)$.

Heat transfer coefficient, $U = \frac{0.050 \, J}{min \cdot cm^2 \cdot °C} \left| \frac{1 \, min}{60 \, s} = 8.33 \times 10^{-4} \frac{J}{s \cdot cm^2 \cdot °C}$.

The heat capacity of the bar: $C_{p_b} = 0.460 \, J/(g°C)$.

Heat conduction in iron is rapid enough for the temperature of the bar to be uniform throughout.

Density of the iron bar: $\rho_b = 11.34 \, g/cm^3$

Volume of the block: $V_b = 2 \times 3 \times 10 \, cm = 60 \, cm^3$

Mass of the block: $m = \rho_b V_b$

Heat transfer area of the block: $A = 2(2 \times 3) + 2(2 \times 10) + 2(3 \times 10) = 112 \, cm^2$

We now rearrange the equation to obtain it in terms of the two variables T and t (T dependent variable, t independent variable)

$$\frac{dT}{dt} = \dot{Q}/mC_p$$

Substituting the heat transferred from the block

$$\frac{dT}{dt} = -UA(T - T_w)/mC_p$$

Solution of the above set of differential and algebraic equations using E–Z solve and Polymath gives the time required to cool the block to 30°C, which is around 2.47 h as shown in Figures E10.8b and E10.8c.

Example 10.9   Heating of solution

An electric heating coil is immersed in a stirred tank. The shaft work of the stirrer is 560 W. A solvent at 15°C with heat capacity 2.1 J/g°C is fed into the tank at a rate of 15 kg/h. Heated solvent is discharged at the same flow rate. The tank is filled initially with 125 kg cold solvent at 10°C. The rate of heating the electric coil is 2000 W. Calculate the time required for the temperature of the solvent to reach 60°C.

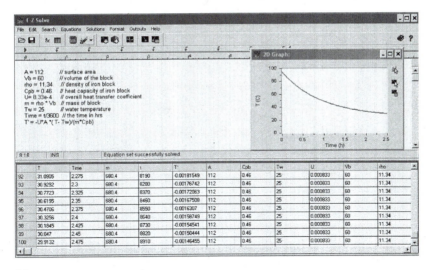

(b)

FIGURE E10.8b
Solution of the differential and algebraic equations using *E–Z* solve.

Solution

Reference: $T_{ref} = T_0 = 15°C$.

The general energy balance equation is simplified to the following equation:

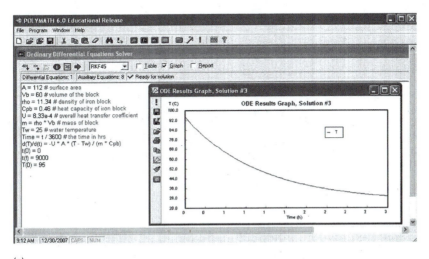

(c)

FIGURE E10.8c
Solution of the set of differential and algebraic equations using Polymath.

$$\frac{d(mC_pT)}{dt} = \sum \dot{m}_{in}\widehat{H}_{in} - \sum \dot{m}_{out}\widehat{H}_{out} + \dot{Q} - \dot{W}_s$$

Rearranging the equation in order to collect and separate variables gives

$$mC_p\frac{d(T)}{dt} = 0 - \dot{m}C_p(T - T_0) + \dot{Q} - \dot{W}_s$$

$$\frac{dT}{dt} = [0 - \dot{m}C_p(T - T_0) + \dot{Q} - \dot{W}_s]/mC_p$$

The schematic diagram is shown in Figure E10.9a.

Solving the above equation necessitates the units to be consistent.

Mass of oil: $m = 125 \text{ kg} \left|\frac{1000 \text{ g}}{1 \text{ kg}}\right. = 1.25 \times 10^5 \text{ g}$

Inlet and exit oil flow rates: $\dot{m} = \frac{15 \text{ kg}}{h}\left|\frac{1000 \text{ g}}{1 \text{ kg}}\right|\frac{h}{3600 \text{ s}} = 4.17\frac{g}{s}$

Heat capacity: $C_p = 2.1\frac{J}{g°C}$

Heat added to the oil from the cooling coil: $Q = 2000\frac{J}{s}$

Work applied on the system from the stirrer: $W_s = -560\frac{J}{s}$

The solution of the set of differential and algebraic equations using $E-Z$ solve and Polymath is shown in Figures E10.9b and E10.9c, respectively.

Example 10.10 Heating glycol solution

An adiabatic stirred tank is used to heat 100 kg of a 45 wt% glycol solution in water (mass heat capacity 3.54 J/g°C). An electrical coil delivers 2.5 kJ/s of

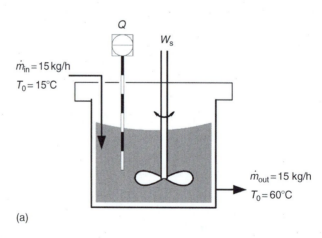

(a)

FIGURE E10.9a
Schematic diagram of heating oil tank using an electrical heater.

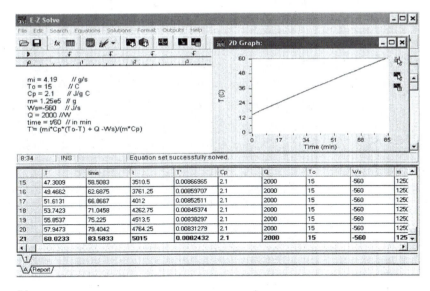

(b)

FIGURE E10.9b
Solution of the set of differential and algebraic equations using *E–Z* solve.

power to the tank; 88% of the energy delivered by the coil goes into heating the vessel contents. The shaft work of the stirrer is 500 W. The glycerol solution is initially at 15°C.

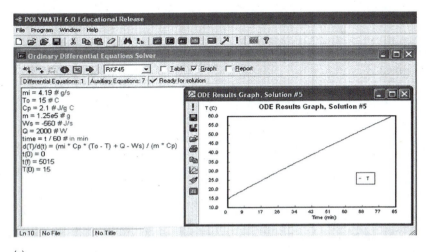

(c)

FIGURE E10.9c
Solution of the set of differential and algebraic equations using Polymath.

  a. Write a differential equation for the energy balance.
  b. Solve the equation using available software package.
  c. How long will the solution take to reach 90°C.

Solution
Reference: $T_{ref} = T_0 = 15°C$.

  (a) The general energy balance equation is simplified to the following equation:

$$\frac{d(u)}{dt} = \sum \dot{m}_{in}\widehat{H}_{in} - \sum \dot{m}_{out}\widehat{H}_{out} + \dot{Q} - \dot{W}_s$$

$$\frac{d(mC_vT)}{dt} = \sum \dot{m}_{in}\widehat{H}_{in} - \sum \dot{m}_{out}\widehat{H}_{out} + \dot{Q} - \dot{W}_s$$

Notice that for solids and liquids C_p and C_v are equal

$$\frac{d(mC_pT)}{dt} = \sum \dot{m}_{in}\widehat{H}_{in} - \sum \dot{m}_{out}\widehat{H}_{out} + \dot{Q} - \dot{W}_s$$

Assumptions
The system is adiabatic, that is, no heat is transferred to or from the surrounding, but still there is heat added by the electrical coil (Figure E10.10a).

$$\dot{Q}_{net} = \dot{Q}_e + \dot{Q}_{sur}$$

Since the system is adiabatic, $\dot{Q}_{sur} = 0$.
  Heat added to the glycol solution is 88% of the heating coil

$$\dot{Q}_e = 0.88 \times 2.5 \text{ kJ/s} = 2.2 \text{ kJ/s}$$

The mass of the glycol solution in the heated tank: $m = 100$ kg
The heat capacity of the glycol solution: $C_p = 3.54$ J/g°C
  Rearranging the equation in order to collect and separate variables gives

$$\frac{dT}{dt} = [\dot{Q} - \dot{W}_s]/mC_p$$

  (b) Results of the above equation solved using E–Z solve and Polymath are shown in Figures E10.10b and E10.10c, respectively.

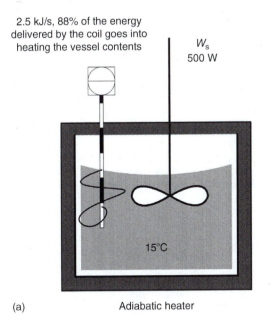

2.5 kJ/s, 88% of the energy delivered by the coil goes into heating the vessel contents

W_s
500 W

15°C

(a)   Adiabatic heater

FIGURE E10.10a
Schematic diagram of adiabatic tank for heating glycol solution.

Use the following data:

$$\dot{Q} = \frac{2.2\ \text{kJ}}{s} \left| \frac{1000\ \text{J}}{\text{kJ}} \right. = 2200\ \frac{\text{J}}{s}$$

$$W_s = -500\ \text{J/s}, \quad m = 100\ \text{kg}$$

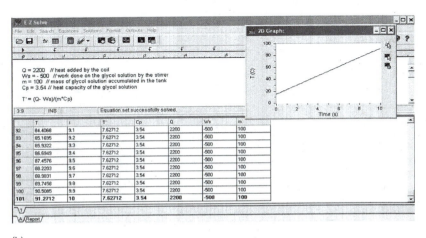

(b)

FIGURE E10.10b
Solution of the model equations using E–Z solve.

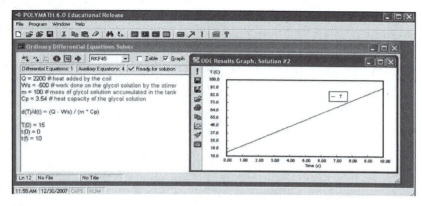

(c)

FIGURE E10.10c
Solution of the model equations using Polymath.

$$C_p = 3.54 \ \text{J/g}^\circ\text{C}$$

(c) As shown in Figures E10.10b and E10.10C below the solution will take around 10 s to reach the temperature of 90°C.

10.3 Problems

10.3.1 Fluid Flow from Storage Tank

A storage tank that is 2.0 m in diameter is filled at the rate of 2.0 m³/min. The exit fluid flow rate is flowing at a rate proportional to the head of fluid (0.5 h m³/min), where h is the height of fluid in meters. Plot the height of the liquid as a function of time. What is the steady-state height of the fluid in the tank?

10.3.2 Boiling of Water

A boiler used to boil water containing 100 L of water at a temperature of 25°C is placed on an electric heater ($Q = 3000$ J/s). Find the time at which water begins to boil. (Hint: Normal boiling point of water is 100°C.)

10.3.3 Heating Using Saturated Steam

A tank containing 1000 kg water at 25°C is heated using saturated steam at 130°C. The rate of heat transfer from the steam is given by the following equation:

$$\dot{Q} = UA(T_s - T)$$

$\dot{Q}$ is the rate of heat transfer to the system. U is the overall heat transfer coefficient, A is the surface area for heat transfer, and T is the temperature. The heat transfer area provided by the coil is 0.3 m^2 and the heat transfer coefficient is 220 (kcal)/m^2 h °C. The condensate leaves the coil as saturated steam.

a. The tank has a surface area of 0.9 m^2 exposed to the ambient air. The tank exchanges heat through this exposed surface at a rate given by an equation similar to that above. For heat transfer to or from the surrounding air, the heat transfer coefficient is 25 (kcal)/m^2 h °C. If the air temperature is 20°C, calculate the time required to heat the water to 80°C.

b. How much time can be saved if the tank is insulated?

10.3.4 Heating a Solvent in a Stirred Tank

A stirred tank is used to heat 100 kg of a solvent (mass heat capacity 2.5 J/g°C). An electrical coil delivers 2.0 kJ/s of power to the tank; the shaft work of the stirrer is 560 W. The solvent is initially at 25°C. The heat lost from the walls of the tank is 200 J/s.

a. Write a differential equation for the energy balance.

b. Solve the equation using the available software package.

c. How long will the solution take to reach 70°C?

10.3.5 Concentration of Reactant as a Function of Time

The following series reaction takes place in a constant volume batch reactor.

$$A \xrightarrow{k_1} B \xrightarrow{k_2} C$$

Each reaction is first order and is irreversible. If the initial concentration of A is 1 mol/L and if only A is present initially, find an expression for the concentrations of A, B, and C as a function of time ($k_1 = 0.1/s$, $k_2 = 0.2/s$).

Further Readings

1. Himmelblau, D.M. (1996) *Basic Principles and Calculations in Chemical Engineering*, 6th edn, Prentice-Hall, New Jersey.
2. Doran, P. (1995) *Bioprocess Engineering Principles*, Academic Press, San Diago.

3. Reklaitis, G.V. (1983) *Introduction to Material and Energy Balances*, John Wiley & Sons, New York.
4. Felder, R.M. and R.W. Rousseau (1999) *Elementary Principles of Chemical Processes*, 3rd edn, John Wiley, New York.
5. Atkinson, B. and F. Mavituna (1991) *Biochemical Engineering and Biotechnology Handbook*, 2nd edn, Macmillan, Basingstoke.

Appendices

Physical properties (Appendix A), heat capacities (Appendix B), and saturated and superheated steam table (Appendix C) are adapted with permission from the following references:

1. Himmelblau, D.M. (1996) *Basic Principles and Calculations in Chemical Engineering*, 6th edn, Prentice-Hall, New Jersey (Appendices A and B).
2. Haywood, R.W. (1968) *Thermodynamic Tables in SI (Metric) Units*, Cambridge University Press, London (Appendix C).

Appendix	Table	Title	Page
A	A.1	Selected physical properties	346
	A.2	Heats of formation and combustion	352
B	B.1	Heat capacities	358
C	C.1	Properties of saturated steam	362
	C.2	Properties of superheated steam	366

Appendix A

TABLE A.1

Physical Properties of Various Organic and Inorganic Substances

No.	Compound	Formula	Molecular Weight	Sp. Gr.	T_m (Melting Temp., K)	$\Delta \hat{H}_m$ (Fusion, kJ/mol)	T_b (Normal b.p., K)	$\Delta \hat{H}_v$ (Vaporization at T_b, kJ/mol)
1	Acetaldehyde	C_2H_4O	44.05	$0.783^{18/4°}$	149.5		293.2	24.4
2	Acetic acid	CH_3CHO_2	60.05	1.049	289.9	12.09	390.4	
3	Acetone	C_3H_6O	58.08	0.791	178.2	5.69	329.2	30.2
4	Acetylene	C_2H_2	26.04		191.7	3.7	191.7	17.5
5	Air			1.00				
6	Ammonia	NH_3	17.03	$0.817^{-79°}$	195.40	5.653	239.73	23.35
7	Ammonium carbonate	$(NH_4)_2CO_3 \cdot H_2O$	114.11			(decomposes at 331 K)		
8	Ammonium chloride	NH_4CL	53.50	$2.53^{17°}$		(decomposes at 623 K)		
9	Ammonium nitrate	NH_4NO_3	80.05	$1.725^{25°}$	442.8	5.4	(decomposes at 483.2 K)	
10	Ammonium sulfate	$(NH_4)_2SO_4$	132.14	1.769	786		(decomposes at 786 K after melting)	
11	Aniline	C_6H_7N	93.12	1.022	266.9		457.4	
12	Benzaldehyde	C_6H_5CHO	106.12	1.046	247.16		452.16	38.40
13	Benzene	C_6H_6	78.11	0.879	278.693	9.837	353.26	30.76
14	Benzoic acid	$C_7H_6O_2$	122.12	$1.316^{28/4°}$	395.4		523.0	
15	Benzyl alcohol	C_7H_8O	108.13	1.045	257.8		478.4	
16	Boron oxide	B_2O_3	69.64	1.85	723	22.0		
17	Bromine	Br_2	159.83	$3.119^{20°}$	265.8	10.8	331.78	31.0
18	1,2-Butadiene	C_4H_6	54.09	$0.652^{20°}$	136.7		283.3	
19	1,3-Butadiene	C_4H_6	54.09	0.621	164.1		268.6	
20	Butane	$n\text{-}C_4H_{10}$	58.12	0.579	134.83	4.661	272.66	22.31
21	iso-Butane	$iso\text{-}C_4H_{10}$	58.12	0.557	113.56	4.540	261.43	21.29
22	1-Butene	C_4H_8	56.10	0.60	87.81	3.848	266.91	21.92

No.	Name	Formula	M	Density	Melting point	ΔH_{fus}	Boiling point	ΔH_{vap}
23	Butyl phthalate	See Dibutyl phthalate						
24	*n*-Butyric acid	*n*-C$_4$H$_8$O$_2$	88.10	0.958	267		437.1	
25	*iso*-Butyric acid	*iso*-C$_4$H$_8$O$_2$	88.10	0.949	226		427.7	
26	Calcium arsenate	Ca$_3$(AsO$_4$)$_2$	398.06		1723			
27	Calcium carbide	Ca$_2$C$_2$	64.10	$2.22^{18°}$	2573			
28	Calcium carbonate	CaCO$_3$	100.09	2.93	(decomposes at 1098 K)			
29	Calcium chloride	CaCl$_2$	110.99	$2.152^{15°}$	1055	28.4		
		CaCl$_2$·H$_2$O	129.01					
		CaCl$_2$·2H$_2$O	147.03					
		CaCl$_2$·6H$_2$O	219.09	$1.78^{17°}$	303.4	37.3 (–6 H$_2$O at 473 K)		
30	Calcium cyanamide	CaCN$_2$	80.11	2.29				
31	Calcium cyanide	Ca(CN)$_2$	92.12					
32	Calcium hydroxide	Ca(OH)$_2$	74.10	2.24	(–H$_2$O at 853 K)			
33	Calcium oxide	CaO	56.08	2.62	2873		3123	
34	Calcium phosphate	Ca$_3$(PO$_4$)$_2$	310.19	3.14	1943	50		
35	Calcium silicate	CaSiO$_3$	117.17	2.915	1803	48.62		
36	Calcium sulfate (gypsum)	CaSO$_4$·2H$_2$O	172.18	2.32	(–1.5 H$_2$O at 301 K)			
37	Carbon	C	12.010	2.26	3873	46.0	4473	
38	Carbon dioxide	CO$_2$	44.01		$217.0^{5.2\,atm}$	8.32	(sublimate at 195 K)	
39	Carbon disulfide	CS$_2$	76.14	$1.261^{22°/20°}$	161.1	4.39	319.41	26.8
40	Carbon monoxide	CO	28.01		68.10	0.837	81.66	6.042
41	Carbon tetrachloride	CCl$_4$	153.84	1.595	250.3	2.5	349.9	30.0
42	Chlorine	Cl$_2$	70.91		172.16	6.406	239.10	20.41
43	Chlorobenzene	C$_6$H$_5$Cl	112.56	1.107	228		405.26	36.5
44	Chloroform	CHCl$_3$	119.39	$1.489^{20°}$	209.5		334.2	
45	Chromium	Cr	52.01	7.1				
46	Copper	Cu	63.54	8.92	1356.2	13.0	2855	305
47	Cumene	C$_9$H$_{12}$	120.19	0.862	177.125	7.1	425.56	37.5
48	Cupric sulfate	CuSO$_4$	159.61	$3.605^{15°}$	(decomposes at 873 K)			
49	Cyclohexane	C$_6$H$_{12}$	84.16	0.779	279.83	2.677	353.90	30.1
50	Cyclopentane	C$_5$H$_{10}$	70.13	0.745	179.71	0.6088	322.42	27.30

(continued)

TABLE A.1 (continued)

Physical Properties of Various Organic and Inorganic Substances

No.	Compound	Formula	Molecular Weight	Sp. Gr.	T_m (Melting Temp., K)	$\Delta\hat{H}_m$ (Fusion, kJ/mol)	T_b (Normal b.p., K)	$\Delta\hat{H}_v$ (Vaporization at T_b, kJ/mol)
51	Decane	$C_{10}H_{22}$	142.28	$0.730^{20°}$	243.3		447.0	
52	Dibutyl phthalate	$C_8H_{22}O_4$	278.34	$1.045^{21°}$			613	
53	Diethyl ether	$(C_2H_5)_2O$	74.12	$0.708^{25°}$	156.86	7.301	307.76	26.05
54	Ethane	C_2H_6	30.07		89.89	2.860	184.53	14.72
55	Ethanol	C_2H_6O	46.07	0.789	158.6	5.021	351.7	38.6
56	Ethyl acetate	$C_4H_8O_2$	88.10	0.901	189.4		350.2	
57	Ethyl benzene	C_8H_{10}	106.16	0.867	178.185	9.163	409.35	36.0
58	Ethyl bromide	C_2H_5Br	108.98	1.460	154.1		311.4	
59	Ethyl chloride	CH_3CH_2Cl	64.52	$0.903^{10°}$	134.83	4.452	285.43	25
60	3-Ethyl hexane	C_8C_{18}	114.22	0.7169			391.69	34.3
61	Ethylene	C_2H_4	28.05		103.97	3.351	169.45	13.54
62	Ethylene glycol	$C_2H_6O_2$	62.07	$1.113^{19°}$	260	11.23	470.4	56.9
63	Ferric oxide	Fe_2O_3	159.70	5.12	1833		(decomposes at 1833 K)	
64	Ferric sulfide	Fe_2S_3	207.90	4.3	(decomposes)			
65	Ferrous sulfide	FeS	87.92	4.84	1466	(decomposes)		
66	Formaldehyde	H_2CO	30.03	$0.815^{-20°}$	154.9		253.9	24.5
67	Formic acid	CH_2O_2	46.03	1.220	281.46	12.7	373.7	22.3
68	Glycerol	$C_3H_8O_3$	92.09	$1.260^{50°}$	291.36	18.30	563.2	
69	Helium	He	4.00		3.5	0.02	4.216	0.084
70	Heptane	C_7H_{16}	100.20	0.684	182.57	14.03	371.59	31.69
71	Hexane	C_6H_{14}	86.17	0.659	177.84	13.03	341.90	28.85
72	Hydrogen	H_2	2.016		13.96	0.12	20.39	0.904
73	Hydrogen chloride	HCl	36.47	1.15	158.94	1.99	188.11	16.15
74	Hydrogen fluoride	HF	20.01		238		293	
75	Hydrogen sulfide	H_2S	34.08		187.63	2.38	212.82	18.67
76	Iodine	I_2	253.8	$4.93^{20°}$	386.5		457.4	
77	Iron	Fe	55.85	7.7	1808	15	3073	353

No.	Name	Formula						
78	Iron oxide	Fe_3O_4	231.55	5.2	1867	138		(decomposes at 1867 K after melting)
79	Lead	Pb	207.21	$11.337^{20°}$	600.6	5.10	2023	180
80	Lead oxide	PbO	223.21	9.5	1159	11.7	1745	213
81	Magnesium	Mg	24.32	1.74	923	9.2	1393	132
82	Magnesium chloride	$MgCl_2$	95.23	$2.325^{25°}$	987	43.1	1691	137
83	Magnesium hydroxide	$Mg(OH)_2$	58.34	2.4			(decomposes at 623 K)	
84	Magnesium oxide	MgO	40.32	3.65	3173	77.4	3873	
85	Mercury	Hg	200.61	$13.546^{20°}$				
86	Methane	CH_4	16.04		90.68	0.941	111.67	8.180
87	Methanol	CH_3OH	32.04	0.792	175.26	3.17	337.9	35.3
88	Methyl acetate	$C_3H_6O_2$	74.08	0.933	174.3		330.3	
89	Methyl amine	CH_5N	31.06	$0.699^{-11°}$	180.5		$266.3^{758\ mm}$	
90	Methyl chloride	CH_3Cl	50.49		175.3		249	
91	Methyl ethyl ketone	C_4H_8O	72.10	0.805	186.1		352.6	
92	Methyl cyclohexane	C_7H_{14}	98.18	0.769	146.58	6.751	374.10	31.7
93	Molybdenum	Mo	95.95	10.2				
94	Napthalene	$C_{10}H_8$	128.16	1.145	353.2		491.0	
95	Nickel	Ni	58.69	$8.90^{20°}$	1725		3173	
96	Nitric acid	HNO_3	63.02	1.502	231.56	10.47	359	30.30
97	Nitrobenzene	$C_6H_5O_2N$	123.11	1.203	278.7		483.9	
98	Nitrogen	N_2	28.02		63.15	0.720	77.34	5.577
99	Nitrogen dioxide	NO_2	46.01		263.86	7.334	294.46	14.73
100	Nitrogen (nitric) oxide	NO	30.01		109.51	2.301	121.39	13.78
101	Nitrogen pentoxide	N_2O_5	108.02	$1.63^{18°}$	303		320	
102	Nitrogen tetraoxide	N_2O_4	92	$1.448^{20°}$	263.7		294.3	
103	Nitrogen trioxide	N_2O_3	76.02	$1.447^{2°}$	171		276.5	
104	Nitrous oxide	N_2O	44.02	$1.226^{-89°}$	182.1		184.4	
105	n-Nonane	C_9H_{20}	128.25	0.718	219.4		423.8	
106	n-Octane	C_8H_{18}	114.22	0.703	216.2		398.7	
107	Oxalic acid	$C_2H_2O_4$	90.04	1.90	(decomposes at 459 K)			

(continued)

TABLE A.1 (continued)

Physical Properties of Various Organic and Inorganic Substances

No.	Compound	Formula	Molecular Weight	Sp. Gr.	T_m (Melting Temp, K)	$\Delta\hat{H}_m$ (Fusion, kJ/mol)	T_b (Normal b.p., K)	$\Delta\hat{H}_v$ (Vaporization at T_b, kJ/mol)
108	Oxygen	O_2	32.00		54.40	0.443	90.19	6.820
109	n-Pentane	C_5H_{12}	72.15	$0.630^{18°}$	143.49	8.393	309.23	25.77
110	iso-Pentane	iso-C_5H_{12}	72.15	$0.621^{19°}$	113.1		300.9	
111	1-Pentane	C_5H_{10}	70.13	0.641	107.96	4.937	303.13	
112	Phenol	C_6H_5OH	94.11	$1.071^{25°}$	315.66	11.43	454.56	
113	Phenyl hydrazine	$C_6H_8N_2$	108.14	$1.097^{23°}$	292.76	16.43	51.66	
114	Phosphoric acid	H_3PO_4	98.00	$1.834^{18°}$	315.51	10.5	($-0.5H_2O$ at 486 K)	
115	Phosphorus (red)	P_4	123.90	2.20	863	81.17	863	41.84
116	Phosphorus (white)	P_4	123.90	1.82	317.4	2.5	553	49.71
117	Phosphorus pentoxide	P_2O_5	141.95	2.387	(sublimes at 523 K)			
118	Propane	C_3H_8	44.09		85.47	3.524	231.09	18.77
119	Propene	C_3H_6	42.08		87.91	3.002	255.46	18.42
120	Propionic acid	$C_3H_6O_2$	74.08	0.993	252.2		414.4	
121	n-Propyl alcohol	C_3H_8O	60.09	0.804	146		370.2	
122	Iso-Propyl alcohol	C_3H_8O	60.09	0.785	183.5		355.4	
123	n-Propyl benzene	C_9H_{12}	120.19	0.862	173.660	8.54	432.38	38.2
124	Silicon dioxide	SiO_2	60.09	2.25	1883	8.54	2503	
125	Sodium bisulfate	$NaHSO_4$	120.07	2.742	455			

No.	Name	Formula						
126	Sodium carbonate (sal soda)	$Na_2CO_3\cdot10H_2O$	286.15	1.46	306.5		(−H_2O at 306.5 K)	171
127	Sodium carbonate (soda ash)	Na_2CO_3	105.99	2.533	1127	33.4	(decomposes)	155
128	Sodium chloride	$NaCl$	58.45	2.163	1081	28.5	1738	
129	Sodium cyanide	$NaCN$	49.01		835	16.7	1770	
130	Sodium hydroxide	$NaOH$	40.00	2.130	592	8.4	1663	
131	Sodium nitrate	$NaNO_3$	85.00	2.257	583	15.9	(decomposes at 653 K)	
132	Sodium nitrite	$NaNO_2$	69.00	2.168^0	544		(decomposes at 593 K)	
133	Sodium sulfate	Na_2SO_4	142.05	2.698	1163	24.3		
134	Sodium sulfide	Na_2S	78.05	1.856	1223	6.7		
135	Sodium sulfite	Na_2SO_3	126.05	2.633^{15}		(decomposes)		
136	Sodium thiosulfate	$Na_2S_2O_3$	158.11	1.667				
137	Sulfur (rhombic)	S_8	256.53	2.07	386	10.0	717.76	84
138	Sulfur (monoclinic)	S_8	256.53	1.96	392	14.17	717.76	84
139	Sulfur chloride (mono)	S_2Cl_2	135.05	1.687	193.0		411.2	36.0
140	Sulfur dioxide	SO_2	64.07		197.68	7.402	263.14	24.92
141	Sulfur trioxide	SO_3	80.07		290.0	24.5	316.5	41.8
142	Sulfuric acid	H_2SO_4	98.08	1.834^{18}	283.51	9.87		
143	Toluene	$C_6H_5CH_3$	92.13	0.866	178.169	6.619	383.78	33.5
144	Water	H_2O	18.016	1.00^4	273.16	6.009	373.16	40.65
145	m-Xylene	C_8H_{10}	106.16	0.864	225.288	11.57	412.26	34.4
146	o-Xylene	C_8H_{10}	106.16	0.880	247.978	13.60	417.58	36.8
147	p-Xylene	C_8H_{10}	106.16	0.861	286.423	17.11	411.51	36.1
148	Zinc	Zn	65.38	7.140	692.7	6.673	1180	114.8
149	Zinc sulfate	$ZnSO_4$	161.44	3.74^{15}	(decomposes at 1013 K)			

TABLE A.2

Heats of Formation and Combustion

No.	Compound	Formula	Molecular Weight	State	$\Delta \hat{H}_f^\circ$ (kJ/mol)	$\Delta \hat{H}_c^\circ$ (kJ/mol)
1	Acetic acid	CH_3COOH	60.05	l	−486.2	−871.69
				g		−919.73
2	Acetaldehyde	CH_3CHO	40.052	g	−166.4	−1192.36
3	Acetone	C_3H_6O	58.08	aq, 200	−410.03	
4	Acetylene	C_2H_2	26.04	g	226.75	−1821.38
5	Ammonia	NH_3	17.032	g	−67.20	−1299.61
				l	−46.191	
6	Ammonium carbonate	$(NH_4)_2CO_3$	96.09	g		−382.58
				c	−941.86	
7	Ammonium chloride	NH_4Cl	53.50	aq	−315.4	
8	Ammonium hydroxide	NH_4OH	35.05	c	−366.5	
9	Ammonium nitrate	NH_4NO_3	80.05	aq	−366.1	
				c	−339.4	
10	Ammonium sulfate	$(NH_4)SO_4$	132.15	aq	−1179.3	
				c	−1173.1	
11	Benzaldehyde	C_6H_5CHO	106.12	aq	−88.83	
				l	−40.0	
12	Benzene	C_6H_6	78.11	g	48.66	−3267.6
				l	82.927	−3301.5
13	Boron oxide	B_2O_3	69.64	g	−1263	
				c	−1245.2	
14	Bromine	Br_2	159.832	l	0	
				g	30.7	
15	n-Butane	C_4H_{10}	58.12	l	−147.6	−2855.6
				g	−124.73	−2878.52
16	Isobutane	C_4H_{10}	58.12	l	−158.5	−2849.0
				g	−134.5	−2868.8

No.	Name	Formula	Molar mass	State	ΔH	ΔH (combustion)
17	1-Butene	C_4H_8	56.104	g	1.172	−2718.58
18	Calcium arsenate	$Ca_3(AsO_4)_2$	398.06	c	−3330.5	
19	Calcium carbide	CaC_2	64.10	c	−62.7	
20	Calcium carbonate	$CaCO_3$	100.09	c	−1206.9	
21	Calcium chloride	$CaCl_2$	110.99	c	−794.9	
22	Calcium cyanamide	$CaCN_2$	80.11	c	−352	
23	Calcium hydroxide	$Ca(OH)_2$	74.10	c	−986.56	
24	Calcium oxide	CaO	56.08	c	−635.6	
25	Calcium phosphate	$Ca_3(PO_4)_2$	310.19	c	−4137.6	
26	Calcium silicate	$CaSiO_3$	116.17	c	−1584	
27	Calcium sulfate	$CaSO_4$	136.15	c	−1432.7	
				aq	−1450.5	
28	Calcium sulfate (gypsum)	$CaSO_4 \cdot 2H_2O$	172.18	c	−2021.1	
29	Carbon	C	12.01	c Graphite (β)	0	−393.51
30	Carbon dioxide	CO_2	44.01	g	−393.51	
				l	−412.92	
31	Carbon disulfide	CS_2	76.14	l	87.86	−1075.2
				g	115.3	−1102.6
32	Carbon monoxide	CO	28.01	g	−110.52	−282.99
33	Carbon tetrachloride	CCl_4	153.838	l	−139.5	−352.2
				g	−106.69	−384.9
34	Chloroethane	C_2H_5Cl	64.52	g	−105.0	−1421.1
35	Cumene (isopropylbenzene)	$C_6H_5CH(CH_3)_2$	120.19	l	−41.20	−5215.44
				g	3.93	−5260.59
36	Cupric sulfate	$CuSO_4$	159.61	c	−769.86	
				aq	−843.12	
				c	−751.4	
37	Cyclohexane	C_6H_{12}	84.16	g	−123.1	−3953.0
38	Cyclopentane	C_5H_{10}	70.130	l	−105.8	−3290.9

(continued)

TABLE A.2 (continued)

Heats of Formation and Combustion

No.	Compound	Formula	Molecular Weight	State	$\Delta \hat{H}_f^\circ$ (kJ/mol)	$\Delta \hat{H}_c^\circ$ (kJ/mol)
39	Ethane	C_2H_6	30.07	g	−77.23	−3319.5
40	Ethyl acetate	$CH_3CO_2C_2H_5$	88.10	g	−84.667	−1559.9
41	Ethyl alcohol	C_2H_5OH	46.068	l	−442.92	−2274.48
				l	−277.63	−1366.91
42	Ethyl benzene	$C_6H_5\cdot C_2H_5$	106.16	g	−235.31	−1409.25
				l	−12.46	−4564.87
43	Ethyl chloride	C_2H_5Cl	64.52	g	29.79	−4607.13
44	Ethylene	C_2H_4	28.052	g	−105	−1410.99
45	Ethylene chloride	C_2H_3Cl	62.50	g	52.283	−1271.5
46	3-Ethyl hexane	C_8H_{18}	114.22	l	31.38	−5470.12
				l	−250.5	−5509.78
47	Ferric chloride	$FeCl_3$		g	−210.9	
48	Ferric oxide	Fe_2O_3	159.70	c	−403.34	
49	Ferric sulfide	FeS_2	See Iron sulfide	See Iron sulfide	−822.156	
50	Ferrosoferric oxide	Fe_3O_4	231.55	c	−1116.7	
51	Ferrous chloride	$FeCl_2$		c	−342.67	−303.76
52	Ferrous oxide	FeO	71.85	c	−267	
53	Ferrous sulfide	FeS	87.92	c	−95.06	
54	Formaldehyde	H_2CO	30.026	g	−115.89	−563.46
55	*n*-Heptane	C_7H_{16}	100.20	l	−224.4	−4816.91
				l	−187.8	−4853.48
56	*n*-Hexane	C_6H_{14}	86.17	g	−198.8	−4163.1
				l	−167.2	−4194.753
57	Hydrogen	H_2	2.016	g	0	−285.84
58	Hydrogen bromide	HBr	80.924	g	−36.23	

	Name	Formula	Molar mass	State		
59	Hydrogen chloride	HCl	36.465	g	−92.311	
60	Hydrogen cyanide	HCN	27.026	g	130.54	
61	Hydrogen sulfide	H$_2$S	34.082	g	−20.15	−562.589
62	Iron sulfide	FeS$_2$	119.98	c	−177.9	
63	Lead oxide	PbO	223.21	c	−219.2	
64	Magnesium chloride	MgCl$_2$	95.23	c	−641.83	
65	Magnesium hydroxide	Mg(OH)$_2$	58.34	c	−924.66	
66	Magnesium oxide	MgO	40.32	c	−601.83	
67	Methane	CH$_4$	16.041	g	−74.84	−890.4
68	Methyl alcohol	CH$_3$OH	32.042	l	−238.64	−726.55
				g	−201.25	−763.96
69	Methyl chloride	CH$_3$Cl	50.49	g	−81.923	−766.63
70	Methyl cyclohexane	C$_7$H$_{14}$	98.182	l	−190.2	−4565.29
				g	−154.8	−4600.68
71	Methyl cyclopentane	C$_6$H$_{12}$	84.156	l	−138.4	−3937.7
				g	−106.7	−3969.4
72	Nitric acid	HNO$_3$	63.02	l	−173.23	
				aq	−206.57	
73	Nitric oxide	NO	30.01	g	90.374	
74	Nitrogen dioxide	NO$_2$	46.01	g	33.85	
75	Nitrous oxide	N$_2$O	44.02	g	81.55	
76	n-Pentane	C$_5$H$_{12}$	72.15	l	−173.1	−3509.5
				g	−146.4	−3536.15
77	Phosphoric acid	H$_3$PO$_4$	98.00	c	−1281	
				aq (H$_2$O)	−1278	
78	Phosphorus	P$_4$	123.90	c	0	
79	Phosphorus pentoxide	P$_2$O$_5$	141.95	c	−1506	
80	Propane	C$_3$H$_8$	44.09	l	−119.84	−2204.0
				g	−103.85	−2220.0
81	Propene	C$_3$H$_6$	42.078	g	20.41	−2058.47
82	n-Propyl alcohol	C$_3$H$_8$O	60.09	g	−255	−2068.6
83	n-Propylbenzene	C$_6$H$_5$·CH$_2$·C$_2$H$_5$	120.19	l	−38.40	−5218.2

(continued)

TABLE A.2 (continued)

Heats of Formation and Combustion

No.	Compound	Formula	Molecular Weight	State	$\Delta \hat{H}_f^\circ$ (kJ/mol)	$\Delta \hat{H}_c^\circ$ (kJ/mol)
84	Silicon dioxide	SiO_2	60.09	g	7.824	−5264.5
85	Sodium bicarbonate	$NaHCO_3$	84.01	c	−851.0	
86	Sodium bisulfate	$NaHSO_4$	120.07	c	−945.6	
87	Sodium carbonate	Na_2CO_3	105.99	c	−1126	
88	Sodium chloride	$NaCl$	58.45	c	−1130	
89	Sodium cyanide	$NaCN$	49.01	c	−411.00	
90	Sodium nitrate	$NaNO_3$	85.00	c	−89.79	
91	Sodium nitrite	$NaNO_2$	69.00	c	−466.68	
92	Sodium sulfate	Na_2SO_4	142.05	c	−359	
93	Sodium sulfide	Na_2S	78.05	c	−1384.5	
94	Sodium sulfite	Na_2SO_3	126.05	c	−373	
95	Sodium thiosulfate	$Na_2S_2O_3$	158.11	c	−1090	
96	Sulfur	S	32.07	c	−1117	
				(rhombic) c	0	
				(monoclinic)	0.297	

No.	Name	Formula	MW	State	$\Delta \hat{H}_f^0$	$\Delta \hat{H}_c^0$
97	Sulfur chloride	S_2Cl_2	135.05	l	−60.3	
98	Sulfur dioxide	SO_2	64.066	g	−296.90	
99	Sulfur trioxide	SO_3	80.066	g	−395.18	
100	Sulfuric acid	H_2SO_4	98.08	l	−811.32	
				aq	−907.51	
101	Toluene	$C_6H_5CH_3$	92.13	l	11.99	−3909.9
				g	50.000	−3947.9
102	Water	H_2O	18.016	g	−285.840	
				l	−241.826	
103	m-Xylene	$C_6H_4(CH_3)_2$	106.16	g	−25.42	−4551.86
				l	17.24	−4594.53
104	o-Xylene	$C_6H_4(CH_3)_2$	106.16	g	−24.44	−4552.86
				l	19.00	−4596.29
105	p-Xylene	$C_6H_4(CH_3)_2$	106.16	g	−24.43	−4552.86
				l	17.95	−4595.25
106	Zinc sulfate	$ZnSO_4$	161.45	c	−978.55	
				aq	−1059.93	

Note: Heats of formation and combustion of compounds at 25°C. Standard states of products for $\Delta \hat{H}_c^0$ are CO_2 (g), H_2O (l), N_2 (g), SO_2 (g), and HCl (aq).

Appendix B

Forms:

(1) C_p J/(g mol)(K or °C). $= a + bT + cT^2 + dT^3$

(2) C_p J/(g mol)(K or °C) $= a + bT + cT^{-2}$

TABLE B.1

Heat Capacity Equations for Organic and Inorganic Compounds (at Low Pressures)

No.	Compound	Formula	Mol. Wt.	State	Form	T	a	B	c	d	Temp. Range (in T)
1	Acetone	CH_3COCH_3	58.08	g	1	°C	71.96	0.201	-1.278×10^{-4}	3.476×10^{-8}	0–1200
2	Acetylene	C_2H_2	26.04	g	1	°C	42.43	0.06053	-5.033×10^{-5}	1.820×10^{-8}	0–1200
3	Air		29.0	g	1	°C	28.94	0.004147	0.3191×10^{-5}	-1.965×10^{-9}	0–1500
				g	1	K	28.09	0.001965	0.4799×10^{-5}	-1.965×10^{-9}	273–1800
4	Ammonia	NH_3	17.03	g	1	K	35.15	0.02954	0.4421×10^{-5}	-6.686×10^{-9}	273–1800
5	Ammonium sulfate	$(NH_4)_2SO_4$	132.15	c	1	K	215.9		0		275–328
6	Benzene	C_6H_6	78.11	l	1	K	-7.27329	0.77054	-1.6482×10^{-3}	1.8979×10^{-6}	279–350
				g	1	°C	74.06	0.3295	-2.520×10^{-4}	7.757×10^{-8}	0–1200
7	Isobutane	C_4H_{10}	58.12	g	1	°C	89.46	0.3013	-1.891×10^{-4}	4.987×10^{-8}	0–1200
8	n-Butane	C_4H_{10}	58.12	g	1	°C	92.30	0.2788	-1.547×10^{-4}	3.498×10^{-8}	0–1200
9	Isobutene	C_4H_8	56.10	g	1	°C	82.88	0.2564	-1.727×10^{-4}	5.050×10^{-8}	0–1200
10	Calcium carbide	CaC_2	64.10	c	2	K	68.62	0.0119	-8.66×10^{-5}	—	298–720
11	Calcium carbonate	$CaCO_3$	100.09	c	2	K	82.34	0.04975	-1.287×10^{-4}	—	273–1033
12	Calcium hydroxide	$Ca(OH)_2$	74.10	c	1	K	89.5				276–373
13	Calcium oxide	CaO	56.08	c	2	K	41.84	0.0203	-4.52×10^{-5}		273–1173
14	Carbon	C	12.01	c#	2	K	11.18	0.01095	-4.891×10^{-5}		273–1373
15	Carbon dioxide	CO_2	44.01	g	1	°C	36.11	0.04233	-2.887×10^{-5}	7.464×10^{-9}	0–1500
16	Carbon monoxide	CO	28.01	g	1	°C	28.95	0.00411	0.3548×10^{-5}	-2.220×10^{-9}	0–1500
17	Carbon tetrachloride	CCl_4	153.84	l	1	K	12.285	0.0001095	-3.1826×10^{-3}	3.4252×10^{-6}	273–343
18	Chlorine	Cl_2	70.91	g	1	°C	33.60	0.01367	-1.607×10^{-5}	6.473×10^{-9}	0–1200
19	Copper	Cu	63.54	c	1	K	22.76	0.0006117			273–1357
20	Cumene	C_9H_{12}	120.19	G	1	°C	139.2	0.5376	-3.979×10^{-4}	1.205×10^{-7}	0–1200
21	Cyclohexane	C_6H_{12}	84.16	G	1	°C	94.140	0.4962	-3.190×10^{-4}	8.063×10^{-8}	0–1200
22	Cyclopentane	C_5H_{10}	70.13	G	1	°C	73.39	0.3928	-2.554×10^{-4}	6.866×10^{-8}	0–1200
23	Ethane	C_2H_6	30.07	G	1	°C	49.37	0.1392	-5.816×10^{-5}	7.280×10^{-9}	0–1200

(continued)

TABLE B.1 (continued)

Heat Capacity Equations for Organic and Inorganic Compounds (at Low Pressures)

No.	Compound	Formula	Mol. Wt.	State	Form	T	a	B	c	d	Temp. Range (in T)
24	Ethyl alcohol	C_2H_6O	46.07	l	1	K	−325.137	0.00041379	-1.4031×10^{-2}	1.7035×10^{-5}	250–400
				G	1	°C	61.34	0.1572	-8.749×10^{-5}	1.983×10^{-8}	0–1200
25	Ethylene	C_2H_4	28.05	G	1	°C	40.75	0.1147	-6.891×10^{-5}	1.766×10^{-8}	0–1200
26	Ferric oxide	Fe_2O_3	159.70	C	2	K	103.4	0.06711	-17.72×10^{-5}	—	273–1097
27	Formaldehyde	CH_2O	30.03	G	1	°C	34.28	0.04268	0.0000	-8.694×10^{-9}	0–1200
28	Helium	He	4.00	G	1	°C	20.8				All
29	n-Hexane	C_6H_{14}	86.17	l	1	K	31.421	0.0097606	-2.3537×10^{-3}	3.0927×10^{-6}	273–400
				G	1	°C	137.44	0.4085	-2.392×10^{-4}	5.766×10^{-8}	0–1200
30	Hydrogen	H_2	2.016	G	1	°C	28.84	0.0000765	0.3288×10^{-5}	-0.8698×10^{-9}	0–1500
31	Hydrogen bromide	HBr	80.92	g	1	°C	29.10	−0.000227	0.9887×10^{-5}	-4.858×10^{-9}	0–1200
32	Hydrogen chloride	HCl	36.47	g	1	°C	29.13	−0.001341	0.9715×10^{-5}	-4.335×10^{-9}	0–1200
33	Hydrogen cyanide	HCN	27.03	g	1	°C	35.3	0.02908	1.092×10^{-5}		0–1200
34	Hydrogen sulfide	H_2S	34.08	g	1	°C	33.51	0.01547	0.3012×10^{-5}	-3.292×10^{-9}	0–1500
35	Magnesium chloride	$MgCl_2$	95.23	c	1	K	72.4	0.0158			273–991
36	Magnesium oxide	MgO	40.32	c	2	K	45.44	0.005008	-8.732×10^{-5}		273–2073
37	Methane	CH_4	16.04	g	1	°C	34.31	0.05469	0.3661×10^{-5}	-1.100×10^{-8}	0–1200
				g	1	K	19.87	0.05021	1.268×10^{-5}	-1.100×10^{-8}	273–1500
38	Methyl alcohol	CH_3OH	32.04	l	1	K	−259.25	0.0003358	-1.1639×10^{-5}	1.4052×10^{-5}	273–400
				g		°C	42.93	0.08301	-1.87×10^{-5}	-8.03×10^{-9}	0–700

39	Methyl cyclohexane	C_7H_{14}	98.18	g	1	°C	121.3	0.5653	-3.772×10^{-4}	1.008×10^{-7}	0–1200
40	Methyl cyclopentane	C_6H_{12}	84.16	g	1	°C	98.83	0.45857	-3.044×10^{-4}	8.381×10^{-8}	0–1200
41	Nitric acid	HNO_3	63.02	l	1	°C	110.0				25
42	Nitric oxide	NO	30.01	g	1	°C	29.50	0.008188	-0.2925×10^{-5}	0.3652×10^{-9}	0–3500
43	Nitrogen	N_2	28.02	g	1	°C	29.00	0.002199	0.5723×10^{-5}	-2.871×10^{-9}	0–1500
44	Nitrogen dioxide	NO_2	46.01	g	1	°C	36.07	0.0397	-2.88×10^{-5}	7.87×10^{-9}	0–1200
45	Nitrogen tetraoxide	N_2O_4	92.02	g	1	°C	75.7	0.125	-1.13×10^{-4}		0–300
46	Nitrous oxide	N_2O	44.02	g	1	°C	37.66	0.04151	-2.694×10^{-5}	1.057×10^{-8}	0–1200
47	Oxygen	O_2	32.00	g	1	°C	29.10	0.01158	-0.6076×10^{-5}	1.311×10^{-9}	0–1500
48	n-Pentane	C_5H_{12}	72.15	l	1	K	33.24	1.9241	-2.3687×10^{-3}	1.7944×10^{-5}	270–350
					1	°C	114.8	0.3409	-1.899×10^{-4}	4.226×10^{-8}	0–1200
49	Propane	C_3H_8	44.09	g	1	°C	68.032	0.2259	-1.311×10^{-4}	3.171×10^{-8}	0–1200
50	Propylene	C_3H_6	42.08	g	1	°C	59.580	0.1771	-1.017×10^{-4}	2.460×10^{-8}	0–1200
51	Sodium carbonate	Na_2CO_3	105.99	c	1	K	121				288–371
52	Sodium carbonate · 10H₂O	$Na_2CO_3 \cdot 10H_2O$	286.15	c	1	K	535.6				298
53	Sulfur	S	32.07	c*	1	K	15.2	0.0268			273–368
				c§	1	K	18.5	0.0184			368–392
54	Sulfuric acid	H_2SO_4	98.08	l	1	°C	139.1	0.1559			10–45
55	Sulfur dioxide	SO_2	64.07	g	1	°C	38.91	0.03904	-3.104×10^{-5}	8.606×10^{-9}	0–1500
56	Sulfur trioxide	SO_3	80.07	g	1	°C	48.50	0.09188	-8.540×10^{-5}	3.240×10^{-8}	0–1000
57	Toluene	C_7H_8	92.13	l	1	K	1.8083	0.81222	-151.27×10^{-5}	1.630×10^{-6}	270–370
				g	1	°C	94.18	0.3800	-27.86×10^{-5}	8.033×10^{-8}	0–1200
58	Water	H_2O	18.016	l	1	K	18.2964	0.47212	-133.88×10^{-5}	1.3142×10^{-6}	273–373
				g	1	°C	33.46	0.00688	0.7604×10^{-5}	-3.593×10^{-9}	0–1500

Graphite

* Rhombic. § Moinoclinik (at 1 atm)

Appendix C: Steam Table

TABLE C.1

Properties of Saturated Steam

P (bar)	T (°C)	$\hat{V}$ (m³/kg)		$\hat{U}$ (kJ/kg)		$\hat{H}$ (kJ/mol)		
		Water	Steam	Water	Steam	Water	Evap.	Steam
0.00611	0.01	0.001000	206.2	0.0	2375.6	0.0	2501.6	2501.6
0.008	3.8	0.001000	159.7	15.8	2380.7	15.8	2492.6	2508.5
0.010	7.0	0.001000	129.2	29.3	2385.2	29.3	2485.0	2514.4
0.012	9.7	0.001000	108.7	40.6	2388.9	40.6	2478.7	2519.3
0.014	12.0	0.001000	93.9	50.3	2392.0	50.3	2473.2	2523.5
0.016	14.0	0.001001	82.8	58.9	2394.8	58.9	2468.4	2527.3
0.018	15.9	0.001001	74.0	66.5	2397.4	66.5	2464.1	2530.6
0.020	17.5	0.001001	67.0	73.5	2399.6	73.5	2460.2	2533.6
0.022	19.0	0.001002	61.2	79.8	2401.7	79.8	2456.6	2536.4
0.024	20.4	0.001002	56.4	85.7	2403.6	85.7	2453.3	2539.0
0.026	21.7	0.001002	52.3	91.1	2405.4	91.1	2450.2	2541.3
0.028	23.0	0.001002	48.7	96.2	2407.1	96.2	2447.3	2543.6
0.030	24.1	0.001003	45.7	101.0	2408.6	101.0	2444.6	2545.6
0.035	26.7	0.001003	39.5	111.8	2412.2	111.8	2438.5	2550.4
0.040	29.0	0.001004	34.8	121.4	2415.3	121.4	2433.1	2554.5
0.045	31.0	0.001005	31.1	130.0	2418.1	130.0	2428.2	2558.2
0.050	32.9	0.001005	28.2	137.8	2420.6	137.8	2423.8	2561.6
0.060	36.2	0.001006	23.74	151.5	2425.1	151.5	2416.0	2567.5
0.070	39.0	0.001007	20.53	163.4	2428.9	163.4	2409.2	2572.6
0.080	41.5	0.001008	18.10	173.9	2432.3	173.9	2403.2	2577.1
0.090	43.8	0.001009	16.20	183.3	2435.3	183.3	2397.9	2581.1
0.10	45.8	0.001010	14.67	191.8	2438.0	191.8	2392.9	2584.8
0.11	47.7	0.001011	13.42	199.7	2440.5	199.7	2388.4	2588.1
0.12	49.4	0.001012	12.36	206.9	2442.8	206.9	2384.3	2591.2
0.13	51.1	0.001013	11.47	213.7	2445.0	213.7	2380.4	2594.0
0.14	52.6	0.001013	10.69	220.0	2447.0	220.0	2376.7	2596.7
0.15	54.0	0.001014	10.02	226.0	2448.9	226.0	2373.2	2599.2
0.16	55.3	0.001015	9.43	231.6	2450.6	231.6	2370.0	2601.6
0.17	56.6	0.001015	8.91	236.9	2452.3	236.9	2366.9	2603.8
0.18	57.8	0.001016	8.45	242.0	2453.9	242.0	2363.9	2605.9
0.19	59.0	0.001017	8.03	246.8	2455.4	246.8	2361.1	2607.9
0.20	60.1	0.001017	7.65	251.5	2456.9	251.5	2358.4	2609.9
0.22	62.2	0.001018	7.00	260.1	2459.6	260.1	2353.3	2613.5
0.24	64.1	0.001019	6.45	268.2	2462.1	268.2	2348.6	2616.8
0.26	65.9	0.001020	5.98	275.6	2464.4	275.7	2344.2	2619.9
0.28	67.5	0.001021	5.58	282.7	2466.5	282.7	2340.0	2622.7
0.30	69.1	0.001022	5.23	289.3	2468.6	289.3	2336.1	2625.4
0.35	72.7	0.001025	4.53	304.3	2473.1	304.3	2327.2	2631.5
0.40	75.9	0.001027	3.99	317.6	2477.1	317.7	2319.2	2636.9
0.45	78.7	0.001028	3.58	329.6	2480.7	329.6	2312.0	2641.7
0.50	81.3	0.001030	3.24	340.5	2484.0	340.6	2305.4	2646.0
0.55	83.7	0.001032	2.96	350.6	2486.9	350.6	2299.3	2649.9

TABLE C.1 (continued)

Properties of Saturated Steam

P (bar)	T (°C)	$\hat{V}$ (m³/kg)		$\hat{U}$ (kJ/kg)		$\hat{H}$ (kJ/mol)		
		Water	Steam	Water	Steam	Water	Evap.	Steam
0.60	86.0	0.001033	2.73	359.9	2489.7	359.9	2293.6	2653.6
0.65	88.0	0.001035	2.53	368.5	2492.2	368.6	2288.3	2656.9
0.70	90.0	0.001036	2.36	376.7	2494.5	376.8	2283.3	2660.1
0.75	91.8	0.001037	2.22	384.4	2496.7	384.5	2278.6	2663.0
0.80	93.5	0.001039	2.087	391.6	2498.8	391.7	2274.1	2665.8
0.85	95.2	0.001040	1.972	398.5	2500.8	398.6	2269.8	2668.4
0.90	96.7	0.001041	1.869	405.1	2502.6	405.2	2265.6	2670.9
0.95	98.2	0.001042	1.777	411.4	2504.4	411.5	2261.7	2673.2
1.00	99.6	0.001043	1.694	417.4	2506.1	417.5	2257.9	2675.4
1.01325	100.0	0.001044	1.673	419.0	2506.5	419.1	2256.9	2676.0
1.1	102.3	0.001046	1.549	428.7	2509.2	428.8	2250.8	2679.6
1.2	104.8	0.001048	1.428	439.2	2512.1	439.4	2244.1	2683.4
1.3	107.1	0.001049	1.325	449.1	2514.7	449.2	2237.8	2687.0
1.4	109.3	0.001051	1.236	458.3	2517.2	458.4	2231.9	2690.3
1.5	111.4	0.001053	1.159	467.0	2519.5	467.1	2226.2	2693.4
1.6	113.3	0.001055	1.091	475.2	2521.7	475.4	2220.9	2696.2
1.7	115.2	0.001056	1.031	483.0	2523.7	483.2	2215.7	2699.0
1.8	116.9	0.001058	0.977	490.5	2525.6	490.7	2210.8	2701.5
1.9	118.6	0.001059	0.929	497.6	2527.5	497.8	2206.1	2704.0
2.0	120.2	0.001061	0.885	504.5	2529.2	504.7	2201.6	2706.3
2.2	123.3	0.001064	0.810	517.4	2532.4	517.6	2193.0	2710.6
2.4	126.1	0.001066	0.746	529.4	2535.4	529.6	2184.9	2714.5
2.6	128.7	0.001069	0.693	540.6	2538.1	540.9	2177.3	2718.2
2.8	131.2	0.001071	0.646	551.1	2540.6	551.4	2170.1	2721.5
3.0	133.5	0.001074	0.606	561.1	2543.0	561.4	2163.2	2724.7
3.2	135.8	0.001076	0.570	570.6	2545.2	570.9	2156.7	2727.6
3.4	137.9	0.001078	0.538	579.6	2547.2	579.9	2150.4	2730.3
3.6	139.9	0.001080	0.510	588.1	2549.2	588.5	2144.4	2732.9
3.8	141.8	0.001082	0.485	596.4	2551.0	596.8	2138.6	2735.3
4.0	143.6	0.001084	0.462	604.2	2552.7	604.7	2133.0	2737.6
4.2	145.4	0.001086	0.442	611.8	2554.4	612.3	2127.5	2739.8
4.4	147.1	0.001088	0.423	619.1	2555.9	619.6	2122.3	2741.9
4.6	148.7	0.001089	0.405	626.2	2557.4	626.7	2117.2	2743.9
4.8	150.3	0.001091	0.389	633.0	2558.8	633.5	2112.2	2745.7
5.0	151.8	0.001093	0.375	639.6	2560.2	640.1	2107.4	2747.5
5.5	155.5	0.001097	0.342	655.2	2563.3	655.8	2095.9	2751.7
6.0	158.8	0.001101	0.315	669.8	2566.2	670.4	2085.0	2755.5
6.5	162.0	0.001105	0.292	683.4	2568.7	684.1	2074.7	2758.9
7.0	165.0	0.001108	0.273	696.3	2571.1	697.1	2064.9	2762.0
7.5	167.8	0.001112	0.2554	708.5	2573.3	709.3	2055.5	2764.8
8.0	170.4	0.001115	0.2403	720.0	2575.5	720.9	2046.5	2767.5
8.5	172.9	0.001118	0.2268	731.1	2577.1	732.0	2037.9	2769.9
9.0	175.4	0.001121	0.2148	741.6	2578.8	742.6	2029.5	2772.1
9.5	177.7	0.001124	0.2040	751.8	2580.4	752.8	2021.4	2774.2
10	179.9	0.001127	0.1943	761.5	2581.9	762.6	2013.6	2776.2
11	184.1	0.001133	0.1774	779.9	2584.5	781.1	1998.5	2779.7
12	188.0	0.001139	0.1632	797.1	2586.9	798.4	1984.3	2782.7
13	191.6	0.001144	0.1511	813.2	2589.0	814.7	1970.7	2785.4

(continued)

TABLE C.1 (continued)

Properties of Saturated Steam

P (bar)	T (°C)	$\hat{V}$ (m³/kg) Water	Steam	$\hat{U}$ (kJ/kg) Water	Steam	$\hat{H}$ (kJ/mol) Water	Evap.	Steam
14	195.0	0.001149	0.1407	828.5	2590.8	830.1	1957.7	2787.8
15	198.3	0.001154	0.1317	842.9	2592.4	844.7	1945.2	2789.9
16	201.4	0.001159	0.1237	856.7	2593.8	858.6	1933.2	2791.7
17	204.3	0.001163	0.1166	869.9	2595.1	871.8	1921.5	2793.4
18	207.1	0.001168	0.1103	882.5	2596.3	884.6	1910.3	2794.8
19	209.8	0.001172	0.1047	894.6	2597.3	896.8	1899.3	2796.1
20	212.4	0.001177	0.0995	906.2	2598.2	908.6	1888.6	2797.2
21	214.9	0.001181	0.0949	917.5	2598.9	920.0	1878.2	2798.2
22	217.2	0.001185	0.0907	928.3	2599.6	931.0	1868.1	2799.1
23	219.6	0.001189	0.0868	938.9	2600.2	941.6	1858.2	2799.8
24	221.8	0.001193	0.0832	949.1	2600.7	951.9	1848.5	2800.4
25	223.9	0.001197	0.0799	959.0	2601.2	962.0	1839.0	2800.9
26	226.0	0.001201	0.0769	968.6	2601.5	971.7	1829.6	2801.4
27	228.1	0.001205	0.0740	978.0	2601.8	981.2	1820.5	2801.7
28	230.0	0.001209	0.0714	987.1	2602.1	990.5	1811.5	2802.0
29	232.0	0.001213	0.0689	996.0	2602.3	999.5	1802.6	2802.2
30	233.8	0.001216	0.0666	1004.7	2602.4	1008.4	1793.9	2802.3
32	237.4	0.001224	0.0624	1021.5	2602.5	1025.4	1776.9	2802.3
34	240.9	0.001231	0.0587	1037.6	2602.5	1041.8	1760.3	2802.1
36	244.2	0.001238	0.0554	1053.1	2602.2	1057.6	1744.2	2801.7
38	247.3	0.001245	0.0524	1068.0	2601.9	1072.7	1728.4	2801.1
36	244.2	0.001238	0.0554	1053.1	2602.2	1057.6	1744.2	2801.7
40	250.3	0.001252	0.0497	1082.4	2601.3	1087.4	1712.9	2800.3
42	253.2	0.001259	0.0473	1096.3	2600.7	1101.6	1697.8	2799.4
44	256.0	0.001266	0.0451	1109.8	2599.9	1115.4	1682.9	2798.3
46	258.8	0.001272	0.0430	1122.9	2599.1	1128.8	1668.3	2797.1
48	261.4	0.001279	0.0412	1135.6	2598.1	1141.8	1653.9	2795.7
50	263.9	0.001286	0.0394	1148.0	2597.0	1154.5	1639.7	2794.2
52	266.4	0.001292	0.0378	1160.1	2595.9	1166.8	1625.7	2792.6
54	268.8	0.001299	0.0363	1171.9	2594.6	1178.9	1611.9	2790.8
56	271.1	0.001306	0.0349	1183.5	2593.3	1190.8	1598.2	2789.0
58	273.3	0.001312	0.0337	1194.7	2591.9	1202.3	1584.7	2787.0
60	275.6	0.001319	0.0324	1205.8	2590.4	1213.7	1571.3	2785.0
62	277.7	0.001325	0.0313	1216.6	2588.8	1224.8	1558.0	2782.9
64	279.8	0.001332	0.0302	1227.2	2587.2	1235.7	1544.9	2780.6
66	281.8	0.001338	0.0292	1237.6	2585.5	1246.5	1531.9	2778.3
68	283.8	0.001345	0.0283	1247.9	2583.7	1257.0	1518.9	2775.9
70	285.8	0.001351	0.0274	1258.0	2581.8	1267.4	1506.0	2773.5
72	287.7	0.001358	0.0265	1267.9	2579.9	1277.6	1493.3	2770.9
74	289.6	0.001364	0.0257	1277.6	2578.0	1287.7	1480.5	2768.3
76	291.4	0.001371	0.0249	1287.2	2575.9	1297.6	1467.9	2765.5
78	293.2	0.001378	0.0242	1296.7	2573.8	1307.4	1455.3	2762.8
80	295.0	0.001384	0.0235	1306.0	2571.7	1317.1	1442.8	2759.9
82	296.7	0.001391	0.0229	1315.2	2569.5	1326.6	1430.3	2757.0
84	298.4	0.001398	0.0222	1324.3	2567.2	1336.1	1417.9	2754.0
86	300.1	0.001404	0.0216	1333.3	2564.9	1345.4	1405.5	2750.9
88	301.7	0.001411	0.0210	1342.2	2562.6	1354.6	1393.2	2747.8
90	303.3	0.001418	0.02050	1351.0	2560.1	1363.7	1380.9	2744.6

TABLE C.1 (continued)

Properties of Saturated Steam

		$\hat{V}$ (m³/kg)		$\hat{U}$ (kJ/kg)		$\hat{H}$ (kJ/mol)		
P (bar)	T (°C)	Water	Steam	Water	Steam	Water	Evap.	Steam
92	304.9	0.001425	0.01996	1359.7	2557.7	1372.8	1368.6	2741.4
94	306.4	0.001432	0.01945	1368.2	2555.2	1381.7	1356.3	2738.0
96	308.0	0.001439	0.01897	1376.7	2552.6	1390.6	1344.1	2734.7
98	309.5	0.001446	0.01849	1385.2	2550.0	1399.3	1331.9	2731.2
100	311.0	0.001453	0.01804	1393.5	2547.3	1408.0	1319.7	2727.7
105	314.6	0.001470	0.01698	1414.1	2540.4	1429.5	1289.2	2718.7
110	318.0	0.001489	0.01601	1434.2	2533.2	1450.6	1258.7	2709.3
115	321.4	0.001507	0.01511	1454.0	2525.7	1471.3	1228.2	2699.5
120	324.6	0.001527	0.01428	1473.4	2517.8	1491.8	1197.4	2689.2
125	327.8	0.001547	0.01351	1492.7	2509.4	1512.0	1166.4	2678.4
130	330.8	0.001567	0.01280	1511.6	2500.6	1532.0	1135.0	2667.0
135	333.8	0.001588	0.01213	1530.4	2491.3	1551.9	1103.1	2655.0
140	336.6	0.001611	0.01150	1549.1	2481.4	1571.6	1070.7	2642.4
145	339.4	0.001634	0.01090	1567.5	2471.0	1591.3	1037.7	2629.1
150	342.1	0.001658	0.01034	1586.1	2459.9	1611.0	1004.0	2615.0
155	344.8	0.001683	0.00981	1604.6	2448.2	1630.7	969.6	2600.3
160	347.3	0.001710	0.00931	1623.2	2436.0	1650.5	934.3	2584.9
165	349.8	0.001739	0.00883	1641.8	2423.1	1670.5	898.3	2568.8
170	352.3	0.001770	0.00837	1661.6	2409.3	1691.7	859.9	2551.6
175	354.6	0.001803	0.00793	1681.8	2394.6	1713.3	820.0	2533.3
180	357.0	0.001840	0.00750	1701.7	2378.9	1734.8	779.1	2513.9
185	359.2	0.001881	0.00708	1721.7	2362.1	1756.5	736.6	2493.1
190	361.4	0.001926	0.00668	1742.1	2343.8	1778.7	692.0	2470.6
195	363.6	0.001977	0.00628	1763.2	2323.6	1801.8	644.2	2446.0
200	365.7	0.00204	0.00588	1785.7	2300.8	1826.5	591.9	2418.4
205	367.8	0.00211	0.00546	1810.7	2274.4	1853.9	532.5	2386.4
210	369.8	0.00220	0.00502	1840.0	2242.1	1886.3	461.3	2347.6
215	371.8	0.00234	0.00451	1878.6	2198.1	1928.9	366.2	2295.2
220	373.7	0.00267	0.00373	1952	2114	2011	185	2196
221.2	374.15	0.00317	0.00317	2038	2038	2108	0	2108

TABLE C.2

Properties of Superheated Steam

P (bar) (T_{sat} °C)		Saturated Water	Saturated Steam	Temperature (°C)							
				50	75	100	150	200	250	300	350
0.0 (—)	$\hat{H}$	—	—	2595	2642	2689	2784	2880	2978	3077	3177
	$\hat{U}$	—	—	2446	2481	2517	2589	2662	2736	2812	2890
	$\hat{V}$	—	—	—	—	—	—	—	—	—	—
0.1 (45.8)	$\hat{H}$	191.8	2584.8	2593	2640	2688	2783	2880	2977	3077	3177
	$\hat{U}$	191.8	2438.0	2444	2480	2516	2588	2661	2736	2812	2890
	$\hat{V}$	0.00101	14.7	14.8	16.0	17.2	9.5	21.8	24.2	26.5	28.7
0.5 (81.3)	$\hat{H}$	340.6	2646.0	209.3	313.9	2683	2780	2878	2979	3076	3177
	$\hat{U}$	340.6	2484.0	209.2	313.9	2512	2586	2660	2735	2811	2889
	$\hat{V}$	0.00103	3.24	0.00101	0.00103	3.41	3.89	4.35	4.83	5.29	5.75
1.0 (99.6)	$\hat{H}$	417.5	2675.4	209.3	314.0	2676	2776	2875	2975	3074	3176
	$\hat{U}$	417.5	2506.1	209.2	313.9	2507	2583	2658	2734	2811	2889
	$\hat{V}$	0.00104	1.69	0.00101	0.00103	1.69	1.94	2.17	2.40	2.64	2.87
5.0 (151.8)	$\hat{H}$	640.1	2747.5	209.7	314.3	419.4	632.2	2855	2961	3065	3168
	$\hat{U}$	639.6	2560.2	209.2	313.8	418.8	631.6	2643	2724	2803	2883
	$\hat{V}$	0.00109	0.375	0.00101	0.00103	0.00104	0.00109	0.425	0.474	0.522	0.571
10 (179.9)	$\hat{H}$	762.6	2776.2	210.1	314.7	419.7	632.5	2827	2	3052	3159
	$\hat{U}$	761.5	2582	209.1	313.7	418.7	631.4	2621	2710	2794	2876
	$\hat{V}$	0.00113	0.194	0.00101	0.00103	0.00104	0.00109	0.206	0.233	0.258	0.282
20 (212.4)	$\hat{H}$	908.6	2797.2	211.0	315.5	420.5	633.1	852.6	2902	3025	3139
	$\hat{U}$	906.2	2598.2	209.0	313.5	418.4	603.9	850.2	2679	2774	2862
	$\hat{V}$	0.00118	0.09950	0.00101	0.00102	0.00104	0.00109	0.00116	0.111	0.125	0.139
40 (250.3)	$\hat{H}$	1087.4	2800.3	212.7	317.1	422.0	634.3	853.4	1085.8	2962	3095
	$\hat{U}$	1082.4	2601.3	208.6	313.0	417.8	630.0	848.8	1080.8	2727	2829
	$\hat{V}$	0.00125	0.04975	0.00101	0.00102	0.00104	0.00109	0.00115	0.00125	0.0588	0.0665

P (T_sat)											
60 (275.6)	$\hat{H}$	1213.7	2785.0	214.4	318.7	423.5	635.6	854.2	1085.8	2885	3046
	$\hat{U}$	1205.8	2590.4	208.3	312.6	417.3	629.1	847.3	1078.3	2668	2792
	$\hat{V}$	0.00132	0.0325	0.00101	0.00103	0.00104	0.00109	0.00115	0.00125	0.0361	0.0422
80 (295.0)	$\hat{H}$	1317.1	2759.9	216.1	320.3	425.0	636.8	855.1	1085.8	2787	2990
	$\hat{U}$	1306.0	2571.7	208.1	312.3	416.7	628.2	845.9	1075.8	2593	2750
	$\hat{V}$	0.00139	0.0235	0.00101	0.00102	0.00104	0.00109	0.00115	0.00124	0.0243	0.0299
100 (311.0)	$\hat{H}$	1408.0	2727.7	217.8	322.9	426.5	638.1	855.9	1085.8	1343.4	2926
	$\hat{U}$	1393.5	2547.3	207.8	311.7	416.1	627.3	844.4	1073.4	1329.4	2702
	$\hat{V}$	0.00145	0.0181	0.00101	0.00102	0.001049	0.00109	0.00115	0.00124	0.00140	0.0224
150 (342.1)	$\hat{H}$	1611.0	2615.0	222.1	326.0	430.3	641.3	858.1	1086.2	1338.2	2695
	$\hat{U}$	1586.1	2459.9	207.0	310.7	414.7	625.0	841.0	1067.7	1317.6	2523
	$\hat{V}$	0.00166	0.0103	0.00101	0.00102	0.00104	0.00108	0.00114	0.00123	0.00138	0.0115
200 (365.7)	$\hat{H}$	1826.5	2418.4	226.4	330.0	434.0	644.5	860.4	1086.7	1334.3	1647.1
	$\hat{U}$	1785.7	2300.8	206.3	309.7	413.2	622.9	837.7	1062.2	1307.1	1613.7
	$\hat{V}$	0.00204	0.005875	0.00100	0.00102	0.00103	0.00108	0.00114	0.00122	0.00136	0.00167
221.2(Pc) (374.15)(Tc)	$\hat{H}$	2108	2108	228.2	331.7	435.7	645.8	861.4	1087.0	1332.8	1635.5
	$\hat{U}$	2037.8	2037.8	206.0	309.2	412.8	622.0	836.3	1060.0	1302.9	1600.3
	$\hat{V}$	0.00317	0.00317	0.00100	0.00102	0.00103	0.00108	0.00114	0.00122	0.00135	0.00163
250 (—)	$\hat{H}$	—	—	230.7	334.0	437.8	647.7	862.8	1087.5	1331.1	1625.0
	$\hat{U}$	—	—	205.7	308.7	412.1	620.8	834.4	1057.0	1297.5	1585.0
	$\hat{V}$	—	—	0.00100	0.00101	0.00103	0.00108	0.00113	0.00122	0.00135	0.00160
300 (—)	$\hat{H}$	—	—	235.0	338.1	441.6	650.9	865.2	1088.4	1328.7	1609.9
	$\hat{U}$	—	—	205.0	307.7	410.8	618.7	831.3	1052.1	1288.7	1563.3
	$\hat{V}$	—	—	0.0009990	0.00101	0.00103	0.00107	0.00113	0.00121	0.00133	0.00155
500 (—)	$\hat{H}$	—	—	251.9	354.2	456.8	664.1	875.4	1093.6	1323.7	1576.3
	$\hat{U}$	—	—	202.4	304.0	405.8	611.0	819.7	1034.3	1259.3	1504.1
	$\hat{V}$	—	—	0.0009911	0.00100	0.00102	0.00106	0.00111	0.00119	0.00129	0.00144
1000 (—)	$\hat{H}$	—	—	293.9	394.3	495.1	698.0	903.5	1113.0	1328.7	1550.5
	$\hat{U}$	—	—	196.5	295.7	395.1	594.4	795.3	999.0	1207.1	1419.0
	$\hat{V}$	—	—	0.0009737	0.0009852	0.00100	0.00104	0.00108	0.00114	0.00122	0.00131

(continued)

TABLE C.2 (continued)

Properties of Superheated Steam

P (bar) (T_{sat} °C)		Temperature (°C)							
		400	450	500	550	600	650	700	750
0.0	$\hat{H}$	3280	3384	3497	3597	3706	3816	3929	4043
	$\hat{U}$	2969	3050	3132	3217	3303	3390	3480	3591
	$\hat{V}$	—	—	—	—	—	—	—	—
0.1 (45.8)	$\hat{H}$	3280	3384	3489	3596	3706	3816	3929	4043
	$\hat{U}$	2969	3050	3132	3217	3303	3390	3480	3571
	$\hat{V}$	21.1	33.3	35.7	38.0	40.3	42.6	44.8	47.2
0.5 (81.3)	$\hat{H}$	3279	3383	3489	3596	3705	3816	3929	4043
	$\hat{U}$	2969	3049	3132	3216	3302	3390	3480	3571
	$\hat{V}$	6.21	6.67	7.14	7.58	8.06	8.55	9.01	9.43
1.0 (99.6)	$\hat{H}$	3278	3382	3488	3596	3705	3816	3928	4042
	$\hat{U}$	2968	3049	3132	3216	3302	3390	3479	3570
	$\hat{V}$	3.11	3.33	3.57	3.80	4.03	4.26	4.48	4.72
5.0 (151.8)	$\hat{H}$	3272	3379	3484	3592	3702	3813	3926	4040
	$\hat{U}$	2964	3045	3128	3213	3300	3388	3477	3569
	$\hat{V}$	0.617	0.664	0.711	0.758	0.804	0.850	0.897	0.943
10 (179.9)	$\hat{H}$	3264	3371	3478	3587	3697	3809	3923	4038
	$\hat{U}$	2958	3041	3124	3210	3296	3385	3475	3567
	$\hat{V}$	0.307	0.330	0.353	0.377	0.402	0.424	0.448	0.472
20 (212.4)	$\hat{H}$	3249	3358	3467	3578	3689	3802	3916	4032
	$\hat{U}$	2946	3031	3115	3202	3290	3379	3470	3562
	$\hat{V}$	0.151	0.163	0.175	0.188	0.200	0.211	0.223	0.235
40 (250.3)	$\hat{H}$	3216	3331	3445	3559	3673	3788	3904	4021
	$\hat{U}$	2922	3011	3100	3188	3278	3368	3460	3554
	$\hat{V}$	0.0734	0.0799	0.0864	0.0926	0.0987	0.105	0.111	0.117
60 (275.6)	$\hat{H}$	3180	3303	3422	3539	3657	3774	3892	4011
	$\hat{U}$	2896	2991	3083	3174	3265	3357	3451	3545

		1	2	3	4	5	6	7	8
80 (295.0)	$\hat{V}$	0.0474	0.0521	0.0566	0.0609	0.0652	0.0693	0.0735	0.0776
	$\hat{H}$	3142	3274	3399	3520	3640	3759	3879	4000
	$\hat{U}$	2867	2969	3065	3159	3252	3346	3441	3537
100 (311.0)	$\hat{V}$	0.0344	0.0382	0.0417	0.0450	0.0483	0.0515	0.0547	0.0578
	$\hat{H}$	3100	3244	3375	3500	3623	3745	3867	3989
	$\hat{U}$	2836	2946	3047	3144	3240	3335	3431	3528
150 (342.1)	$\hat{V}$	0.0264	0.0298	0.0328	0.0356	0.0383	0.0410	0.0435	0.0461
	$\hat{H}$	2975	3160	3311	3448	3580	3708	3835	4962
	$\hat{U}$	2744	2883	2999	3105	3207	3307	3407	3507
200 (365.7)	$\hat{V}$	0.0157	0.0185	0.0208	0.0229	0.0249	0.0267	0.6286	0.0304
	$\hat{H}$	2820	3064	3241	3394	3536	3671	3804	3935
	$\hat{U}$	2622	2810	2946	3063	3172	3278	3382	3485
221.2(Pc) (374.15)(Tc)	$\hat{V}$	0.009950	0.0127	0.0148	0.0166	0.0182	0.0197	0.0211	0.0225
	$\hat{H}$	2733	3020	3210	3370	3516	3655	3790	3923
	$\hat{U}$	2553	2776	2922	3045	3157	3265	3371	3476
250 (↑)	$\hat{V}$	0.008157	0.0110	0.0130	0.0147	0.0162	0.0176	0.0190	0.0202
	$\hat{H}$	2582	2954	3166	3337	3490	3633	3772	3908
	$\hat{U}$	2432	2725	2888	3019	3137	3248	3356	3463
300 (↑)	$\hat{V}$	0.006013	0.009174	0.0111	0.0127	0.0141	0.0143	0.0166	0.0178
	$\hat{H}$	2162	2826	3085	3277	3443	3595	3740	3880
	$\hat{U}$	2077	2623	2825	2972	3100	3218	3330	3441
500 (↑)	$\hat{V}$	0.002830	0.006734	0.008680	0.0102	0.0114	0.0126	0.0136	0.0147
	$\hat{H}$	1878	2293	2723	3021	3248	3439	3610	3771
	$\hat{U}$	1791	2169	2529	2765	2946	3091	3224	3350
1000 (↑)	$\hat{V}$	0.001726	0.002491	0.003882	0.005112	0.006112	0.007000	0.007722	0.008418
	$\hat{H}$	1798	2051	2316	2594	2857	3105	3324	3526
	$\hat{U}$	1653	1888	2127	2369	2591	2795	2971	3131
	$\hat{V}$	0.001446	0.00162	0.001893	0.00224	0.00266	0.00310	0.003536	0.00395

Index